ALLE ZEIT WACH
1842

Springer Series in Synergetics

Editor: Hermann Haken

Synergetics, an interdisciplinary field of research, is concerned with the cooperation of individual parts of a system that produces macroscopic spatial, temporal or functional structures. It deals with deterministic as well as stochastic processes.

1 **Synergetics** An Introduction 3rd Edition
By H. Haken

2 **Synergetics** A Workshop
Editor: H. Haken

3 **Synergetics** Far from Equilibrium
Editors: A. Pacault, C. Vidal

4 **Structural Stability in Physics**
Editors: W. Güttinger, H. Eikemeier

5 **Pattern Formation by Dynamic Systems and Pattern Recognition**
Editor: H. Haken

6 **Dynamics of Synergetic Systems**
Editor: H. Haken

7 **Problems of Biological Physics**
By L. A. Blumenfeld

8 **Stoachastic Nonlinear Systems**
in Physics, Chemistry, and Biology
Editors: L. Arnold, R. Lefever

9 **Numerical Methods in the Study of Critical Phenomena**
Editors: J. Della Dora, J. Demongeot, B. Lacolle

10 **The Kinetic Theory of Electromagnetic Processes** By Yu. L. Klimontovich

11 **Chaos and Order in Nature**
Editor: H. Haken

12 **Nonlinear Phenomena in Chemical Dynamics**
Editors: C. Vidal, A. Pacault

13 **Handbook of Stochastic Methods**
for Physics, Chemistry, and the Natural Sciences
By C. W. Gardiner

14 **Concepts and Models of a Quantitative Sociology**
The Dynamics of Interacting Populations
By W. Weidlich, G. Haag

15 **Noise-Induced Transitions** Theory and Applications in Physics, Chemistry, and Biology
By W. Horsthemke, R. Lefever

16 **Physics of Bioenergetic Processes**
By L. A. Blumenfeld

17 **Evolution of Order and Chaos** in Physics, Chemistry, and Biology Editor H. Haken

18 **The Fokker-Planck Equation**
By H. Risken

19 **Chemical Oscillations, Waves, and Turbulence**
By Y. Kuramoto

20 **Advanced Synergetics**
By H. Haken

21 **Stoachastic Phenomena and Chaotic Behaviour in Complex Systems**
Editor: P. Schuster

22 **Synergetics - From Microscopic to Macroscopic Order** Editor: E. Frehland

23 **Synergetics of the Brain**
Editors: E. Başar, H. Flohr, H. Haken, A. J. Mandell

24 **Chaos and Statistical Methods**
Editor: Y. Kuramoto

25 **Dynamics of Hierarchical Systems**
By J. S. Nicolis

26 **Self-Organization and Management of Social Systems**
Editors: H. Ulrich, G. J. B. Probst

27 **Non-Equilibrium Dynamics in Chemical Systems** Editors: C. Vidal, A. Pacault

28 **Self-Organization.** Autowaves and Structures Far from Equilibrium
Editor: V. I. Krinsky

Non-Equilibrium Dynamics in Chemical Systems

Proceedings of the International Symposium,
Bordeaux, France, September 3–7, 1984

Editors: C. Vidal and A. Pacault

With 137 Figures

Springer-Verlag
Berlin Heidelberg New York Tokyo 1984

Professeur Dr. *Christian Vidal*
Professeur Dr. *Adolphe Pacault*
Centre de Recherche Paul Pascal, Domaine Universitaire
F-33405 Talence Cédex, France

Series Editor:
Professor Dr. Dr. h. c. Hermann Haken
Institut für Theoretische Physik der Universität Stuttgart, Pfaffenwaldring 57/IV,
D-7000 Stuttgart 80, Fed. Rep. of Germany

ISBN-13: 978-3-642-70198-6 e-ISBN-13: 978-3-642-70196-2
DOI: 10.1007/978-3-642-70196-2

Softcover reprint of the hardcover 1st edition 1984

2153/3130-543210

Preface

Markedly apart from elementary particle physics, another current has been building up and continuously growing within contemporary physics for several decades, and even expanding into many other disciplines, especially chemistry, biology and, quite recently, economics. Several reasons account for this: presumably the most important one lies in the fact that, whatever the specific problem, model or material concerned, the same basic mathematical features are always involved. In this way, a general phenomenology has emerged which, unlike thermodynamics, is no longer dependent upon the details or specifics: what largely prevails is the nonlinear character of the underlying dynamics. Perhaps we are witnessing the emergence of a "nonlinear physics" - in a way similar to the birth of "quantum physics" in the twenties - a physics which deals with the general behaviour of systems, whatever they are or may be.

Over the past fifteen years, chemical systems evolving sufficiently far from equilibrium have proved to be particularly well fitted to experimental research on nonlinear behaviour: oscillation, multistability, birhythmicity, chaotic evolution, spatial self-organization and hysteresis are displayed by chemical reactions whose number is growing each year. In this volume are collected the lectures, communications and posters (abstracts) presented at an international meeting entitled: "Non-Equilibrium Dynamics in Chemical Systems", held in Bordeaux (France), September 3rd-7th, 1984. Papers have been grouped into five parts devoted to what appear to be up-to-date trends in this field. By comparing these proceedings to those of the previous Bordeaux meetings [Springer Series in Synergetics, Vol. 3 (1978), Vol. 12 (1981)], one can get a good idea about the evolution of research on nonlinear chemical systems. Though things appear to be more complex than at first believed, as finer investigations are performed - a quite trivial conclusion -, it is striking to note how fertile are the concepts of chemical oscillators, waves and chaos, of non-equilibrium evolution and of dissipative structures.

This research domain is so full of life that it seems henceforth advisable to hold an annual discussion meeting. At the same time, the need for putting a little bit of order into a perfectly unscheduled set of conferences has become evident. Suggestions for achieving such an international settlement were presented at the meeting (see Introduction by A. Pacault). It is hoped that non-attendees will agree, as attendees did.

For the third time the generous sponsorship of the French Centre National de la Recherche Scientifique made this conference possible, and it is greatly acknowledged. We are also indebted to the Université de Bordeaux I, Mairie de Bordeaux and Conseil Général de la Gironde for their hospitality and financial support. Last, but not least, we wish to address special thanks to our secretary, Mrs. Maurat, who once again did a lot of work before, during and after the meeting, without losing her good humour.

Bordeaux
October 1984

C. Vidal
A. Pacault

Contents

Part I General Introduction

Part II Oscillating Reactions and Modelling Problems

Part III Spatial Structures and Chemical Waves

Part IV Chemical Chaos

Part V Noise Effects

Part VI Stochastic Analysis

Part VII Posters

Part I

General Introduction

Introduction to the Meeting
"Non-Equilibrium Dynamics in Chemical Systems"

A. **Pacault** *(Chairman of the Organizing Committee)*
Centre de Recherche Paul Pascal-Domaine universitaire
F-33405 Talence Cédex, France

The field of science with which we are concerned is fascinating because it covers a much wider domain of knowledge than traditional topics do. It requires difficult experimental studies, full of surprises, in which the results of mechanics, thermodynamics, hydrodynamics and kinetics must be taken into account, only to mention the main subjects. In the course of translating experimental results into mathematical language, real objects, i.e. a set of variables with which we choose to describe a part of the universe are compared with mathematical objects with similar properties.

Our field of science requires talented experimentalists, familiar with various domains of knowledge, able to converse with those accustomed to the mathematical language.

For a better understanding of the surrounding world, we have to build interdisciplinary connections, but without forgetting that *"there is nothing interdisciplinary in the meeting of twelve experts around a table : cross-fertilization between domains arises only through the penetration of each domain by the same field of conscience"**. Precisely, collaborations between laboratories exist, which contribute to creating this *field of conscience*. Nevertheless, these usually distant collaborations can only exist between small groups of researchers. It is desirable that a large community be able to meet periodically and communicate research results.

Such is the goal of the interdisciplinary meeting that we have organized. But, since 1967, over 26 meetings have been held, and their yearly number is growing. So many meetings cause unnecessary repetition? Couldn't we take advantage of this meeting in Bordeaux to try to harmonize the organization of these necessary interdisciplinary meetings?

This harmonization has been achieved in other fields. The advances in carbon research, to which I have been contributing for many years, are reported every year during an international conference that is held in odd years in the U.S.A., and in even years successively in Germany, Great Britain and France. Last July we organized the "CARBON 84" conference in Bordeaux. Over 350 industrial and academic participants were present. For ten years now this modus vivendi has been satisfactory, each country being entirely and solely responsible for the organization of the conference.

The scientific community that is concerned with the phenomena for whose study we are together today is now sufficiently structured to plan an annual conference that can be successively focoused on the various topics of interest, in order to review their achievements and to discuss their present developments. Each theme need not be considered every year.

I know, although I don't know everything, that:

- from September 17th to 22nd, 1984, we shall talk about *"Temporal Order"* in Bremen (a symposium on Mechanisms, Models, and Significance of Oscillations in Heterogeneous Chemical and Biological Systems),

---

*A. Moles: *Une science de l'imprécis,* Le Monde aujourd'hui *XIII, (12-13)-8-84.*

- that from October 15th to 17th, 1984, a meeting on *"Dynamically Organized Systems"* will be held at Schloss Elmau, Germany,

- that on December 3rd and 4th, 1984, in Brussels, we shall deal with *"The Physics and Chemistry of Complex Phenomena - an interface between pure and applied research"*,

- that from July 22nd to 26th, 1985, a Gordon conference in Plymouth will deal with *"Oscillations and Dynamic Instabilities"*.

Would it not be convenient to decide now what should be dealt with, and where, in 1986 ?

Now I must thank all those who have contributed to the organization of this meeting:

First the C.N.R.S. for financial support, the University of Bordeaux I, the Mayor or Bordeaux who will receive us tonight in the magnificent Rohan Palace. I should thank also the Organizing Committee* and, more specifically, Professor Vidal.

Special thanks should be addressed to those who agreed to deliver a plenary lecture, a difficult pedagogical task but so necessary in view of the interdisciplinary aspect of our field.

I must also thank those who present their most recent and original research results.

I am sure that when we depart at the end of the week, we shall have deepened our comprehension of the world, for such is the recent scientific revolution that has emerged from our studies: now we know that a deterministic world is not necessarily predictable, and we begin to measure its degree of unpredictability more knowledgeably.

Science joining Philosophy: Isn't this stimulating?

Thursday September 6th Session, 14.30

Discussions seem to have led to a general agreement on the following points:

1. the number of meetings is too high;

2. it is necessary to inform the scientific community interested in the subject that could be called *"Nonlinearities and Instabilities in Chemistry"* (N.I.C.) well in advance of conference planning. This information could be delivered through the scientific associations of each country;

3. such information would lead to more coherence, avoiding overlap and promoting interdisciplinary complementarities;

4. each conference could have a different style - summer school, workshop, conference - insisting either on pedagogical and synthetic aspects or on original results and breakthroughs;

5. this is no attempt to create an *"institution"*, but rather a friendly cooperation which allows each country to take part in the international concert.

We know that a Gordon conference will take place in 1985. It would be interesting if an Eastern European contry could be involved in 1986. Hungary could consider such an opportunity. In 1987, Great Britain together with Belgium, in 1988 Germany, in 1989 the U.S.A., could organize such an N.I.C. conference. We have time to think of 1990!

Of course, all these are only proposals, and correspondence to exchange points of view will confirm or modify this very informal attempt at harmonization.

---

* *F. Argoul, J.M. Bodet, J. Boissonade, P. De Kepper, P. Hanusse, A. Pacault, P. Richetti A. Rossi, J.C. Roux, C. Vidal.*

Introduction au Colloque

Le domaine scientifique qui nous occupe est passionnant parce qu'il recouvre un champ de connaissances bien plus vaste que celui des disciplines traditionnelles. Il requiert des études expérimentales difficiles et pleines d'embûches prenant en compte les acquis de la mécanique, de la thermodynamique, de l'hydrodynamique, de la cinétique pour ne citer que quelques grandes rubriques. La traduction des résultats expérimentaux en langage mathématique rapproche les objets matériels - ensemble des variables que nous choisissons pour décrire un morceau d'Univers - d'objets mathématiques ayant des propriétés semblables.

Notre domaine scientifique mobilise donc des expérimentateurs de talent, familiers de disciplines différentes, dialoguant avec les habitués du langage mathématique.

Pour mieux comprendre le monde qui nous entoure, il nous faut donc construire des réseaux interdisciplinaires mais en notant *"qu'il n'existe pas d'interdisciplinarité dans l'assemblage de douze spécialistes autour d'une table : il n'y a de fécondation réciproque d'une discipline par une autre qu'à l'intérieur d'un même champ de conscience passé successivement par des disciplines différentes"**. Or, justement, des collaborations existent entre laboratoires qui permettent d'établir ce *"champ de conscience"*. Cependant ces collaborations lointaines ne peuvent avoir lieu qu'entre un petit nombre de chercheurs et il est souhaitable que le plus grand nombre puisse se retrouver périodiquement pour confronter ses résultats.

Tel est l'objectif des colloques interdisciplinaires que nous organisons. Cependant, depuis 1967, ont eu lieu plus de 26 colloques dont le nombre annuel a cru durant ces dernières années. N'est-ce pas trop si on désire éviter les répétitions ? Ne pourrait-on profiter de cette rencontre de Bordeaux pour tenter d'harmoniser l'organisation de ces nécessaires réunions interdisciplinaires ?

Cette harmonisation a été possible dans d'autres domaines. Les recherches sur les carbones, auxquelles j'ai contribué depuis longtemps, sont exposées chaque année dans une conférence internationale qui a lieu les années impaires aux U.S.A. et les années paires successivement en Allemagne, en Angleterre et en France. En Juillet dernier nous avons organisé à Bordeaux "CARBONE 84" qui rassemblait 350 participants industriels et universitaires. Depuis dix ans ce modus vivendi donne satisfaction, chaque pays étant seul entièrement responsable de l'organisation de la conférence.

La communauté scientifique qui s'intéresse aux phénomènes pour lesquels nous sommes aujourd'hui réunis est maintenant suffisamment constituée pour organiser une conférence annuelle qui ordonnerait les sujets de manière telle que le point soit fait régulièrement à la fois sur l'acquis et sur le présent. Les mêmes thèmes pourraient n'être pas traités chaque année.

Je sais - mais je ne sais pas tout -,

- que du 17 au 22 Septembre 1984, on parlera à Brême de *"Temporal order"* (A Symposium on Mechanisms, Models, and Significance of Oscillations in Heterogeneous Chemical and Biological Systems),

---

* A. Moles: *Une science de l'imprécis*, Le Monde aujourd'hui *XIII, (12-13)-8-84.*

- que du 15 au 17 Octobre 1984 à Schloss Elmaü est organisée une rencontre sur *"Dynamically Organized Systems"*,

- que les 3 et 4 Décembre 1984 à Bruxelles on parlera de *"The Physics and Chemistry of Complex Phenomena - an interface between pure and applied research"*,

- que du 22 au 26 Juillet 1985, on traitera à Plymouth dans une Gordon Conference de *"Oscillations and Dynamic Instabilities"*.

Peut-être serait-il agréable de décider maintenant de quoi s'entretenir et où en 1986 ?

Je dois maintenant remercier tous ceux qui ont participé à l'organisation de cette réunion.

D'abord le C.N.R.S. qui l'a financée, l'Université, le Maire de Bordeaux qui nous recevra ce soir dans le magnifique Palais Rohan, ensuite le Comité d'Organisation* et plus particulièrement le Professeur Vidal.

Des remerciements particuliers doivent être adressés à tous ceux qui ont accepté de faire une conférence plénière, lourde tâche pédagogique justement si nécessaire compte tenu de la pluridisciplinarité de notre champ d'étude.

Je remercie enfin tous ceux qui présentent leurs derniers résultats.

Je suis sûr que nous nous quitterons, en fin de semaine, ayant encore approfondi notre compréhension du Monde car telle est la récente révolution scientifique qui est née des études qui nous occupent : maintenant nous savons qu'un monde déterministe peut n'être pas prévisible, et mieux encore nous commençons à mesurer son degré d'imprévisibilité.

N'est-il pas stimulant que la science rejoigne enfin la philosophie ?

Séance du jeudi 6 Septembre à 14 h 30

Des conversations ont permis de dégager une sorte de consensus général sur les points suivants :

1. le nombre des conférences est trop grand;

2. il est nécessaire d'informer suffisamment longtemps à l'avance la communauté scientifique intéressée par le sujet dont le titre général pourrait être *Non-linéarités et instabilités en chimie*. Cette information pourrait être donnée par les sociétés scientifiques des différents pays ;

3. une telle information conduirait à une cohérence permettant d'éviter les recouvrements et de susciter les complémentarités interdisciplinaires ;

4. chaque conférence pourrait avoir un style différent - école, atelier, colloque - l'une insistant plus sur l'aspect pédagogique et synthétique, l'autre sur les résultats originaux et les découvertes par exemple ;

5. il n'est pas question d'institutionnaliser mais d'entretenir une aimable coopération qui permette à chaque pays de prendre une place dans notre concert international.

On sait qu'en 1985 une Gordon Conference aura lieu aux U.S.A. Il serait agréable qu'en 1986 les pays de l'Est interviennent, et la Hongrie pourrait envisager une telle éventualité.

En 1987, l'Angleterre et la Belgique réunies, en 1988 l'Allemagne, et en 1989 les U.S.A. pourraient être les organisateurs de cette conférence. Attendons pour penser à 1990.

Bien entendu il ne s'agit que de points de vue échangés et des correspondances permettront de consolider ou de modifier cette tentative très informelle d'harmonisation.

---

* *F. Argoul, J.M. Bodet, J. Boissonade, P. De Kepper, P. Hanusse, A. Pacault, P. Richetti A. Rossi, J.C. Roux, C. Vidal.*

Spatial and Temporal Patterns Formed by Systems Far from Equilibrium

H. Haken

Institut für Theoretische Physik, Universität Stuttgart, Pfaffenwaldring 57/IV
D-7000 Stuttgart 80, Fed. Rep. of Germany

Over the past years enormous progress has been made in chemistry in the experimental and theoretical study of temporal and spatio-temporal patterns formed in systems far from thermal equilibrium. In this paper I discuss why quite different systems can show similar behavior and how this behavior can be adequately described by evolution equations or by discrete maps. Examples for the formation of spatial patterns in fluids and flames are provided. The problem of chaos and routes to it, including that via quasi-periodicity, are discussed in the framework described above. Particular attention is paid to the relation between discrete maps and description via trajectories in a phase space.

1. Introduction

Over the past years enormous progress has been made in making experiments on macroscopic spatial or temporal structures formed in chemical reactions and interpreting them theoretically. In this way chemistry has given an outstanding contribution to the study of systems driven far from thermal equilibrium. It is a particular pleasure for me to present this talk in Bordeaux where important contributions to this new field were given which is also witnessed by the two volumes on nonlinear chemical dynamics edited by A. Pacault and C. Vidal [1].

The systematic study of systems driven far from thermal equilibrium is a rather new field of science. At least two features are most surprising. When we think of systems in thermal equilibrium, we all admire the great power of thermodynamics with its universal laws. But for a long time it was unclear how to extend thermodynamics in an adequate way to systems far from thermal equilibrium. Furthermore it seemed quite counter-intuitive to expect ordered structures to occur when systems are driven far from thermal equilibrium. Rather, one would expect wild fluctuations. As we now know, well ordered patterns appear and even seemingly chaotic phenomena can obey laws of order. Furthermore, strikingly analogous phenomena are found in seemingly quite different systems, such as lasers, fluids, electronic devices, solids, in acoustics, and other fields.

Personally, I must confess that I was not very surprised by this development, because about 15 years ago I stressed that far reaching analogies of systems driven far from thermal equilibrium can be expected, and I suggested studying these phenomena under unifying concepts, within an interdisciplinary field of research I called "Synergetics" [2].

In my present paper I would like to show why quite different systems may show precisely the same phenomena, and should like also to include some of the more recent methods of treating both ordered and chaotic states.

2. Modelling Processes by Differential Equations

Let us consider two chemicals A and B which by their interaction produce a third type of chemical C and let us further assume that chemical C may decay into a further chemical D. Denoting the corresponding concentrations by n_1, n_2, n_3, n_4, respectively, we then have the scheme

$$\underset{n_1}{A} + \underset{n_2}{B} \Rightarrow \underset{n_3}{C}, \quad C \rightarrow \underset{n_4}{D} \qquad (2.1)$$

It is quite simple to write down equations describing the processes (2.1), namely the differential equation

$$\dot{n}_3 = k_1 n_1 n_2 - k_1' n_3 \quad . \qquad (2.2)$$

As it has turned out, the theoretical treatment of temporal or spatial structures formed by chemical reactions requires kinetic equations of which (2.2) is an example, rather than it is sufficient to use any concepts of thermodynamics, e.g. entropy, or quantities related to it.

When we have a network of chemicals reacting with each other we have to introduce the corresponding concentrations $n_j(t)$ which we lump together into a state vector

$$(n_1, n_2, \ldots\ldots) = \underset{\sim}{n} \quad . \qquad (2.3)$$

The corresponding equations can be written in the form

$$\dot{\underset{\sim}{n}} = \underset{\sim}{R}(\underset{\sim}{n}) + D \, \Delta \, \underset{\sim}{n} \quad , \qquad (2.4)$$

where R represents the reactions and D represents diffusion. Thus (2.4) are the well known reaction diffusion equations.

I am sure that also in years to come other transport effects, such as convection, must be considered in the study of pattern formation which is already done e.g. in pattern formation of flames. In such a case convection terms of the type

$$v \frac{\partial v}{\partial x} \tag{2.5}$$

must be incorporated into eqs.(2.4). Furthermore not only velocity fields as in (2.5) but also temperature fields must be taken into account. Describing all these different variables by means of a state vector $\tilde{q}$ the equations to be studied acquire the general form

$$\dot{\tilde{q}} = \tilde{N}(\tilde{q},\alpha) \tag{2.6}$$

where α is an abbreviation for the control parameters by which we control the system from the outside, e.g. by matter flux, or energy flux into the system. Even if the system can be considered as homogeneous, e.g. in a well-stirred tank reactor, typically 30 variables describing also intermediate products must be considered, at least in principle. If the system is not well stirred, patterns can evolve and to their description many more variables are needed. But as is shown in synergetics, at the onset of spatio-temporal patterns only a few degrees of freedom or variables dominate the system and this fact lies at the origin of the far reaching analogies found between quite different systems.

3. Many or few variables?

The access to the reduction of many to few variables was provided in particular by a study of nonequilibrium phase transitions [3] but the general principle is valid also for many other types of formation of macroscopic structures in systems far from equilibrium. Since I have described the whole idea and the detailed methods at various instances, I just want to make a few comments [4] and present the special case of a single order parameter.

We first assume that for a given control parameter α a state described by $\tilde{q}_0$ is established which, from a mathematical point of view, obeys the equations (2.6). When the control parameter is changed, this state, e.g. a homogeneous and quiescent state, can become unstable and a small spatial structure $\tilde{v}(\tilde{x})$, superimposed on that homogeneous state, can grow. Thus a new state evolves which is mathematically described by a superposition of three terms

$$\tilde{q} = \tilde{q}_0 + \xi\,(t)\;\tilde{v}(x) + \tilde{r} \tag{3.1}$$

where the rest term contains all other configurations of the system which interact with the growing mode, and eventually serve to stabilize it or to cause some kind of oscillation. All what matters for the time being is to note that the multivariable dynamic system is entirely governed by the behavior of the amplitude $\xi(t)$ in

(3.1) which is called an order parameter. A typical equation for the order parameter reads

$$\dot{\xi} = \lambda \xi - \xi^3 \quad . \tag{3.2}$$

By means of (3.1) and (3.2) it becomes possible to treat mathematically evolving patterns. When the system is driven farther and farther away from the instability point, the newly formed structure is deformed but does not change qualitatively, i.e. for instance, a periodic oscillation remains periodic. The whole dynamics and pattern is still governed by the order parameter. But eventually with further increase of control parameters the pattern can become unstable again and can be replaced by a qualitatively new pattern, e.g. a spatial pattern can be replaced by a spatio-temporal oscillation.

4. Typical order parameter equations. Spatial Patterns

Order parameter equations for spatio-temporal patterns developing in chemical reactions are treated in a recent book by Kuramoto [5]. In this paper I should rather like to present two other examples, namely fluid dynamics and flames. Just to illustrate how such equations in extended media may look, I present a describing pattern formation in the convection instability. The order parameter $\Psi(x,t)$ can be essentially considered as the deviation of the temperature field from a constant gradient. The corresponding patterns can be directly measured optically. The order parameter equation reads [6]

$$\dot{\Psi} = [\varepsilon - (1-\Delta)^2]\, \Psi + \delta\, \Psi^2 - \Psi^3 \tag{4.1}$$

Figs. 1 to 4 show typical examples of evolving structures [7]. Corresponding results have been obtained also experimentally [8] so that there exists good qualitative agreement. As another example we consider pattern formation in fluids where the plane flame front of a plane burner becomes unstable. The vertical deviation Ψ of the flame front from the horizontal plane can be considered as an order parameter. It obeys an equation derived by Sivashinsky [9] to which we have added the buoyancy term, which serves for the stabilization of the evolving pattern

$$\dot{\phi} = \frac{1}{Le}\,[1+\beta(Le-1)]\,\nabla^2\phi - \frac{1}{2}(\nabla\phi)^2 + \frac{(1-\sigma)}{8\pi^2}\iint \| dk \|\, \phi(\underset{\sim}{x}',t)\, e^{i\underset{\sim}{k}(\underset{\sim}{x}-\underset{\sim}{x}')}\, d\underset{\sim}{k}\, d\underset{\sim}{x}' \, . \tag{4.2}$$

Typical evolving patterns calculated by methods of synergetics by Schnaufer and myself [10] are shown in Figs.5 and 6. I personally believe that the study of spatial and temporal patterns of flames still provides a vast field of further experimental and theoretical research where new interesting results can be gained.

Fig.1 Evolution of roll pattern of the convection instability. Delta = δ in eq.(4.1). T = time, L = Lyapunov functional. The initial state, which is not shown here, is a random field. (After Bestehorn and Haken, to be published)

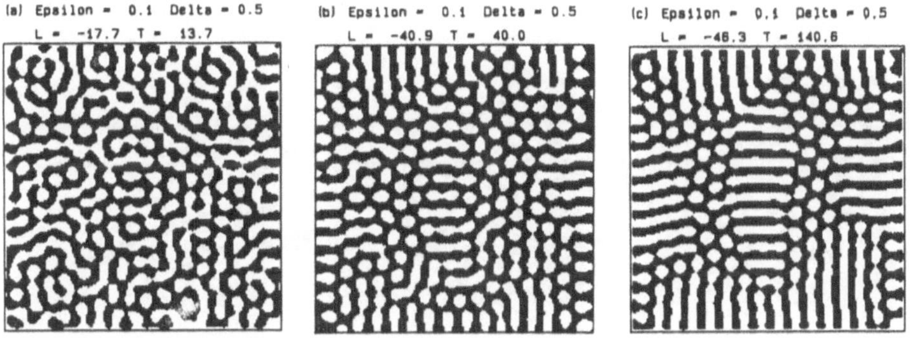

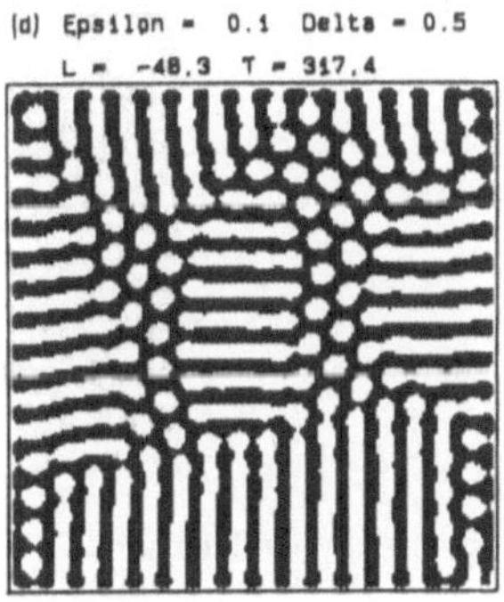

Fig. 2 Same as Fig.1, but δ is increased. Hexagons occur in addition to the rolls.

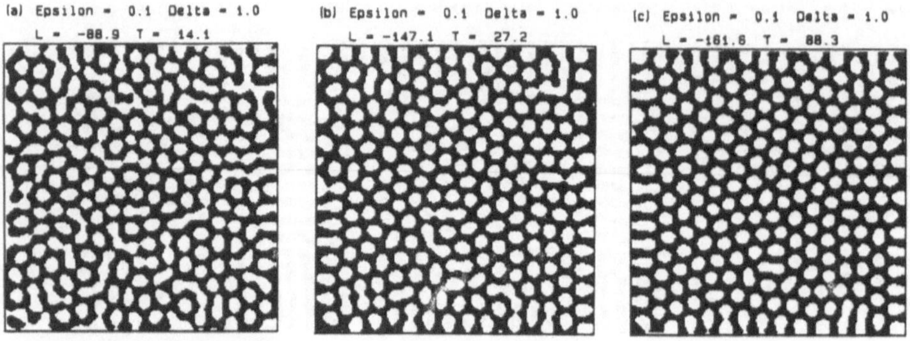

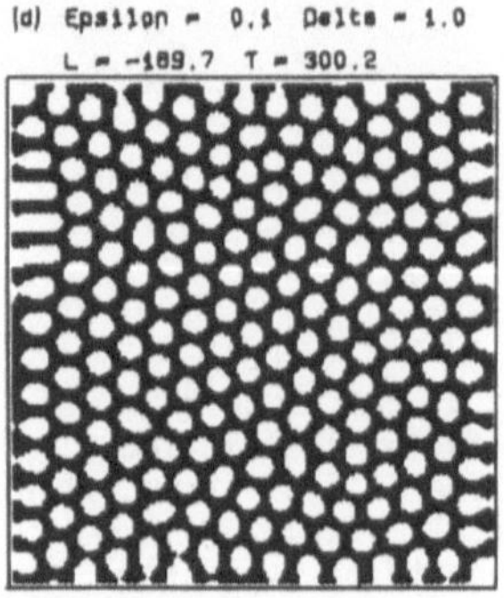

Fig. 3 Same as Figs.1 and 2, but δ is further increased. Only hexagons are formed now.

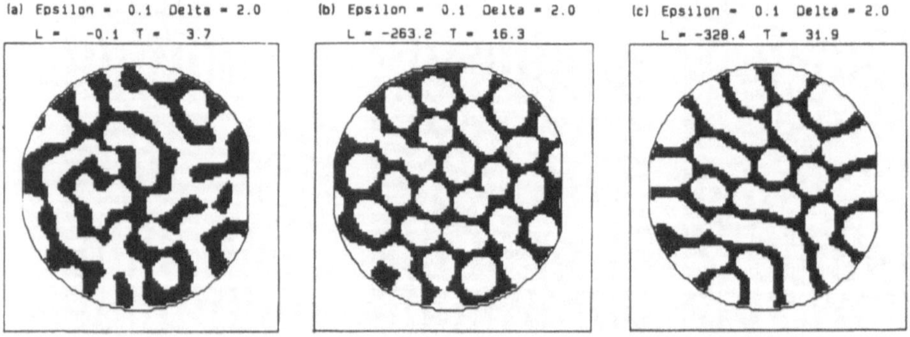

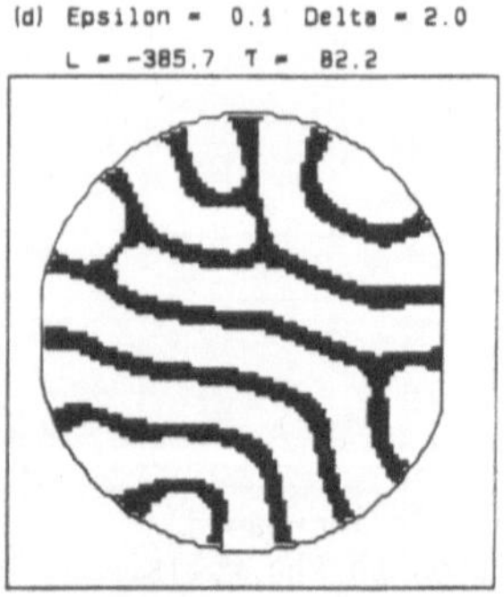

Fig. 4 Same as Fig.3, but other geometry and other aspect ratio

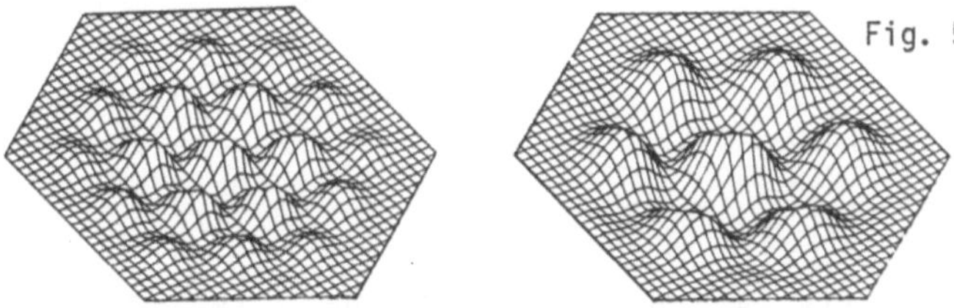

Fig. 5 Pattern formation in flames (After [10])

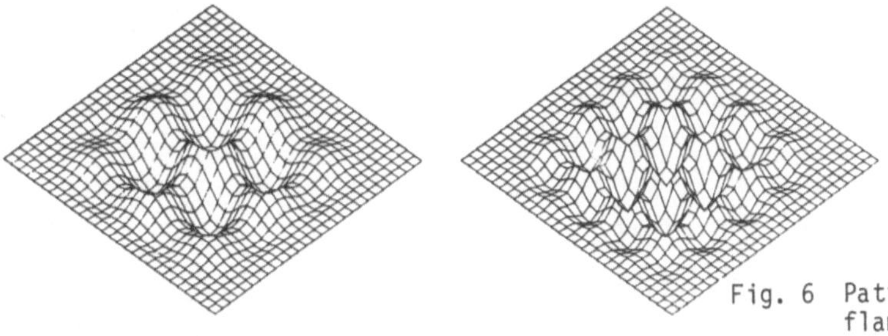

Fig. 6 Pattern formation in flames (After [10])

5. Temporal Patterns. Chaos

For simplicity we shall ignore from now on the spatial dependence in order to facilitate our presentation, so that in the realm of chemistry we deal with well-stirred tank reactors. We shall assume that the order parameter equations have been derived, and we should like to remind the reader of a few typical examples. The order parameter equation

$$\dot{\xi} = \lambda \xi - \xi^3 \tag{5.1}$$

describes the relaxation of the system towards a steady state, while the order parameter equation

$$\dot{z} = \lambda z - z|z|^2 , \quad z = \xi + i \eta \tag{5.2}$$

describes a limit cycle, i.e. periodic oscillations, and the relaxation of the system towards a limit cycle. The really exciting phenomena occur for order parameter equations which contain at least three variables. Surprisingly enough, a single nonlinearity is sufficient to produce an irregular motion, called deterministic chaos. Such an equation is of the Roessler type [11]

$$\dot{x} = -y - z$$

$$\dot{y} = x + ay$$

$$\dot{z} = b + z(x-c) \quad .$$

By now famous model equations are those of Lorenz [12], derived first in the context of fluid dynamics:

$$\dot{x} = \sigma\,(y-x)$$

$$\dot{y} = x(r-z) - y$$

$$\dot{z} = xy - bz \quad .$$

Other equations, in which typical phenomena of the formation of temporal patterns can be studied (mostly nowadays by computers), are the Helmholtz equation

$$\ddot{x} + \gamma\dot{x} + ax + bx^2 = A\sin\omega t$$

and the Duffing equation

$$\ddot{x} + \gamma\dot{x} + ax + bx^3 = A\sin\omega t$$

which both describe damped nonlinear driven oscillators. One can easily convince oneself that these equations are equivalent to three coupled first order differential equations which are autonomous.

When the amplitude of the driver A is increased, various temporal patterns can be found, such as oscillations at the driver frequency, but also oscillations at fractions of ω, i.e. at multiples of the fundamental period. One may find sequences of period doubling, but also, other multiples of the fundamental period can occur. Furthermore irregular oscillations, i.e. chaotic oscillations are obtained. The occurrence of such specific regions of oscillations does not only depend on A but also on other parameters, e.g. the damping γ. In this way one is led to study regions in parameter space referring to specific kinds of oscillations. However, an important point must not be overlooked. Even for the same set of parameters quite different kinds of oscillations can be found depending on the initial conditions. In more technical terms, that means that in the corresponding phase space of the variables, different basins of attraction may coexist.

Therefore in particular one can study how a specific basin of attraction is changed when control parameters are changed. In such a case, the system may run through a hierarchy of instability points, at each of which a new oscillation sets in, e.g. with periods 2T, 4T, 8T, etc.

In order to get some insight into the behavior of the solutions of such equations, even within a limited range of control parameters and initial conditions, a good deal of effort must be spent in solving the equations numerically. Fortunately quite a different approach has evolved over the past decade which is from a computational point of view much more accessible, namely the study of discrete maps.

6. Discrete Maps

This idea, which was developed by Poincaré around the turn of the century, is as follows. Instead of following up the whole path of trajectories in phase space we consider the crossing points of these trajectories with a plane (or a corresponding hyperplane). Quite often one finds from numerical studies or experimentally that the crossing points can be connected, at least to some approximation, by a line, so that we can label the crossing points, $x_1, x_2, x_3, \ldots$ In the next step one studies how the point x_{n+1} is connected with its previous point x_n. This is described by $x_{n+1} = f(x_n)$. A typical and by now wellknown example is the logistic map

$$x_{n+1} = \alpha x_n (1 - x_n) \tag{6.1}$$

which is shown in Fig.7. When the control parameter α is changed, a specific sequence of stationary solution x_n $n = 1,2,\ldots$ is reached, namely a steady state, a periodic state, period 2, period 4 etc. The parameter α, at which bifurcations from period n to period 2n occur, obey a simple law as was first shown by Grossmann and Thomae [13]. Feigenbaum [14] observed that such a law is universal, i.e. that it is valid for a hole class of one-dimensional maps which are similar in shape to Fig.7. For a proof of this conjecture see Eckmann and Coullet [15]. Of course, certain assumptions must be fulfilled. One of such assumptions is that the Schwartzian derivative is positive. If this requirement is violated, new kinds of patterns can occur, violating the period doubling sequence (see [16]). While the logistic map has been studied by a number of authors purely as a model which actually has highly interesting properties, from a more fundamental point of view two questions arise: 1) How can we construct such maps from experimentally observed data, 2) How are such maps connected with trajectories which result as solutions of differential equations?

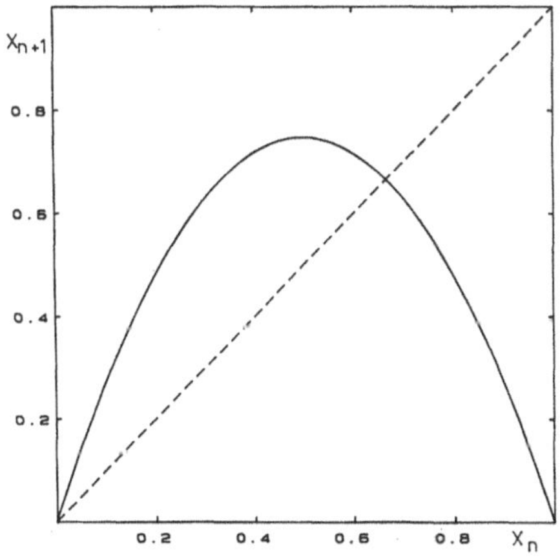

Fig. 7 The logistic maps (eq.(6.1))

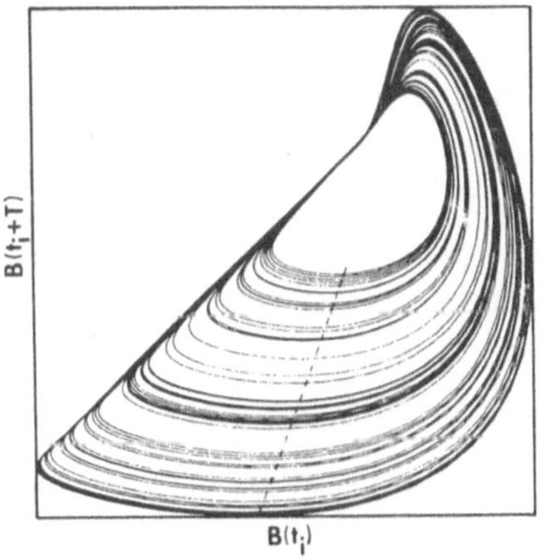

Fig. 8 Strange attractor, constructed from experimental data [19]

1) For the construction of discrete maps and even trajectories from experimental data, the by now well-known method introduced by Ruelle [17] and by Shaw et al [18] is used. Let us assume that the continuous time sequence x(t) of an observable has been measured. Then one introduces x(t+T), x(t+2T)... as additional variables and plots the trajectory within the space spanned by these variables. A typical example is shown in Fig.8 for a chemical reaction [19]. As is seen, in this case the trajectories are well resolved in a three dimensional space. Then one takes a cross-section of the trajectories, e.g. as shown by the dotted line and constructs the Poincaré return map which in the present case is shown in Fig.9. This procedure works for almost all times T and can be based on rigorous grounds by means of embedding theorems [20].

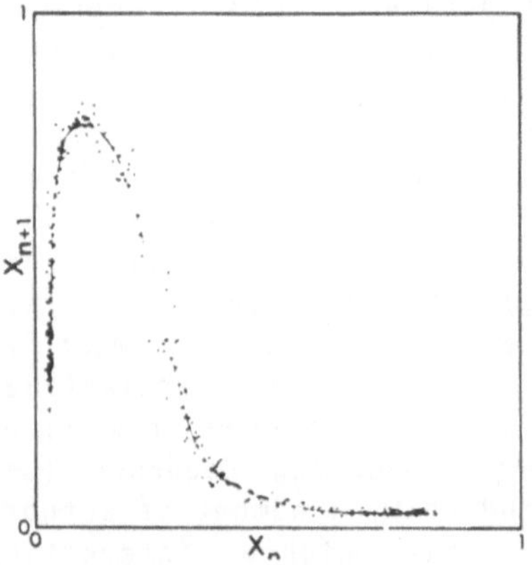

Fig. 9 Return map belonging to Fig. 8 [19]

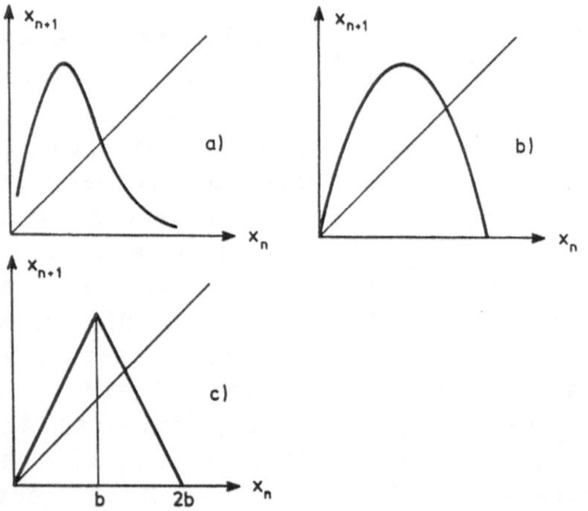

Fig. 10 Examples of one-humped return maps

To my knowledge the point 2, namely the question of how to reconstruct trajectories from the return map, is less well studied. In the following I want to show that a whole class of return maps as shown in Fig.10 are caused by topologically equivalent trajectories. In the following I want to construct a typical representative of such a class of trajectories and also derive the differential equation they obey. Clearly the whole flow of trajectories can be deformed so that different differential equations can give rise to the same class of trajectories. But as I want to show, these trajectories are topologically uniquely determined.

7. Construction of suspensions of discrete maps.

In this section we shall deal with one-humped maps giving rise to chaos. Examples are provided by Fig.10. Since all the essential features of the construction can be studied by means of Fig. 10c we shall deal with this problem. In general each of these maps belongs to a family of maps parametrized by a control parameter. It will be our goal to search for an autonomous system yielding the return maps under discussion. Since in most, if not all cases known in the family of maps, there is at least one in which a periodic oscillation occurs, we shall first assume that the trajectories lie in a plane, say the xy-plane. The part of the map from x = 0...b then appears as a spiral (Fig.11). The non-trivial part of the problem starts when we look at the map from x = b,...,x = 2b. The point x = 2b is mapped onto the origin x = 0. Thus in some way or another it must cross the spiraling trajectories, which is forbidden for autonomous systems. Thus we find the well-known effect that chaos can be produced only by at least three dimensions (a well-known theory states that the only singular points in a plane can be fixed points or limit cycles). Therefore we have to lift the trajectories which originate from the interval $b<x<2b$ into the 3rd dimension. When these trajectories return to the xy-plane they must smoothly join the trajectories starting from the same plane but from the interval $0<x<b$. This requires that the trajectories come in more or less parallel to the y-direction. Since the trajectories stemming from $b<x<2b$ and from $0<x<b$ cannot end up in the same xy-plane (because otherwise the trajectories would merge, in contrast to the assumption of an autonomous system) the trajectories stemming from $0<x<b$ and $b<x<2b$, respectively, must end up in different sections of the xz-plane. We now make the additional, but usual, assumption that the flow is a continuous function of x, y and z (except for one cut line as we shall see below). Up to deformations which retain the topological connections, we choose the part of the xy-plane with $z<0$ as ending part for the trajectories starting from $0<x<b$ and the upper part, $z>0$, for those trajectories starting from $b<x<2b$. Thus we are led to the initial and final locations of the trajectories as indicated in Fig.12. The orientation of the end points which is necessitated by the discrete map and the continuity requirement is indicated by the arrows. The mapping of A onto A′ is generated by trajectories which rotate around the z-axis and describe a compression in the z-direction and an expansion within the xy-plane. The map of B onto B′ can be generated by a continuous

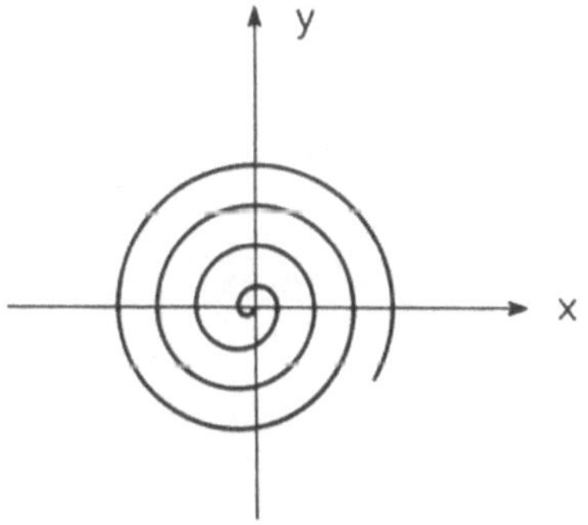

Fig. 11 Compare text

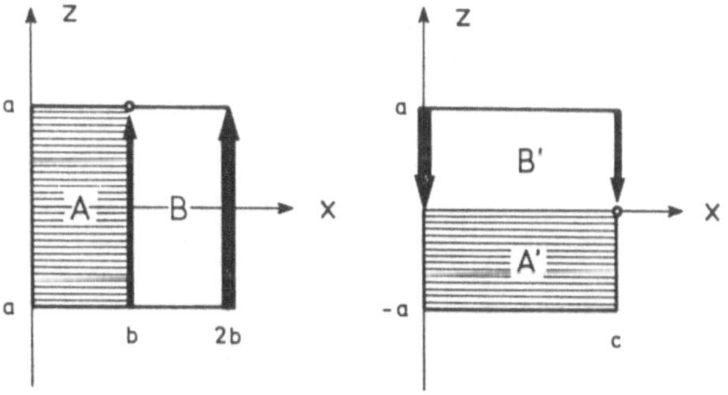

Fig. 12

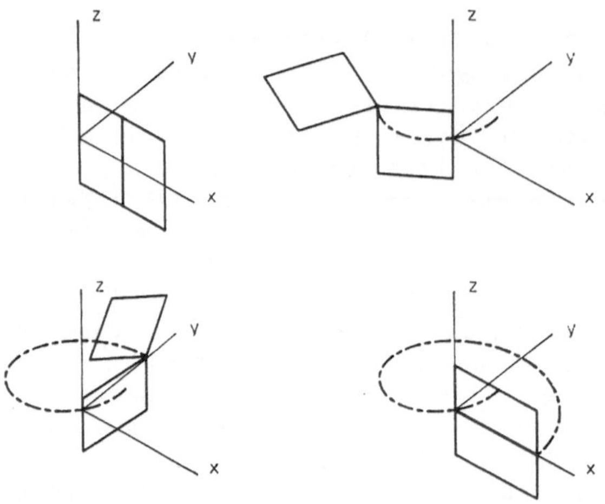

Fig. 13 Compare text

sequence of rotations parallel to the y-axis, by parallel shifts and by rotations around an axis parallel to the z-axis. Also, mixed rotations and parallel shifts are possible, and there exists a huge variety, even including rotations several times. This is possible provided the cut between A and B is total. However, if we require that the point generated from the trajectory starting at (a,b) is transferred along a unique trajectory, the only possible connection under the assumptions made above can be achieved by a sequence of rotations of B as shown in Fig. 13. In this way B is mapped onto B´ via trajectories within a Moebius strip, whose edges are the trajectories shown in Fig. 14. These considerations lead us in a natural way to the Baker´s transformation and its suspension. Eqs. (7.1) describe the flow (cf. also Fig. 15). Here we have assumed expansion factors, i.e. a spreading of the trajectories which is not reflected by the discrete map. Therefore our example shows that the calculation of Ljapunov exponents from discrete maps may give quite different results from the calculation of these exponents using the trajectories. For sake of completeness we mention that chaotic attractors may be also characterized by various kinds of

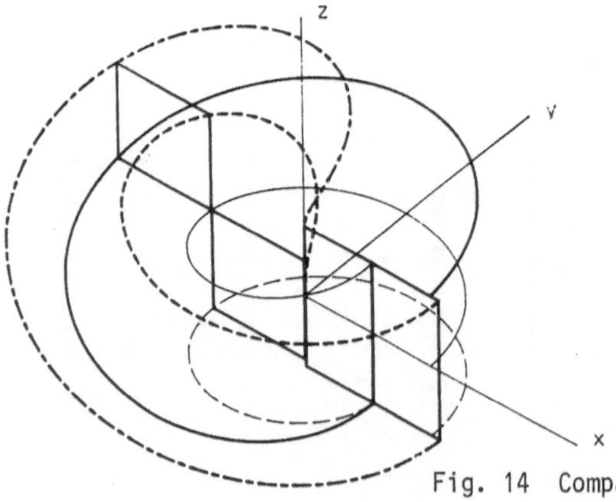

Fig. 14 Compare text

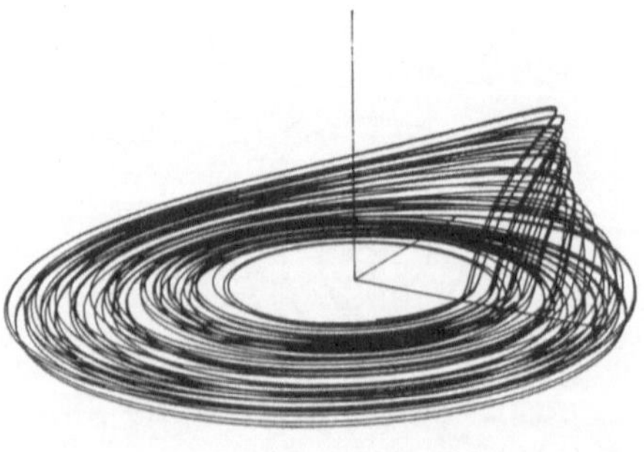

Fig. 15 Compare text

"dimensions" and the Kolmogorov entropy. Unfortunately space does not allow me to discuss these problems here, but these concepts play an important role in studying strange attractors.

8. Routes to Chaos

As the quadratic map teaches us, before the onset of chaos a sequence of period doublings occurs. This has become one famous route to chaos and has been found in quite different kinds of systems such as in fluids, lasers, and electronic devices [21]. Another way is called the Ruelle-Takens route in which a system first undergoes an oscillation at a frequency ω_1; a further oscillation at a second frequency ω_2 occurs in addition, and afterwards chaos should set in in the "generic" case. I have pointed out for a number of years [23] that the question of whether chaos occurs or whether still more incommensurate frequencies in a quasiperiodic motion occur, is not a question of "genericity", but is a quantitative question, so that under suitable circumstances 4, 5 or even more incommensurate frequencies should be found. My prediction, which is based on a detailed theory [23], was in the meantime confirmed by experiments done on the Rayleigh-Bénard instability [24]. I personally expect that similar things may happen in chemical reactions also, for instance in tank reactors, which are not well stirred so that the different oscillators are essentially spatially separated, and interact only weakly with each other. But in principle such phenomena should occur also in well stirred tank reactors. A third way, which has become of great interest, is that via intermittency [25]. In this case chaotic outbursts vary periodically with quiescent behavior of a system. Over the past years great and quite successful efforts have been devoted to finding such routes in various systems. But it seems to me that the universality of some routes is sometimes overstressed. I rather think that the situation is nowadays similar to one before chaotic oscillations came into the focus of research interest. Before that, any records of nonperiodic oscillations were thrown into the waste paper box because the experimenters had not found the fine beautiful periodic oscillations they had been looking for. A similar thing may happen nowadays. Many important results are probably thrown away because they do not show the expected route to turbulence. I think one rather should keep one's eyes open and also keep track of any other routes to chaos.

9. Concluding Remarks

Over the past years it has become possible to classify quite a number of chaotic oscillations and the routes leading to them. Such analysis is to a great deal based on discrete maps. Therefore quite generally one may ask what the relation of discrete maps to differential equations is. The situation if as follows. In a number of concrete examples, depending on the basin of attraction and the routes of parameter changes followed up, specific "universal" discrete maps can be found, which mirror the behavior of the trajectories very well. But at the same time discrete maps can belong to the same original differential equation. Conversely different original differential equations in their specific ranges of parameters and initial conditions can give rise to the same class of discrete maps, e.g. the logistic map. I think at the present moment we must be fully aware of the puzzling variety of phenomena which is still ahead of us. A good deal of work is still to be done to clarify the relations between time-continuous processes decribed by differential equations and the more model-types of discrete maps.

References

1 Pacault, A., and C. Vidal, eds. SYNERGETICS: FAR FROM EQUILIBRIUM. Springer, Berlin, Heidelberg, New York, 1974

Vidal, C. and A. Pacault, eds. NONLINEAR PHENOMENA IN CHEMICAL DYNAMICS, Springer, Berlin, Heidelberg, New York, 1981

For a recent review cf. Vidal, C., in SYNERGETICS. FROM MICROSCOPIC TO MACROSCOPIC ORDER, ed. Frehland, E., Springer, Berlin, Heidelberg, New York, 1984

2 The term "Synergetics" was coined in my lecture at Stuttgart University 1970. See also Haken, H.,and R. Graham, Umschau 6, 191 (1971)

3 Graham, R. and H. Haken, Z.Physik 213, 420 (1968); 237, 31 (1970), DeGiorgio, V., and M.O.Scully, Phys.Rev.A2, 117a (1970). These papers are based on a theoretical treatment of the laser transition (Haken, H. Z.Physik 181, 96 (1964))

4 Haken, H., SYNERGETICS: AN INTRODUCTION. 3rd edition, Springer, Berlin, Heidelberg, New York, 1983
Haken, H., ADVANCED SYNERGETICS, Springer, Berlin, Heidelberg, New York, 1983

5 Kuramoto, Y. CHEMICAL OSCILLATIONS, WAVES, AND TURBULENCE, Springer, Berlin, Heidelberg, New York, 1983

6 see [4]

7 Bestehorn, M. and H. Haken, Phys.Lett. 99A, 265 (1983) and to be published

8 Bergé, P., in CHAOS AND ORDER IN NATURE, ed. H.Haken, Springer, Berlin, Heidelberg, New York, 1981

Gollub, J.P., Communication at the Nobel Symposium on "The Physics of Chaos and Related Phenomena", Graftevallen, Sweden, 1984

9 Sivashinsky, G.I., Acta Astronautica 4, 1177 (1977)

10 Schnaufer, B., and H. Haken, to be published

11 Roessler, O.E., e.g. in SYNERGETICS. A WORKSHOP, ed. H. Haken, Springer, Berlin, Heidelberg, New York, 1977

12 Lorenz, E.N., J.Atmos.Sci. 20, 130 (1963)

13 Grossmann, S., and S. Thomae, Z.Naturfrschg. A32, 1353 (1977)

14 Feigenbaum, M.J., J. Stat.Phys. 19, 25 (1978); Phys.Lett. A74, 375 (1979)

15 Collet, P., and J.P. Eckmann, Iterated Maps on the Interval as Dynamical System, Birkhäuser, Boston 1980

16 Mayer-Kress, G., and H. Haken, Physica D, to be published

17 Ruelle, D., Ann.N.Y.Acad.Sci. 316, 408 (1979)

18 Packard, N.H., J.P. Crutchfield, J.D. Farmer and R.S.Show, Phys.Rev.Lett. 45, 712 (1980)
Crutchfield, J.P., and N.H. Packard, Physica 7D, 201 (1983)

19 Roux, J.C., and H.L. Swinney, in NONLINEAR PHENOMENA IN CHEMICAL DYNAMICS, Vidal, C. and A. Pacault, eds., Springer, Berlin, Heidelberg, New York, 1981

20 Whitney, H., Differentiable Manifolds, Ann.Math.37, 645 (1936)
Takens, F., LECTURE NOTES IN MATHEMATICS, 898, D.A.Rand and L . Young, eds. Springer, Berlin, Heidelberg, New York, 1981

21 see e.g. the workshop "Testing Nonlinear Dynamics", 6.-9.June 1983, at Haverford College

22 Ruelle, D., and F. Takens, Commun. math.Phys. 20, 167 (1971)

23 Haken, H., ADVANCED SYNERGETICS, l.c. [4]

24 Walden, R.W., P.Kolodner, A. Passner, and C.M. Surko, Phys.Rev.Lett. 53, 3, 242 (1984)

25 Pomeau, Y., and P. Manneville, Commun.Math.Phys. 74, 189 (1980)

Part II

Oscillating Reactions and Modelling Problems

New Chemical Oscillators

Irving R. Epstein

Department of Chemistry, Brandeis University, Waltham, MA 02254, USA

1. Introduction

Any systematic analysis of the literature of oscillating chemical reactions (see., e.g., BURGER and BUJDOSO [1]), would show at least two clear trends. First, there has been a rapid expansion in both the number of papers published annually and in the number of groups working on such problems. Second, and more recent has been a diversification in the set of chemical systems studied; an ever increasing fraction of work is being devoted to systems other than the classic Belousov-Zhabotinskii (BZ) reaction.

We present here a brief survey of some of these new chemical oscillators. Somewhat arbitrarily, we exclude biological and heterogeneous systems and define "new" oscillators as those discovered subsequent to the first of these conferences in 1978. We may also refer to "newer" oscillators as those found since the second Bordeaux meeting in 1981.

We first list these new systems, dividing them between those based on halogen chemistry and those in which non-halogen elements play the central role. Next we discuss briefly the search procedures which have been used to develop new oscillators. Some comments on mechanisms for new oscillators are then followed by a summary of a few of the more exotic dynamical phenomena which have been seen in these systems. We conclude with some cautions to seekers of new oscillating systems and a few predictions of things to come.

2. Reactions

A division of oscillating reactions into halogen based and non-halogen is, at least historically speaking, a natural one. All of the "old" oscillators - the BZ [2,3], Bray-Liebhafsky [4], Briggs-Rauscher [5] reactions and their variants - have either bromate or iodate as the indispensable ingredient. The earliest and to date the majority of the new oscillators are also halogen-based, though breakthroughs into other regions of the Periodic Table have recently been made and are discussed below.

2.1 Halogen Based

While the early oscillators were based upon bromate or iodate chemistry, the largest group of new oscillators contains a third oxyhalogen ion, chlorite. Table 1 lists the known chlorite oscillators, all of which were discovered in a stirred tank reactor (CSTR).

The chlorite-iodate-arsenite system marks the breakthrough [6] from accidentally discovered to systematically designed chemical oscillators. The subsequently discovered ClO_2^--I^- system respresents the "minimal" [7] or simplest of what has proved to be a rather large family of iodine-containing chlorite oscillators. The systems with A = BrO_3^- might equally or better be thought of as a family of new bromate oscillators. Some of the wide variety of behavior exhibited by these systems is discussed in Section 5.

Table 1. Chlorite Oscillators in a CSTR

A	Additional Species B	C	Notes	Reference
I-			"Minimal", Subcritical Hopf Bifurcation	[8]
I^-	IO_3^-, MnO_4^- or $Cr_2O_7^-$		Nucleation Induced Transitions	[9]
I^-	Malonic Acid		Batch Oscillations, Spatial Waves	[10]
IO_3^-	H_3AsO_3		First Chlorite Oscillator	[6]
IO_3^-	$Fe(CN)_6^{4-}$, SO_3^{2-}, Ascorbic Acid or $CH_2O{\cdot}SO_2$			[11]
IO_3^-	$S_2O_3^{2-}$		Batch Oscillation	[10]
I_2	$Fe(CN)_6^{4-}$, SO_3^{2-}			[11]
IO_3^-	I^-	Malonic Acid	Batch Oscillations	[10]
IO_3^-	I^-	H_3AsO_3	Tristability	[9]
I^-	BrO_3^-		Birhythmicity, Compound Oscillation, Chaos	[12,13]
BrO_3^-	SO_3^{2-}, $Fe(CN)_6^{4-}$, H_3AsO_3 or Sn^{2+}			[14]
$S_2O_3^{2-}$			Complex Oscillation, Chaos	[15,16]
$(NH_2)_2CS$			Birhythmicity, Complex Oscillation, Chaos	[17]
I^-	I_2	$S_2O_3^{2-}$	Birhythmicity, Tristability, Complex Oscillation, Chaos	[18]
Br^-				[19]
SCN^-			No bistability	[19]

The range of bromate oscillators has expanded considerably since NOYES' [20] proposal of a mechanistically based classification scheme in 1980. While such phenomena as the reaction of bromate and cerium with oxalic acid in the presence of a purging gas stream [21] or the uncatalyzed bromate oscillators [22] just meet our chronological criterion for novelty, they are more in the spirit of the older BZ type oscillators.

The new bromate oscillators really date from the experimental fulfillment of BAR-ELI's [23] prediction that a system consisting of bromate, bromide and cerous (or manganous) ions would show sustained oscillations in a CSTR. This "minimal bromate oscillator" was soon found by ORBÁN et al. [24] and later independently by GEISELER [25]. Figure 1 shows the excellent agreement between the calculations based on the NFT mechanism [23,26] and the actual experimental conditions for oscillation.

From the ability of the minimal bromate system to oscillate in the CSTR, one may infer that the malonic acid in the BZ system could be replaced by an input flow of bromide or indeed of any other species capable of generating bromide from bromate at an appropriate rate. This insight led to the discovery of the bromate-chlorite-reductant [14] and bromate-manganous-reductant [27] families of new bromate oscillators. A similar system [28] consists of bromate , manganous and hypophosphite ions in a batch reactor with a stream of N_2 gas to remove the product bromine.

Another, apparently quite different, bromate oscillator is the bromate-iodide reaction in a CSTR [29]. This reaction poses fascinating mechanistic

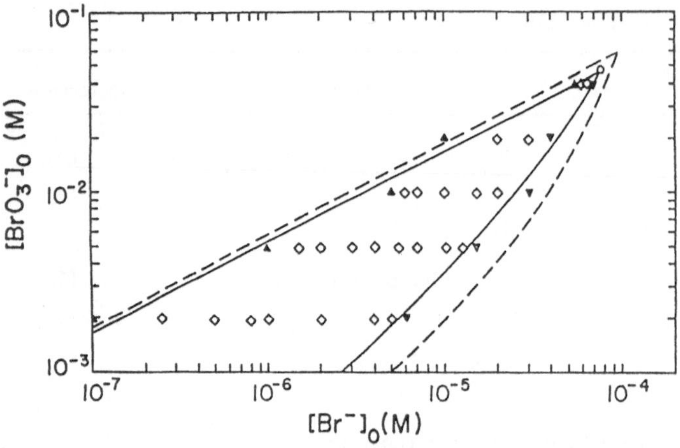

Fig.1 Calculated (---) and experimental (——) phase diagrams for the minimal bromate oscillator in the $[BrO_3^-]_0$-$[Br^-]_0$ plane with $k_0 = 0.0135\ s^{-1}$, $[Mn^{2+}]_0 = 1.02 \times 10^{-4}$ M, $[H_2SO_4]_0 = 1.5$ M. Symbols for experimental points; ▲ , high potential (low bromide) steady state; ▼ , low potential (high bromide) steady state; ◇ , bistability; ○ , oscillation.

problems, and forms part of several coupled oscillator systems which are discussed below.

While no confirmed reports of new iodate oscillators have yet appeared, it seems inevitable that systems based on the bistable arsenite-iodate [30,31] or related Landolt-type [32] reactions will soon be made to oscillate. The iodide-peroxydisulfate-oxalate reaction discussed by Chopin-Dumas elsewhere in this volume may be the first of the new iodate oscillators.

2.2 Non-halogen

All homogeneous non-halogen oscillators fit into our "newer" category, being of quite recent vintage. JENSEN [33] discovered the first non-biological organic-based oscillator, the air oxidation of benzaldehyde catalyzed by cobalt and bromide, in 1982. Although the system contains Br^-, we classify it as "organic-based" because the mechanism [34] revolves primarily about the changing oxidation states of carbon, while bromide plays a peripheral role. The Dupont group [35] has recently reported another system of this type, with cyclohexanone in place of benzaldehyde.

Sulfur chemistry has provided the other source of new non-halogen oscillators. BURGER and FIELD [36] found oscillations in the reaction of methylene blue, sulfide, sulfite and dissolved oxygen. The system, as illustrated in the phase diagram of Fig. 2, does not appear to exhibit bistability or a cross-shaped phase diagram, though it may instead possess a continuous range of steady states [37].

A second sulfur-based oscillator, the hydrogen peroxide-sulfide reaction recently discovered by ORBÁN and EPSTEIN [38] shows, in contrast, an almost classic cross-shaped phase diagram, Fig. 3. The Burger-Field system is the first to oscillate in basic solution, while, as we see in Fig. 4, the H_2O_2-S^{2-} oscillations go from acidic to basic pH, making possible dramatic color effects with acid-base indicators.

3. Search Proceedures

Recent progress in discovering new chemical oscillators has been aided immeasurably by the development and use of systematic search procedures. By far the most productive such technique has the "cross-shaped phase diagram"

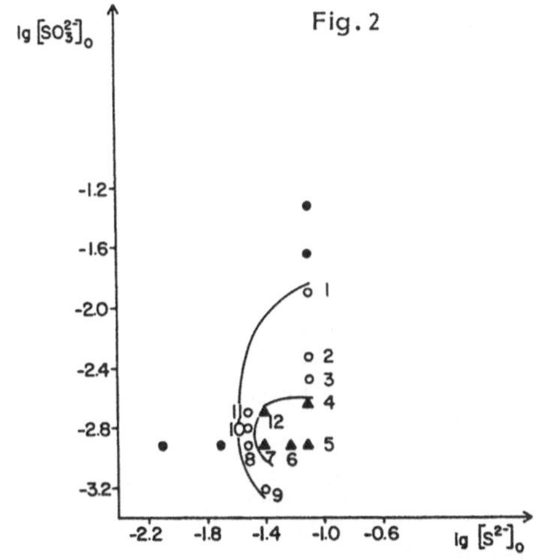

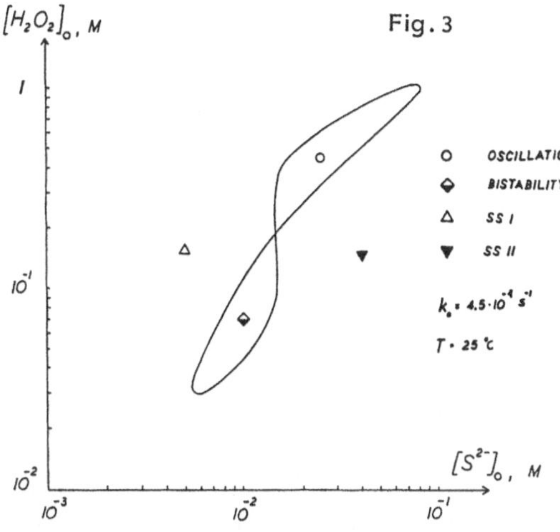

Fig. 2 Experimental phase diagram for the methylene blue oscillator in the $[S^{2-}]_0 - [SO_3^{2-}]_0$ plane. $[MB^+]_0 = 3.0 \times 10^{-6}$ M, $[OH^-]_0 = 5 \times 10^{-2}$ M, $[O_2]_0 = 1.5 \times 10^{-4}$ M, temperature:20 ± 0.1°C. ●, stable steady state; O, transient oscillations; ▲ sustained oscillations. Reciprocal residence times k_0 ($\times 10^3$ s^{-1}) (1) 1.9; (2) 3.5 and 3.1; (3) 3.1; (4) 2.3 and 1.9; (5) 2.0: (6) 2.0; (7) 1.6 and 1.3; (8) 1.3 and 1.0; (9) 1.5; (10) 1.5; (11) 1.3; (12) 1.3.

Fig. 3 Experimental phase diagram for the H_2O_2-S^{2-} system.

approach based on model calculations of BOISSONADE and DE KEPPER [39] and first brought to fruition in the discovery of the chloride-iodate-arsenite oscillator [6].

In this method [40], one starts from an autocatalytic reaction and seeks conditions under which the system shows bistability in a CSTR. The key step is then to find a "feedback species" which modifies the effective value of one of the system constraints (usually an input concentration) by quite different amounts on the two bistable branches. If this effect is in the right

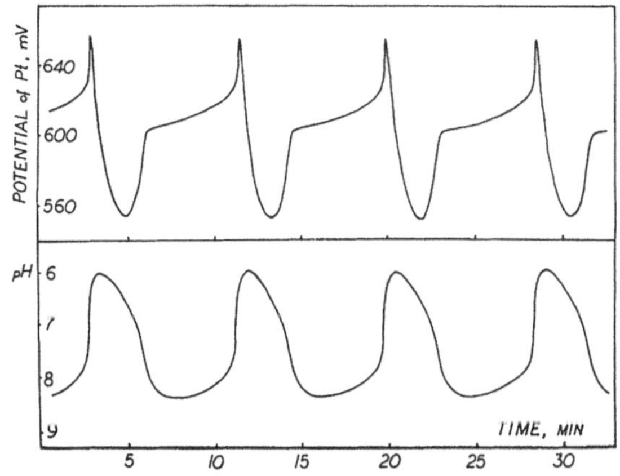

Fig. 4 Oscillations in the pH and redox potential of a sulfide-hydrogen peroxide system with $[H_2O_2]_0 = 0.4$ M, $[S^{2-}]_0 = 0.0167$ M, $[H_2SO_4]_0 = 0.001$ M, $k_0 = 6 \times 10^{-4}$ s^{-1}, T = 25°C.

direction, and if the feedback reaction is slow compared with the relaxation of the unperturbed system to the steady states, then adding enough of the feedback species to the input flow will destabilize the steady states, leading to oscillation. Such systems are characterized by cross-shaped phase diagrams like that of Fig. 3.

In particularly fortunate cases the feedback species is generated internally, and one need only manipulate the constraints of the bistable system (e.g., the flow rate) until the bistable region closes and oscillation appears beyond the critical point. The chlorite-iodide oscillator [8] is one example of such a system.

An alternative procedure, which was exploited by GEISELER [25] in his search for the minimal bromate oscillator, is the linear gradient technique. In this method, one input concentration is varied continuously in time by employing two reservoirs, one containing solution and the other pure solvent.

Still another approach, to which the discovery of the methylene blue oscillator [36] may be attributed, is a combination of careful thought, chemical intuition and perseverance. Such tactics have probably become more attractive with the growing awareness that so many oscillators exist.

Finally, chance, which played such a major role in the early development of the field is still an important source of new oscillators. Like the BZ [2] and BL [4] reactions before it, the JENSEN [33] oscillator was discovered accidentally.

4. Mechanism

Perhaps the major factor which led to the dominant role played by the BZ reaction was the success of the FIELD-KÖRÖS-NOYES [41] mechanism in

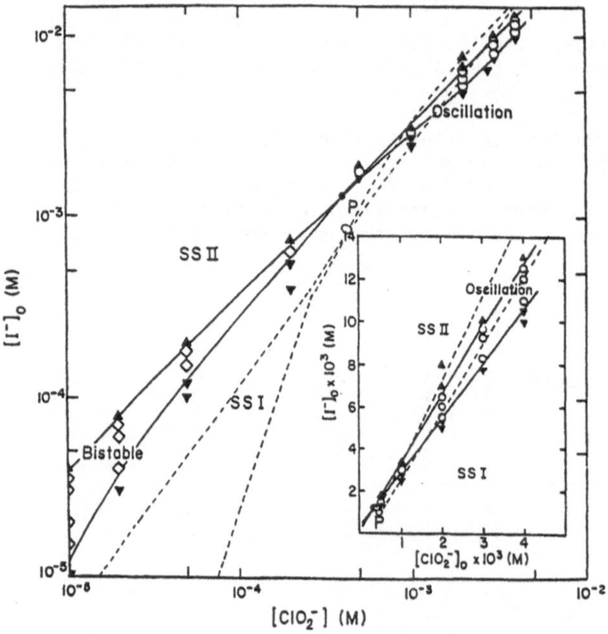

Fig. 5 Phase diagram in the $(ClO_2^-)_0$ - $(I^-)_0$ plane for the chlorite-iodide reaction with pH = 2.04, k_0 = 1.1 x 10^{-3} s^{-1}, T = 25°C. ▲, high iodide steady state; ▼, low iodide steady state; ○, oscillatory state. Solid lines are boundaries of experimentally determined regions; dashed lines show calculated boundaries. Inset shows data near cross-point P on a linear scale.

demonstrating that chemical oscillation could be explained in terms of a reasonable set of elementary steps and associated rate constants. It is therefore appropriate to give a brief summary of the status of the new oscillators with respect to mechanism.

With the single exception of the chlorite-bromate-reductant systems [14], for which a mechanism has been developed by joining the NFT model [26] with THOMPSON'S [42] mechanism for the BrO_3^--ClO_2^- reaction, no mechanism has been published for any chlorite oscillator. In Table 2, we give a recently developed mechanism [43] for the minimal chlorite-iodide oscillator. Of special significance is the fact that it contains no radical species, but rather the binuclear intermediate ClO_2. Calculations with this mechanism give excellent agreement with a wide variety of experimental results. One example is given in Fig. 5.

The mechanisms of sulfur oscillators remain a mystery at this time. Even the autocatalytic species, if any, has yet to be identified.

The Dupont group [34] has developed a radical chain mechanism for the Jensen oscillator which gives good agreement with the observed oscillation. This and work on related organic systems is discussed elsewhere in this volume.

Table II. A Mechanism for the Chlorite-Iodide Reaction

Number	Reaction	k_j^a	k_{-j}^a
(M1)[b]	$H^++ClO_2^-+I^- = HOCl+OI^-$	$1x10^3$	c
(M2)	$HOCl+I^- = Cl^-+HOI$	$2x10^8$	c
(M3)	$H^++HOI+I^- = I_2+H_2O$	$3.1x10^{13}$	2.2
(M4)	$H^++HOI+ClO_2^- = HIO_2+HOCl$	$1x10^6$	c
(M5)	$HIO_2+I^-+H^+ = 2HOI$	$2x10^9$	c
(M6)	$HOI+HOCl= HIO_2+Cl^-+H^+$	$2x10^8$	c
(M7)	$I_2+ClO_2^- = IClO_2+I^-$	$1.1x10^1+\frac{1.0x10^{-2}}{[H^+]_0}$	$1x10^5\ k_{M7}$
(M8)	$IClO_2+H_2O = HIO_2+HOCl$	$1x10^3$	c
(M9)[b]	$IClO_2+I^-+H_2O = HOI+OI^-+HOCl$	$1.3x10^5$	c
(M10)	$HOCl+HIO_2 = Cl^- + IO_3^-+2H^+$	$2x10^8$	c
(M11)	$2H^++IO_3^-+I^- = I_2O_2+H_2O$	$4x10^9$	$4x10^4$
(M12)	$I_2O_2+H_2O = HOI+HIO_2$	$8x10^{-2}$	$1x10^5$
(M13)[b]	$I_2O_2+I^-+H_2O = 2HOI+OI^-$	$6x10^3$	c
Rapid equilibria[d]			
(M1a)	$H^++OI^- = HOI$		
(M4b)	$HClO_2 = H^++ClO_2^-$		

[a] All concentrations in M, times in s.

[b] Followed immediately by (M1a)

[c] Set to zero in the calculation

[d] Assumed instantaneous

5. New Phenomena

One might well ask whether the advent of new types of oscillators has brought with it the discovery of new dynamical phenomena. Of course, the BZ reaction itself exhibits an enormous variety of behavior including chaos, spatial waves, multistability and complex oscillation. Therefore there may not be many new phenomena left to discover (this of course, is always a dangerous view to take).

The new systems have, however, shown that none of these phenomena is unique to the BZ reaction - a logically trivial, but psychologically significant point. Furthermore, in some cases they provide variants or extensions of phenomena which also appear in the BZ reaction. For example, though the BZ system exhibits bistability, it has not been found to give tristability, a behavior which is illustrated for the AsO_3^{3-} - ClO_2^- - I^- - IO_3^- system [9] in Fig. 6. The period adding sequence shown in Fig. 7 for the ClO_2^--$S_2O_3^{2-}$ reaction [16] is similar to, but more intricate than one reported earlier [47] for the BZ reaction.

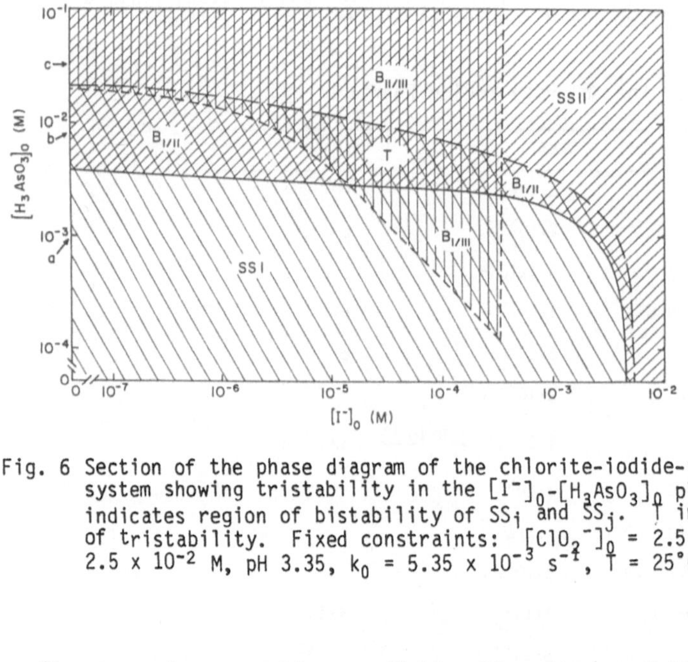

Fig. 6 Section of the phase diagram of the chlorite-iodide-iodate-arsenite system showing tristability in the $[I^-]_0$-$[H_3AsO_3]_0$ plane. $B_{i/j}$ indicates region of bistability of SS_i and SS_j. T indicates region of tristability. Fixed constraints: $[ClO_2^-]_0$ = 2.5×10^{-3} M, $[IO_3^-]_0$ = 2.5×10^{-2} M, pH 3.35, k_0 = 5.35×10^{-3} s^{-1}, T = 25°C.

The range of new reactions available offers the possibility of creating systems of chemically coupled oscillators [48]. For example, Fig. 8 illustrates the phenomenon of birhythmicity in the ClO_2^--BrO_3^--I^- reaction [12], where the ClO_2^--I^- and BrO_3^--I^- oscillatory subsystems are coupled via the common species I^-. The phase diagram in Fig. 9 demonstrates some of the complexity possible in the dynamics of such systems.

6. Some Cautions

One should not be misled from what appears above into thinking that any chemical reaction of some complexity will oscillate when run in a CSTR if only one is patient and careful enough. We have spent considerable time trying to design new oscillators based on the bistable Fe(II)-HNO_3 reaction [49]. While we have learned a great deal about nitrate chemistry and now believe we understand the problem, we have not been able to produce a genuine nitrate

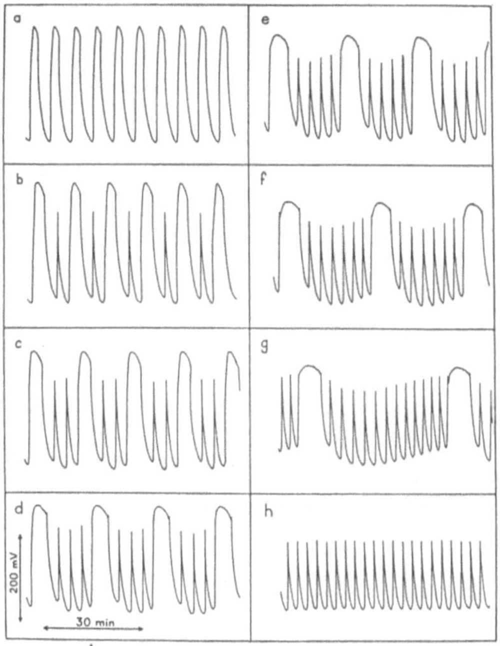

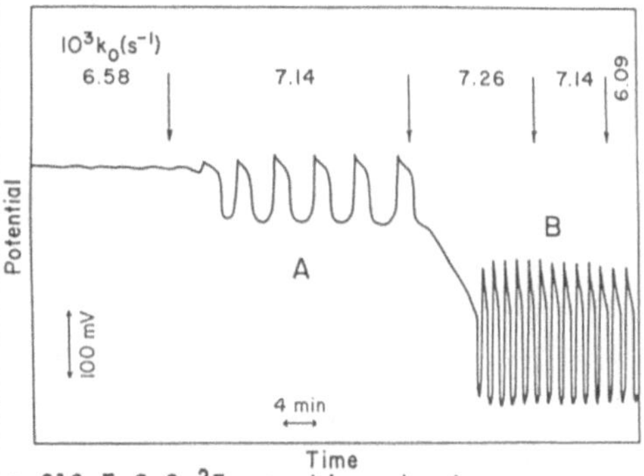

Fig. 7 Complex periodic oscillations in the $ClO_2^- - S_2O_3^{2-}$ reaction showing one large (L) and n small (S) amplitude oscillations (L + nS) per cycle with $[ClO_2^-]_0 = 5 \times 10^{-4}$ M, $[S_2O_3^{2-}]_0 = 3 \times 10^{-4}$ M, pH = 4, T = 25°C, residence time τ. (a) L, τ = 5.9 min; (b) L + S, τ = 9.5 min; (c) L +2S, τ = 10.8 min; (d) L + 3S, τ = 13.5 min; (e) L+4S, τ = 15.8 min; (f) L + 6S, τ = 20.6 min; (g) L + 12S, τ = 26.3 min; (h) S, τ = 47.3 min.

Fig. 8 Birhythmicity in the chlorite-bromate-iodide system. Potential is that of Pt electrode vs. Hg/Hg_2SO_4 reference electrode. At times indicated by the arrows, flow rate is changed. Flow rate in each time segment (measured as reciprocal residence time k_0) is shown at top. Note that A state and B state are both stable at $k_0 = 7.14 \times 10^{-3}$ s^{-1}. Fixed constraints: T = 25°C, $[I^-]_0 = 6.5 \times 10^{-4}$ M, $[BrO_3^-]_0 = 2.5 \times 10^{-3}$ M, $[ClO_2^-]_0 = 1.0 \times 10^{-4}$ M, $[H_2SO_4]_0$ = 0.75 M.

oscillator. We have, however, had several tantalizing failures. In one case, using H_2O_2 as the feedback species, we observed somewhat irreproducible large amplitude oscillations caused by O_2 bubbles produced by peroxide decomposition catalyzed by the walls of the inlet tubes. Other false alarms have resulted from "pump oscillations", i.e., extreme sensitivity to and amplification by the system of small volume changes due to the peristaltic pumping. Pump oscillations are quite reproducible, but characteristically show a frequency which is directly proportional to the pump speed.

Another factor to keep in mind is the stirring rate. Although one generally assumes that mixing is essentially perfect in the CSTR, recent experiments by ROUX et. al. [50] on the chlorite-iodide reaction have shown marked dependence of the stability of steady states on the rate of stirring even at such "high" speeds as 700 rpm. Similar effects are to be expected for oscillatory states.

7. The Future

Where can one expect new advances to be made in the next several years? It seems likely that new oscillators involving elements other than halogens,

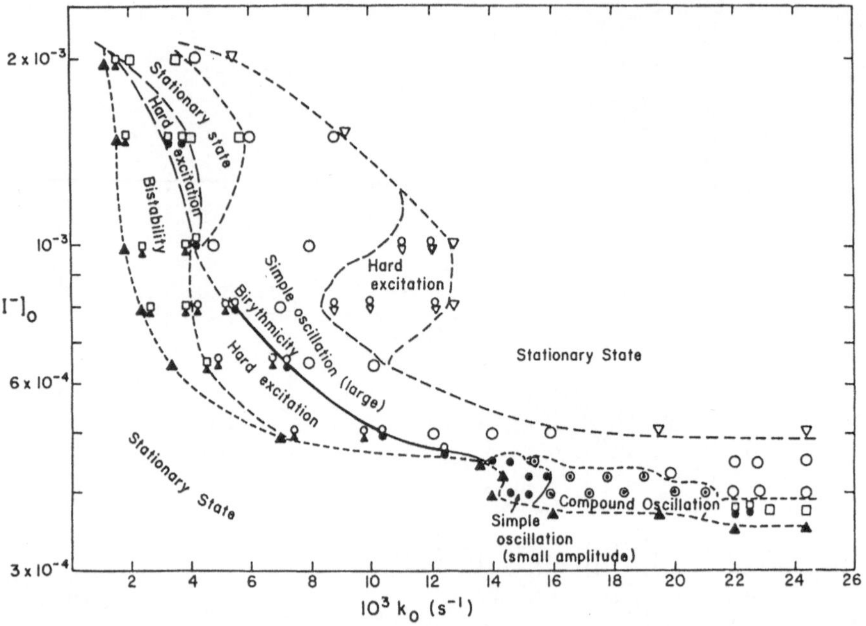

Fig. 9 Phase diagram in the k_0-$[I^-]_0$ plane (fixed constraints as in Fig. 8): ● oscillatory state A; ○, oscillatory state B; ◉, compound oscillation C; 8, birhythmicity, A and B both stable; ▲, high-potential stationary-state; ▽, low-potential stationary state; □, intermediate-potential steady state. Combinations of two symbols imply bistability between the corresponding states.

carbon or sulfur will be sought and found. The transition metals and nitrogen seem like the most promising places to look. One may expect further progress in the area of mechanisms, particularly for sulfur-based oscillators and for the more complex chlorite systems which undoubtedly involve ClO_2 and associated radicals. Experiments involving coupled oscillators will produce increasingly complex dynamical phenomena to ponder. So, we can expect many more new systems showing more complex phenomena, and a deeper understanding of them.

Acknowledgments

My own work on new chemical oscillators has benefited immeasurably from the collaboration of a host of able co-workers, most notably Mohamed Alamgir, György Bazsa, Christopher Dateo, Patrick De Kepper, Kenneth Kustin, Jerzy Masełko and Miklós Orbán. Our work has been supported by grants from the National Science Foundation and by international grants from the NSF, NATO and the Hungarian Academy of Sciences.

References

1. M. Burger and E. Budjoso, in R. J. Field and M. Burger, eds.: Oscillations and Traveling Waves in Chemical Systems (Wiley, New York, 1984).

2. B. P. Belousov: Sb. Ref. Radiats. Med., Medgiz, Moscow, 1959, p. 145.

3. A. M. Zhabotinskii: Biofizika 9, 30 (1964).

4. W. C. Bray: J. Am. Chem. Soc. 43, 1262 (1921).

5. T. S. Briggs and W. C. Rauscher: J. Chem. Educ. 50, 496 (1973).

6. P. De Kepper, I. R. Epstein and K. Kustin: J. Am. Chem. Soc. 103, 2133 (1981).

7. I. R. Epstein and M. Orbán, in R. J. Field and M. Burger, eds.: Oscillations and Traveling Waves in Chemical Systems (Wiley, New York, 1984).

8. C. E. Dateo, M. Orbán, P. De Kepper and I. R. Epstein: J. Am. Chem. Soc. 104, 504 (1982).

9. M. Orbán, C. E. Dateo, P. De Kepper and I. R. Epstein: J. Am. Chem. Soc. 104, 5911 (1982).

10. P. De Kepper, I. R. Epstein, K. Kustin and M. Orbán: J. Phys. Chem. 86, 170 (1982).

11. M. Orbán, P. De Kepper, I. R. Epstein and K. Kustin: Nature 292, 816 (1981).

12. M. Alamgir and I. R. Epstein: J. Am. Chem. Soc. 105, 2500 (1983).

13. J. Maselko, M. Alamgir and I. R. Epstein: submitted for publication.

14. M. Orbán and I. R. Epstein: J. Phys. Chem. 87, 3212 (1983).

15. M. Orbán, P. De Kepper and I. R. Epstein: J. Phys. Chem. 86, 431 (1982).

16. M. Orbán and I. R. Epstein: J. Phys. Chem. 86, 3907 (1982).

17. M. Alamgir and I. R. Epstein: submitted for publication.

18. J. Maselko and I. R. Epstein: J. Phys. Chem., in press.

19. M. Alamgir and I. R. Epstein: to be published.

20. R. M. Noyes: J. Am. Chem. Soc. 102 4644 (1980).

21. Z. Noszticzius and J. Bodiss, J. Am. Chem. Soc. 101, 3177 (1979).

22. M. Orban and E. Koros: J. Phys. Chem. 82, 1672 (1978).

23. K. Bar-Eli, in C. Vidal and A. Pacault, eds.: Nonlinear Phenomena in Chemical Dynamics (Springer, Berlin, 1981), pp 228-239.

24. M. Orbán, P. De Kepper and I. R. Epstein: J. Am. Chem. Soc. 104, 2657 (1982).

25. W. Geiseler: J. Phys. Chem. 86, 4394 (1982).

26. R. M. Noyes, R. J. Field and R. C. Thompson: J. Am. Chem. Soc. 93, 7315 (1971).

27. M. Alamgir, M. Orbán and I. R. Epstein: J. Phys. Chem. 87, 3725 (1983).

28. L. Adamčiková and P. Ševčík: Int. J. Chem. Kinet. 14, 735 (1982).

29. M. Alamgir, P. De Kepper, M. Orbán and I. R. Epstein: J. Am. Chem. Soc. 105 2641 (1983).

30. P. De Kepper, I. R. Epstein and K. Kustin: J. Am. Chem. Soc. 103, 6121 (1981).

31. G. A. Papsin, A. Hanna and K. Showalter: J. Chem. Phys. 85, 2575 (1981).

32. J. Eggert and B. Scharnow: Elektrochem. Z. 27, 455 (1921).

33. J. H. Jensen: J. Am. Chem. Soc. 105, 2639 (1983).

34. M. G. Roelofs, E. Wasserman, J. H. Jensen and A. E. Nader: J. Am. Chem. Soc. 105, 6329 (1983).

35. J. D. Druliner and E. Wasserman: presented at American Chemical Society meeting, Philadelphia, August 27-31, 1984.

36. M. Burger and R. J. Field: Nature, 307, 720 (1984).

37. R. J. Field: private communication.

38. M. Orbán and I. R. Epstein: to be published.

39. J. Boissonade and P. De Kepper: J. Phys. Chem. 84, 501 (1980).

40. I. R. Epstein, K. Kustin, P. De Kepper and M. Orbán: Scientific American, 248(3), 112 (1983).

41. R. J. Field, E. Körös and R. M. Noyes: J. Am. Chem. Soc. 94, 8649 (1972).

42. R. C. Thompson: Inorg. Chem. 12, 1905 (1973).

43. I. R. Epstein and K. Kustin, submitted to J. Am. Chem. Soc.

44. Z. Noszticzius, H. Farkas and Z. A. Schelly: J. Chem. Phys. 80, 6062 (1984).

45. R. M. Noyes: J. Chem. Phys. 80, 6071 (1984).

46. J. J. Tyson: J. Chem. Phys. 80, 6079 (1984).

47. J. S. Turner, J.-C. Roux, W. P. McCormick and H. L. Swinney: Phys. Lett. 85A, 9 (1981).

48. M. Alamgir and I. R. Epstein: J. Phys. Chem. 88, 2848 (1984).

49. M. Orbán and I. R. Epstein: J. Am. Chem. Soc. 104, 5918 (1982).

50. J.-C. Roux, P. De Kepper and J. Boissonade: Phys. Lett. 97A, 168 (1983).

The Mechanism of the Oscillating Air Oxidation of Benzaldehyde

James H. Jensen[1], Mark G. Roelofs[2], and E. Wasserman[2]

E.I. du Pont de Nemours and Company, Inc., Experimental Station
Wilmington, DE 19898, USA

During the oxygen or air oxidation of benzaldehyde catalyzed by cobalt and bromide several measureable parameters oscillate including: redox potential, absorbance at 620 nm, oxygen concentration, rate of benzaldehyde disappearance and free radical concentration. At 55-100°C in a 90/10 acetic acid/water mixture the oscillations will last for many hours[1]. A typical cycle is shown in Figure 1.

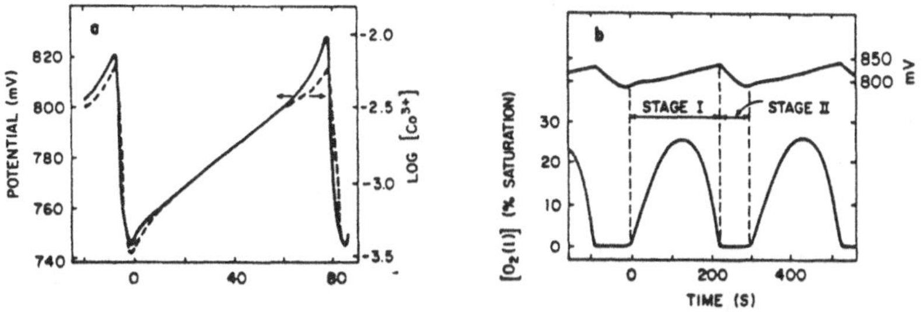

Fig. 1. (a) Oscillations of the platinum electrode potential vs. Ag/AgCl reference electrode (——) and log [Co(III)] determined by optical absorption at 620 nm (---). A 150-mL solution contained initially 500 mM PhCHO, 10 mM $Co(AcO)_2$, 5 mM NaBr, and 90/10 w/w $AcOH/H_2O$ as solvent. The temperature was 70°C, an O_2 pressure of 580 torr was maintained above the solution, and the magnetic stirring speed was 400 rpm. (b) A comparison of the Pt electrode response (top trace) with the dissolved oxygen concentration (bottom trace) expressed in terms of the percentage of saturation with O_2 at a partial pressure of 580 torr. We estimate that 100% on this scale corresponds to $[O_2(l)] = 5 \pm 2$ mM. Conditions are as in (a), but the solution contained initially 750 mM pHCHO, 20 mM $Co(AcO)_2$, and 2 mM NaBr; O_2 was introduced at a flow rate of 20 mL/min through a frit immersed in the liquid. (Figure from [3].)

Oscillations of the bromide ion concentration are barely detectable, ($\Delta[Br^-]<0.2[Br^-]_{average}$), in contrast to the Belousov-Zhabotinsky and other bromate-driven oscillators where changes in [Br-] of several orders of magnitude are an essential feature of the mechanism[2]. Therefore this reaction appears to belong to a different class of oscillating reactions. A second major difference is that the reaction is both homogeneous and heterogeneous; for the latter, a major feature is the dissolution of a gas in the liquid.

We have proposed an 11 step mechanism[3] which accounts for the oscillations, and provides reasonable agreement with all of the experimental observations.

---

[1]Agricultural Chemicals Department
[2]Central Research and Development Department

Initiation steps R1 and R2 use a Co(III) dimer, $Co(III)_2$, to produce a benzoyl radical at a rate proportional to the product of [PhCHO], [Br^-] and [$Co(III)_2$]. During Stage I, benzoyl radical is rapidly intercepted by oxygen in R3 to form the perbenzoyl radical. R6 continues the radical chain, while R7 terminates it. R3, R6 and R7 are well established[4,5]. The rapid oxidation of Co(II) by perbenzoic acid (R9) then completes the autocatalytic generation of $Co(III)_2$.

It is only when oxygen becomes severely depleted that Stage II begins, the benzoyl radical lifetime increases, and its concentration rises sharply. During Stage II, $Co(III)_2$ is now reduced by benzaldehyde through the intermediacy of benzoyl radicals in R4 and R8. The sequence R5 followed by 2xR8 provides an alternate path for the perbenzoxy radicals during Stage II which does not generate more Co(III).

Although the radicals have been written in the standard form, they may exist in solution complexed with Co ion, leading to unusual chemical behavior, e.g. (R5).

$$Co(III)_2 + Br^- \rightleftharpoons Co(III)_2Br^- \quad (R1)$$

$$Co(III)_2Br^- + PhCHO \rightarrow PhCO^\bullet + Br^- + CO(III) + Co(II) + H^+ \quad (R2)$$

$$PhCO^\bullet + O_2(l) \rightarrow PhCO_3^\bullet \quad (R3)$$

$$PhCO^\bullet + Co(III)_2 + H_2O \rightarrow PhCO_2^\bullet + 2Co(II) + 2H^+ \quad (R4)$$

$$PhCO^\bullet + PhCO_3^\bullet \rightarrow 2PhCO_2^\bullet \quad (R5)$$

$$PhCO_3^\bullet + PhCHO \rightarrow PhCO_3H + PhCO^\bullet \quad (R6)$$

$$PhCO_3^\bullet + Co(II) + H^+ \rightarrow PhCO_3H + Co(III) \quad (R7)$$

$$PhCO_2^\bullet + PhCHO \rightarrow PhCO_2H + PhCO^\bullet \quad (R8)$$

$$PhCO_3H + 2Co(II) + 2H^+ \rightarrow PhCO_2H + Co(III)_2 + H_2O \quad (R9)$$

$$Co(III) + Co(III) \rightarrow (Co(III))_2 \quad (R10)$$

$$O_2(g) \rightleftharpoons O_2(l) \quad (R11)$$

If [$O_2(l)$] is always appreciably less than saturation, the rate of mass transfer of oxygen from the gas phase into the liquid determines the average rate of oxidation of PhCHO. The average rate of disappearance of PhCHO during a batch reaction should be, and is, zero order in [PhCHO]. Within a given cycle; however, [PhCHO] shows a rapid decrease during Stage II, supporting a reduction of $Co(III)_2$ by benzaldehyde, R4 and R8.

The rate equations for R1-R11 were integrated using a Gear algorithm[6], with the results shown in Figure 2.

These results were sufficiently encouraging that a sensitivity analysis for this model was done. We chose to compare the calculated period of oscillation in a CSTR reactor to experimentally measured periods as a function of cobalt, bromide and benzaldehyde concentrations.

The effect of varying bromide concentration is shown in Figure 3. The proposed mechanism fits the experimental data quite well at high concentrations of bromide. The experimental points are within the uncertainty of the calculated line. However, experimentally the oscillations stop below about 3 mM bromide, whereas the calculated line predicts that the oscillations should continue down to about 0.02 mM Br and the period should go through a maximum and decrease.

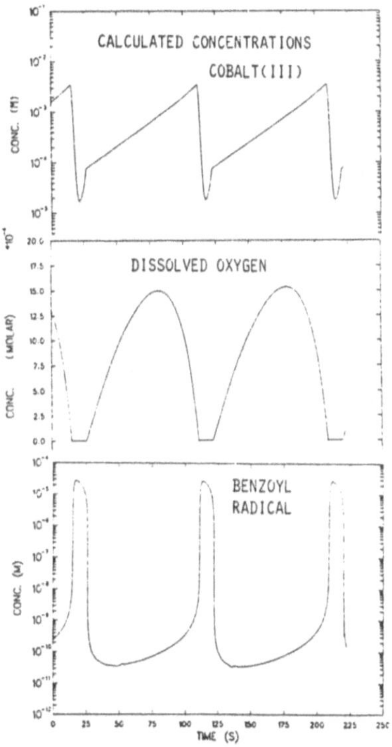

Fig. 2. Calculated concentration oscillations in (a) $[Co(III)_2]$, (b) $[O_2(l)]$, and (c) $[PhCO^{\bullet}]$.

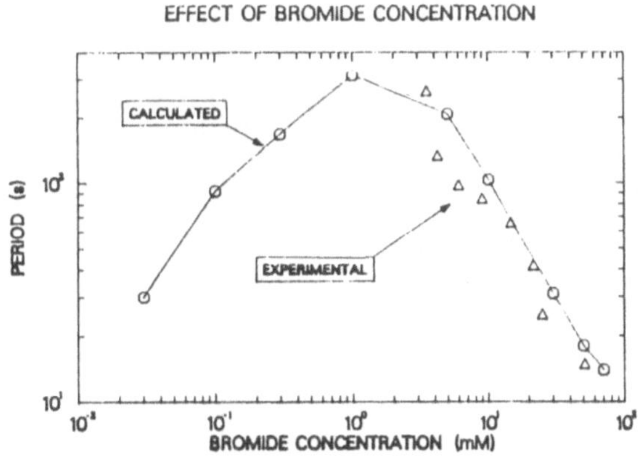

Fig. 3. Calculated, O, and experimental, + (ramp), period of oscillation as a function of bromide ion concentration.

Stage II of the model does not involve Br^-; however, experimentally Stage II is lengthened at low $[Br^-]$. This observation suggests that at low $[Br^-]$ additional reactions are required in the model.

The model predictions for variations of cobalt concentration are shown in Figure 4. The experimentally determined periods do not fit the calculated line although maxima occur in both curves. This difference may be due to complexes of Co(III) and Co(II) particularly at low cobalt which we have not taken into account in the model.

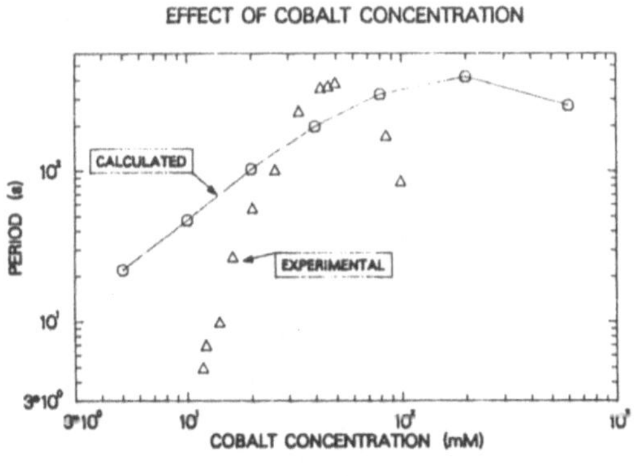

Fig. 4. Calculated, O, and experimental, + (ramp), period of oscillation as a function of cobalt concentration.

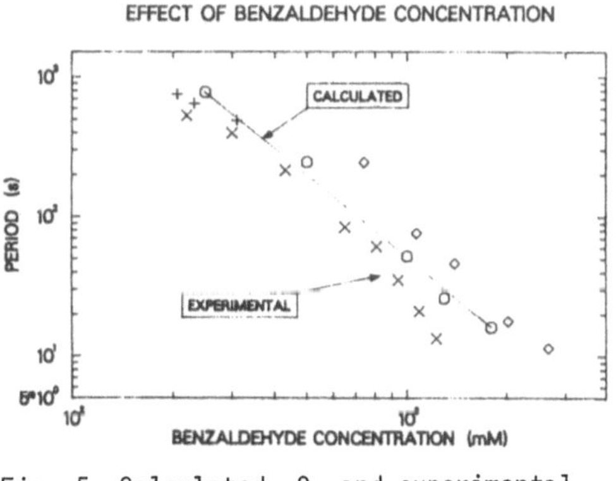

Fig. 5. Calculated, O, and experimental, + 0-400 mM, ◇ 200-1500 mM, x 0-3000 mM PhCHO (ramp), period of oscillation as a function of benzaldehyde concentration.

The agreement between the calculated and measured periods for benzaldehyde concentration variations is satisfactory (Figure 5). The boundaries between oscillation and steady state occur quite close to the prediction at both high and low concentrations. The slope of the log-log plot is approximately -2, which is correct for the kinetic model.

The kinetic model predicts that radicals are produced in an oscillatory manner. We have detected two radicals in Stage II by EPR using a flow system. This appears to be the first time that organic radicals have been detected in an oscillating chemical reaction, although they have been proposed for other systems. We have not been able to confirm the identity of the radicals, but reasonable conjectures are possible.

In conclusion, it appears that the original 11 step mechanism explains the gross oscillatory behavior of this system reasonably well. Many fine details remain to be filled in, especially the role of bromide as a ligand, the effect of Co complexes, and the relatively narrow range of water concentration (5-20%) in which the reaction shows oscillatory behavior.

REFERENCES

1. J. H. Jensen, J. Am. Chem. Soc. 1983, 105, 2639-2641.

2. R. J. Noyes, M. Am. Chem. Soc. 1980, 102, 4644-4649.

3. M. G. Roelofs, E. Wasserman, J. H. Jensen and A. E. Nader, J. Am. Chem. Soc. 1983, 105, 6329-6330.

4. R. A. Sheldon, J. K. Kochi, "Metal-Catalyzed Oxidations of Organic Compounds", Academic Press: New York, 1981.

5. C. F. Hendriks, H. C. A. van Beek, P. M. Heetjes, Ind. Eng. Chem. Prod. Res. C. F. Dev. 1978, 17, 260-264.

6. R. N. Stabler, J. Chesick, Int. J. Chem. Kinet. 1978, 10, 461-469.

Analysis of Elementary Steps in Oscillating Chemical Reactions

Donald W. Boyd, Irving R. Epstein, Kenneth Kustin, and Oscar Valdes-Aguilera
Department of Chemistry, Brandeis University, Waltham, MA 02254, USA

1. Introduction

The recently discovered chlorine(III) (chlorite-based) oscillators show a wide range of dynamical phenomena in "batch" and "flow." In addition to oscillations, these systems give rise to chaos, birhythmicity, spatial inhomogeneity, and other unusual effects [1]. To uncover the chemical origins of these phenomena, we construct multi-step mechanisms, and test them out by modeling the system's observed behavior with them. However, the validity of a mechanism may not be judged fairly, if insufficient rate data exist for elementary steps in the mechanism.

We are therefore attempting to elucidate the kinetics of elementary steps relevant to chlorite-based oscillators when this information is not available, and report several such analyses. In addition to the application to modeling studies, the results are also of general chemical interest. Moreover, under certain conditions, these so-called elementary processes can become complex, leading to the discovery of new phenomena. We begin by reporting on our studies of initial steps in the reaction between MnO_4^- and Cl(III). In later sections we discuss the reactions of Cl(III) with other halogen species.

2. Chlorine(III)-Permanganate

This study was undertaken because the Cl(III)-$\bar{I}$-MnO_4^- system appeared to be a new oscillator, and not simply a perturbation on Cl(III)-I^-[2]. The Cl(III)-MnO_4^- reaction is rapid and complex; consequently, we studied both stoichiometry and kinetics for reaction times less than one minute by conventional and rapid mixing spectrophotometry. Under these conditions the reaction stoichiometry is straightforward,

$$MnO_4^- + 5ClO_2^- + 8H^+ = Mn^{2+} + 5ClO_2 + 4H_2O \qquad (1)$$

with rate law

$$-d[MnO_4^-]/dt = k[MnO_4^-][Cl(III)] \qquad (2)$$

$$k = (k_1 + k_2[H^+]/K_a)(1 + [H^+]/K_d) \qquad (3)$$

where k_1 and k_2 are rate constants for the rate-determining steps $MnO_4^- + ClO_2^- \longrightarrow MnO_4^{2-} + ClO_2$ ($k_1=24.4 \pm 2.0\ M^{-1}\ s^{-1}$), and $MnO_4^- + HClO_2 \longrightarrow MnO_4^{2-} + ClO_2 + H^+$ ($k_2 = 92 \pm 29\ M^{-1}\ s^{-1}$); and $K_a = [H^+][ClO_2^-]/[HClO_2] = (2.70 \pm 0.45) \times 10^{-3}M$ at 25.0 ± 0.5 °C and ionic strength 1.0M ($NaClO_4$) [3].

The unusual reactivity pattern $HClO_4 > ClO_2^-$ explains the inhibition of the Cl(III)-I^- oscillator by permanganate at pH 1.

The I^- ion cannot compete with MnO_4^- for Cl(III). Oscillation occurs in the Cl(III)-I^--MnO_4^- system at pH 2-3.5 because the Cl(III)-I^- reaction competes favorably with the slowed-down Cl(III)-MnO_4^- reaction in this pH range. Other features of the Cl(III)-MnO_4^- study are unusual as well. The value of K_a is smaller than that previously determined, presumably due to avoidance of $HClO_2$ decomposition under rapid-mixing conditions. Neither the values of k_1 and k_2, nor the greater reactivity of $HClO_2$ compared with ClO_2^- are consistent with an outer-sphere electron transfer mechanism, in contrast with most ClO_2^-/ClO_2 reactions [4]; it seems likely that the chlorine(III)-permanganate reaction mechanism is inner-sphere electron transfer.

3. Evidence for $XClO_2$ Intermediates in Chlorite-Based Oscillators

Earlier studies on oxyhalogen kinetics implicated the existence of X_2O_2 (X = halogen) species as transient intermediates in a variety of net and exchange reactions. Species proposed include Cl_2O_2 [5], I_2O_2 [6] and $IClO_2$ [7]. The mixed halogen species $IClO_2$ plays a key role in the mechanism of the fundamental Cl(III)-I^- oscillator [8]. In batch, this system "clocks"; autocatalytic buildup of the brown color due to iodine is followed by an abrupt fadeout. The fadeout is due to the oxidation of iodine by unreacted chlorine(III). The kinetics of this reaction were determined by stopped-flow [7].

The rate law for the reaction

$$5ClO_2^- + 2I_2 + 2H_2O \longrightarrow 5Cl^- + 4IO_3^- + 4H^+ \qquad (4)$$

consists of three terms, one of which is Cl(III)-independent. We have assigned the two Cl(III)-dependent terms to ClO_2^- reactions with I_2 and I_2OH^- leading to $IClO_2$ formation. The Cl(III)-independent term is represented by the process

$$I_2 + H_2O \underset{k_{-5}}{\overset{k_5}{\rightleftharpoons}} IOH_2^+ + I^- \qquad (5)$$

$$IOH_2^+ + ClO_2^- \xrightarrow{k_6} IClO_2 + H_2O \qquad (6)$$

In our experiments, $k_6\,[ClO_2^-] \gg k_{-5}\,[I^-]$, and the Cl(III)-dependence expected when the rate expression for (5) + (6) is derived is not observed.

The kinetics of the batch reaction between chlorine(III) and iodine species has now been documented. It is known that chloride ion accelerates the decomposition of Cl(III) in acid media. It is natural to speculate on the effect bromine species would have on Cl(III). Would the system Cl(III)-Br^- oscillate? Would it "clock"? What reaction would occur when Cl(III) and Br_2 are mixed? Would mechanistic analysis require postulation of $BrClO_2$-type intermediates? To find out, we mixed Cl(III) with Br^- in flow and batch, and reacted Br_2 (and Br_3^-) with Cl(III) in batch.

4. Chlorine(III)-Bromide Ion and Br_2 Reactions

The system Cl(III)-Br^- oscillates in flow [9]. However, guidance from the better described Cl(III)-I^- oscillatory system may be misleading, because bromine species usually possess greater oxidizing power than analogous iodine species, and bromine species show comparable or even greater kinetic lability. We have therefore adopted an empirical approach, and determined that, (a) Br_2 oxidizes Cl(III) to Cl(IV) cleanly, (b) depending on pH and con-

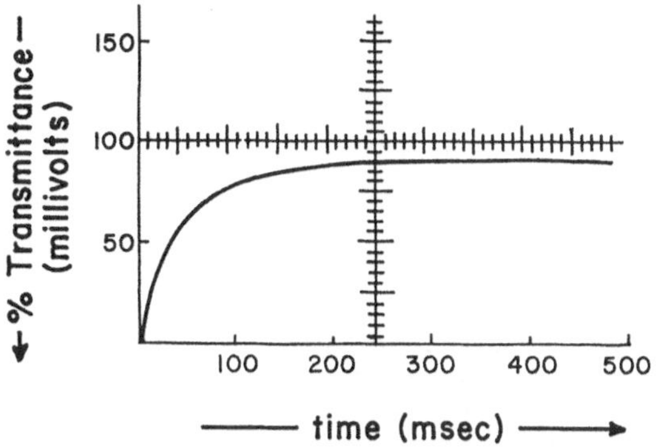

Figure 1. Stopped-flow oscillograph of the reaction Br_2 + Cl(III). Ordinate (50 mv per major division) is proportional to decreasing % transmittance. Initial concentrations in mixing chamber: $[Br_2] = 3.98 \times 10^{-4}M$, $[Cl(III)] = 1.06 \times 10^{-2}M$, pH 3.0, ionic strength = 0.6M [$NaClO_4$], temperature = 25.0 ± 0.5 °C; wavelength = 380 nm

centration, Cl(III) can oxidize Br^- to Br_2, (c) Br^- can catalyze the decomposition of Cl(III). We shall report on the Cl(III)-Br_2, and Cl(III)-Br^- batch reactions.

At pH 3 the oxidation of Cl(III) by Br_2 is capable of spontaneously producing one or more of the products ClO_2, ClO_3^-, ClO_4^-. Within experimental error, we detect only ClO_2; the reaction is therefore

$$2ClO_2^- + Br_2 \longrightarrow 2ClO_2 + 2Br^- \qquad (7)$$

The rate law, determined by stopped-flow spectrophotometry (Figure 1) depends on $[H^+]$, $[Br^-]$, $[Br_2]$ and $[Cl(III)]_o$ (initial [Cl(III)]). At moderate acidities the rate law expression is

$$(\tfrac{1}{2})\frac{d[ClO_2]}{dt} = \frac{(a_1 + a_2[Br^-])[Br_2][Cl(III)]_o^2}{b[Cl(III)]_o + c[Br^-]^2 + d(1 + e[Cl(III)]_o)[Br^-]} \qquad (8)$$

Under conditions where $b[Cl(III)]_o \gg c[Br^-]^2 + d[Br^-]$, and $a_1 \gg a_2[Br^-]$ the rate depends on a_1/b. The remaining parameter a_1/b is influenced by $[H^+]$, increasing with decreasing $[H^+]$, reaching a maximum, then declining to a lower value at pH 7.41.

Several differences exist between the rate law for this system, and the analogous Cl(III)-I_2 [7] and Cl(III)-Cl_2 [10] reactions. The mechanism most consistent with the rate data involves <u>rapid</u> equilibration between Br_2 and Cl(III) to form the mixed halogen intermediate; for example

$$Br_2 + ClO_2^- \rightleftharpoons BrClO_2 + Br^- \qquad (9)$$

(A complete mechanism requires inclusion of reaction between Cl(III) and Br_3^-.) This step is followed by rate-determining Cl(III) attack; e.g.

$$BrClO_2 + ClO_2^- \longrightarrow 2ClO_2 + Br^- \qquad (10)$$

The complex pH dependence in acidic media can be explained by protonation of $BrClO_2$ and greater reactivity of ClO_2^- than $HClO_2$. In alkaline media, the reaction with HOBr should be considered.

The dynamical behavior of the system Cl(III)-Br^- is complex; product formation depends on conditions. We summarize our observations as follows. Br⁻ in excess: the observed reaction is

$$ClO_2^- + 6Br^- + 4H^+ \longrightarrow Cl^- + 2Br_3^- + 2H_2O \quad (11)$$

The rate law is

$$(½)d[Br_3^-]/dt = k[Cl(III)]_o[Br^-][H^+] \quad (12)$$

where $k = (9.5 \pm 0.5) \times 10^{-2}\ M^{-2}\ s^{-1}$ at 25° C and ionic strength 1.0M ($NaClO_4$). This rate law can be explained by rate determining formation of a protonated mixed halogen intermediate, or hydrolyzed mixed halogen intermediate; e.g.

$$H^+ + Br^- + HClO_2 \longrightarrow BrClO + H_2O \quad (13)$$

followed by rapid Br^- attack; viz.

$$Br^- + BrClO \longrightarrow Br_2 + ClO^- \quad (14)$$

Cl(III) in excess: ClO_2 is observed; Br^- appears to catalyze the decomposition of $HClO_2$. The stoichiometry and rate law are difficult to define. However, at concentrations in the narrow range $[Cl(III)]_o \approx 2 \times 10^{-3}M$, $[Br^-] \approx 5 \times 10^{-4}\ M$, a clock reaction occurs (Figure 2a), the lag time of which decreases upon addition of small ($< 2 \times 10^{-4}M$) amounts of added Br_2 (Figure 2b). The mechanistic steps relevant to the Cl(III)-Br^- system should therefore be similar to those of the Cl(III)-I^- system, with appropriate rate constants.

5. Conclusion

The studies we report have provided rate data for the modeling of complex dynamical systems. The importance of intermediates

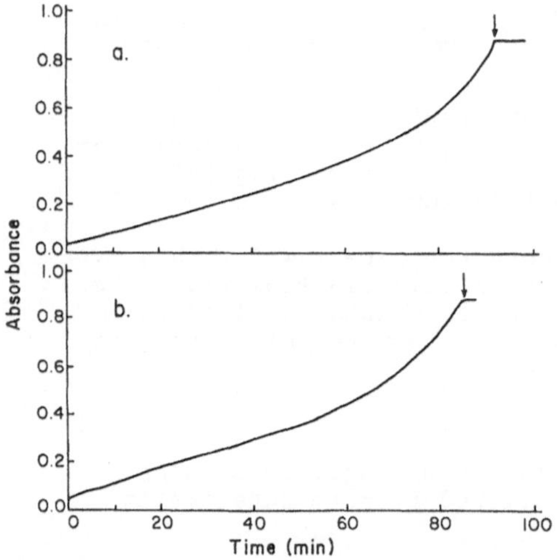

Figure 2. The Cl(III)-Br^- clock reaction. (a) $[Br^-] = 5 \times 10^{-4}M$, $[Cl(III)]_o = 1.65 \times 10^{-3}M$, $[HClO_4] = 0.2M$, ionic strength = 0.5M $NaClO_4$, temperature = 25.0 ± 0.5 °C; (b) Same conditions with $[Br_2] = 2.26 \times 10^{-5}M$ initially present; wavelength of recording = 390 nm

of the form $XClO_2$ for chlorite-based oscillators has been established. This conclusion is strengthened by the finding that, even in the permanganate system, an inner-sphere mechanism predominates. Further modeling studies of other chlorite oscillators, such as Cl(III)-thiosulfate, will determine whether direct attachment of Cl(III) by reductant or oxidant is a common feature of chlorite-based oscillators.

6. Acknowledgments

We thank the National Science Foundation for research grant CHE 8204085, and the National Institutes of Health for research grant GM 08893, which have provided financial support for these studies.

7. References

1. I.R. Epstein and K. Kustin: Structure and Bonding 56, 1 (1984)
2. M. Orbán, C. Dateo, P. DeKepper, and I.R. Epstein: J. Am. Chem. Soc. 104, 5911 (1982)
3. C. Ahlstrom, D.N. Boyd, I.R. Epstein, K. Kustin and J. Romanow: Inorg. Chem. 23, 2185 (1984)
4. L.A. Lednicky and D.M. Stanbury: J. Am. Chem. Soc. 105, 3098 (1983)
5. H. Dodgen and H. Taube: J. Am. Chem. Soc. 71, 2501 (1949)
6. K.J. Morgan, M.G. Peard, and C.F. Cullis: J. Chem. Soc. 1865 (1951); a hydrated version of the I_2O_2 intermediate is given in H. Liebhafsky and G.M. Roe: Int. J. Chem. Kinetics 11, 693 (1979)
7. J.L. Grant, P. DeKepper, I.R. Epstein, K. Kustin, and M. Orbán: Inorg. Chem. 21, 2192 (1982)
8. I.R. Epstein and K. Kustin: J. Am. Chem. Soc., submitted
9. M. Alamgir and I.R. Epstein, to be published
10. F. Emmenegger and G. Gordon: Inorg. Chem. 6, 633 (1967)

Bistability in the Reduction of Permanganate by Hydrogen Peroxide in a Stirred Tank Flow Reactor

P. De Kepper, Q. Ouyang, and E. Dulos

Centre de Recherche Paul Pascal, Domaine universitaire
F-33405 Talence Cédex, France

In recent years, the number of isothermal homogeneous chemical oscillating reactions has considerably increased, but all the chemical oscillators are based on halogen chemistry [1,2] with the exception of the methylene blue-sulfide-sulfite air oxidation [3] and its simplified version : the hydrogen peroxide oxydation of sulfide in neutral or basic solution in flow reactor [4]. The recent cobaltbromide catalized oxidation os some aromatic compounds by air [5] could also be regarded as a non-halogen oscillator if the role of bromide is only to form a cobalt (III)-bromide complex more active than $C_o{}^{3+}$.

Sulfur and carbon chemistry are certainly very rich areas for complex feedback reactions and will probably produce in the near future a wealth of chemical oscillators comparable to that found in halogen chemistry.

In the following, we present our efforts to produce oscillatory dynamics in another area of chemistry : the manganese chemistry. It is based on the cross-shaped phase diagram technique [6,7]. It starts with the quest for bistable systems and then continues with the search of unstable states at the boundaries of the bistability domain. Isothermal chemical bistabilities can be obtained when autocatalytic reactions are performed in flow reactor mode. A number of autocatalytic reactions is reported in manganese chemistry. Among the best documented, let us mention the reduction of permanganate (MnO_4^-) by oxalic acid [8], malonic acid [9] or hydrogen peroxide (H_2O_2) [10,11,12] and the oxidation of manganeous ions (Mn^{2+}) by periodic acid [13] or chloroperbenzoïc acid [14].

Previous studies have shown that the permanganate-oxalic acid reaction can be bistable when performed in a C.S.T.R. [15,16]. This reaction is relatively slow at room temperature. We focus here on the faster MnO_4^- - H_2O_2 reaction.

1. The MnO_4^- - H_2O_2 - Mn^{2+} - $HClO_4$ system

In batch reactor

In excess H_2O_2 and at low enough pH, the reduction of MnO_4^- to Mn^{2+} is quantitative with the following stoichiometry :

$$2MnO_4^- + 5H_2O_2 + 6H^+ = 2Mn^{2+} + 5O_2 + 8H_2O$$

The reaction generally presents an induction period with a slow decrease in MnO_4^- concentration, followed by an abrupt drop off with disappearance of all trace of permanganate color. Initial addition of Mn^{2+}, a species involved in the autocatalytic loop, reduces dramatically the induction period. A strange aspect of this reaction is the *oscillatory* dependence of its half-life time on initial $[H_2O_2]$ or $[Mn^{2+}]$ as already mentioned by several authors [12].

C.S.T.R. results

In a thermally regulated flow reactor, the steady state values of the absorbance at 525 nm (maximum absorbance of MnO_4^-) can take two different values as a function of the flow rate k_0 (k_0 is the reciprocal of the residence time τ) as shown in figure 1.

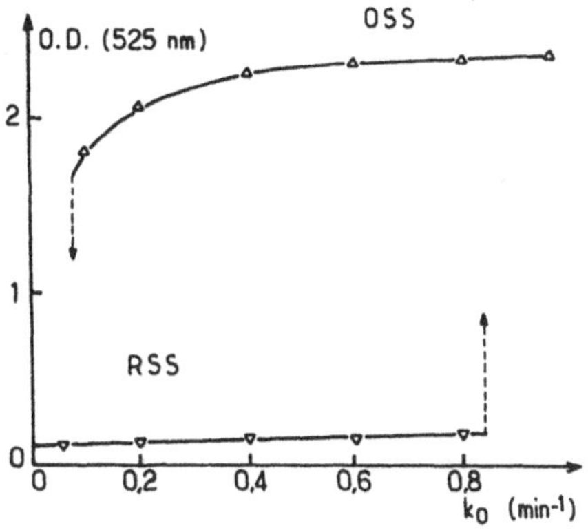

Fig. 1 Steady state bistability in the optical density at 525 nm as a function of flow rate, all other constraints kept constant :

$[MnO_4^-]_0$ = 3x10^{-4} M ; $[H_2O_2]_0$ = 0.033 M ;

$[Mn^{2+}]_0$ = 6.7x10^{-6} M ; $[HClO_4]_0$ = 1.33 M ;

T = 22°C

The set of state with low optical density (OD) is referred to as the reduced steady state (RSS) because manganese is mainly in its reduced Mn(II) form. It corresponds to the extension of the equilibrium state at k_0=0 and is the so called *thermodynamic branch*. The high OD set of state is referred to as the oxydized steady state (OSS), for manganese is mainly oxidized to Mn(VII). This latter state corresponds to the *flow branch* where species concentration is primarily determined by the input flow, because the chemical processes are slow relative to the flow. The two branches overlap over a large range of k_0. Arrows pointing respectively up and down indicate k_0 limiting values for stability on the RSS and the OSS branches. Transition time from OSS to RSS is typically 10 seconds whereas the reverse transition (RSS → OSS) is of the order of the residence time τ. The determination of the stability limit of OSS and RSS can be repeated for a range of input hydrogen peroxide concentration ($[H_2O_2]_0$)* and this for different $[HClO_4]_0$.

Figure 2 represents the bistability limits in the (k_0, $[H_2O_2]_0$) plane for three different $[HClO_4]_0$, all the other constraints being kept constant. The width of the bistable region decreases with increasing $[H_2O_2]_0$, at any $[HClO_4]_0$ and beyond a critical $[H_2O_2]_0$ it reduces to zero at low flow rate. The vanishing regions of bistability are difficult to establish because of considerable slowing down in the transition between steady states. Beyond these critical regions, changes from OSS to RSS are continuous, no oscillatory behaviour is observed as in the case of minimal oscillators [17,18]. The whole pattern is consistent with a cusp point behaviour, in the vicinity of which relaxation oscillations are generally excluded. The other end of the bistable domain, at high flow rate, is not be observed.

Figure 3 shows a typical dependence of the bistability range in the (k_0, $[Mn^{2+}]_0$) plane. Increasing $[Mn^{2+}]_0$ enlarges the bistability range as a function of k_0. But above a limiting value of $[Mn^{2+}]_0$, only the RSS is observed. Below $[Mn^{2+}]_0 \sim 10^{-6}$ M experiments are not easily reproducible, probably due to uncontrolled trace impurities in the permanganate solutions which are buffered at higher $[Mn^{2+}]_0$. When $[H_2O_2]_0$ is increased, the bistability region in the (k_0, $[Mn^{2+}]$) plane shrinks. Here again, only the low flow cusp point can be studied. The high flow end of the bistable domain was not observed for any of the presently investigated conditions. Very short residence time, beyond the reach of our present experimental set up, seems to be required.

2. Perturbed MnO_4^- - H_2O_2 reactions

In the next step, we modified our initial reaction system by introducing additional reagents in order to generate appropriate feedbacks which would destabilize the previous steady states [6,1b]. We mainly focused our attention on persulfate ($S_2O_8^{2-}$).

*Input concentration of any A chemical species is denoted by $[A]_0$.

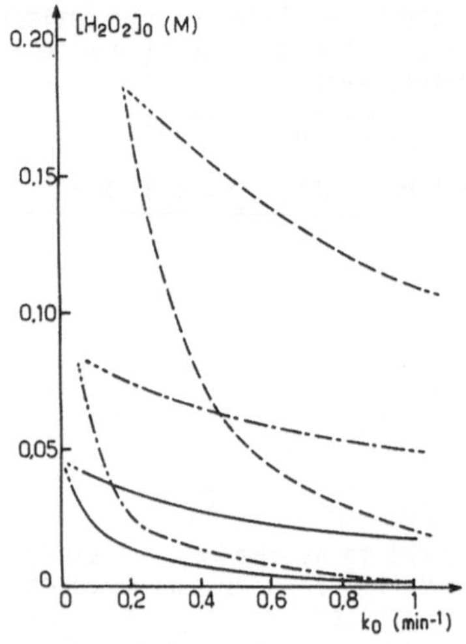

Fig. 2 Sections of the bistability domain in the ($[H_2O_2]_0$, k_0) plane for different $[HClO_4]_0$: 0.045 M ---, 0.16 M -.--.-, 1.33 M ——

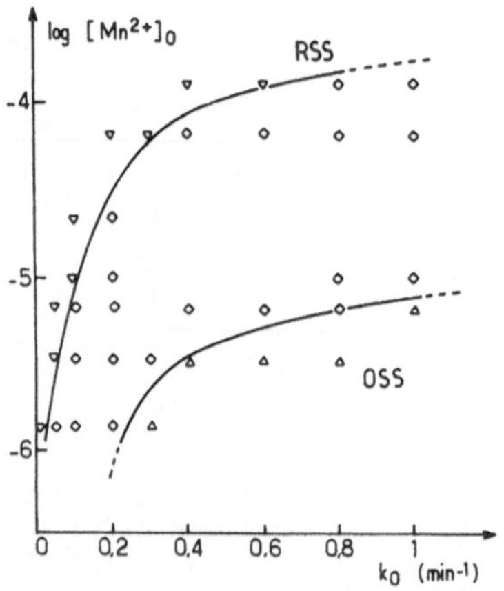

Fig. 3 Section of the phase diagram in the ($[Mn^{2+}]_0$, k_0) plane, with all other constraints as in fig. 1. Symbols : ∇ RSS, Δ OSS, ◇ bistability RSS/OSS

Indeed, persulfate is known to oxidize Mn^{2+} to MnO_4^- [20] with the following stoïchiometry :

$$5S_2O_8^{2-} + 2Mn^{2+} + 8H_2O = 2MnO_4^- + 10\ SO_4^{2-} + 16\ H^+$$

This reaction is very slow at room temperature but its rate can be considerably increased by introducing silver ions (Ag^+) as catalyst [21].

Preliminary experiments have shown that Ag^+ also catalyzes the reduction of MnO_4^- by H_2O_2. We thus first studied the effect of Ag^+ on the MnO_4^- - H_2O_2 bistability.

The MnO_4^- - H_2O_2 - Ag^+ - $HClO_4$ system

Using Ag^+ instead of Mn^{2+} gives similar results. As shown in fig. 4 the bistability range of flow rate is enlarged. Above a critical $[Ag^+]_0$ only the RSS is observed. However this $[Ag^+]_0$ value is an order of magnitude larger than the critical $[Mn^{2+}]_0$ value for similar conditions. An increase in $[H_2O_2]_0$ (Fig. 5) can considerably reduce the $[Ag^+]_0$ range of bistability.

The MnO_4^- - H_2O_2 - $S_2O_8^{2-}$ - Ag^+ - $HClO_4$ system

An additional input flow of persulfate, up to $[S_2O_8^{2-}]_0$ = 0.1 M, in experimental conditions otherwise similar to those used in figure 5, produces only very slight shift in the previous phase diagram. Again only the low flow rate end of the bistable domain can be probed for oscillations. No oscillatory behaviour is observed in these experimental conditions.

Nevertheless, a *different bistable* region is found for lower values of $[H_2O_2]_0$ and $[MnO_4^-]_0$ and at much larger residence time (τ = 30 min). The Pt-potential and the optical density hysteresis as a function of k_0 is shown in figure 6. In this case, the steady states are the OSS corresponding now to the thermodynamic branch and an oxidized intermediate steady state (ISS) characterized by a potential plateau at 200 mV (the RSS corresponds to potential values which range from 600 to 700 mV). Though it is not visible on the absorption measurements at 525 nm, there is a very definite color difference between the OSS and of the ISS, the latter is reddish-purple, probably due to significant amounts of Mn(III) and Mn(IV), while the for-

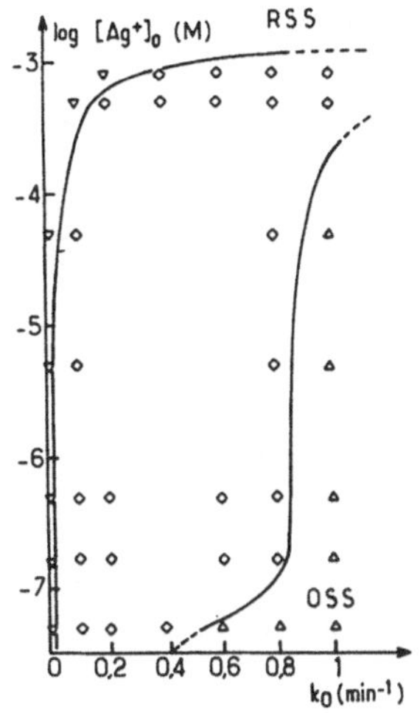

Fig. 4 Dependence of the bistability region versus k_0 as a function of $[Ag^+]_0$. All other constraints and symbols as in fig. 1 and 3

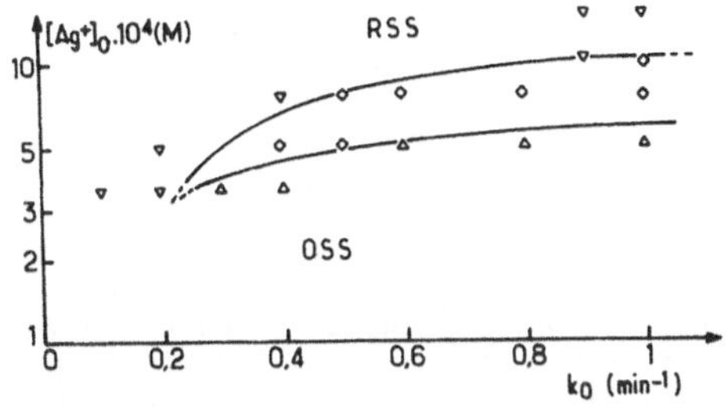

Fig. 5 Dependence of the bistability region versus k_0 as a function of $[Ag^+]_0$, with $[H_2O_2]_0$ = 0.05 M, $[HClO_4]_0$ = 2 M. All other constraints and symbols as in fig. 1 and 3

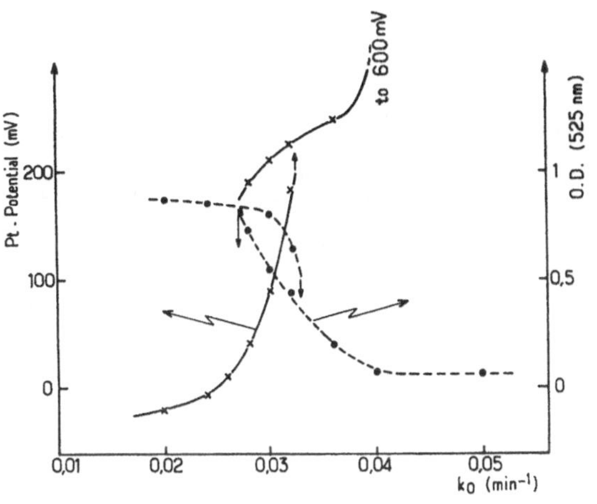

Fig. 6 Steady state bistability in the optical density (- -●-) and the Pt-potential (—x—) versus mercurous sulfate reference electrode as a function of flow rate. All other constraints kept constant : $[MnO_4^-]_0$ = $1x10^{-3}$M, $[H_2O_2]_0$ = $2x10^{-3}$M, $[S_2O_8^{2-}]_0$ = 0.075 M, $[Ag^+]_0$ = $3x10^{-3}$M, $[HClO_4]_0$ = 1.25 M, T = 22°C

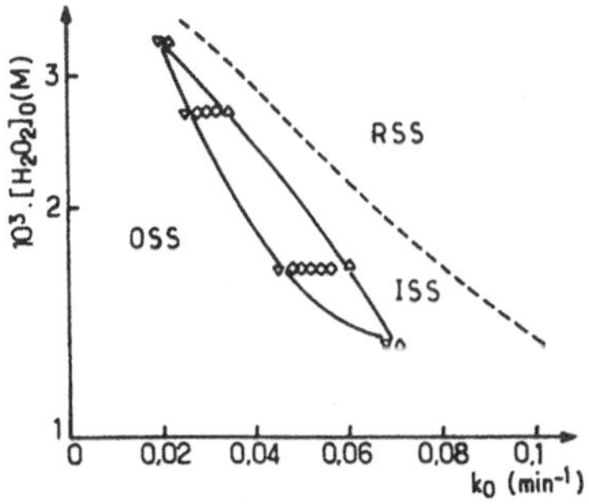

Fig. 7 Section of the new bistability domain in the ($[H_2O_2]_0$, k_0) plane with all other constraints as in figure 6. Symbols ▼ OSS, ▲ ISS, ◇ bistability OSS/ISS

mer is deep purple. Figure 7 displays a section of this new bistability in the (k_0, $[H_2O_2]_0$) phase plane. The dashed line corresponds to the locus of the steep inflexion in the steady state potential which separates the ISS from the OSS. Both ends of the bistable region were studied but only cusp point behaviour has been observed.

3. Concluding remarks

Despite the fact that no oscillatory behaviour has yet been found, the MnO_4^- - H_2O_2 - $S_2O_8^{2-}$ - Ag^+ - $HClO_4$ system seems to be a quite promising system since it produces two different types of bistability. They are actually widely separated in the constraint space, which suggests little coupling in the chemical processes from which they are issued. The point now is to see if they can be brought closer, in other experimental conditions, to produce more significant coupling between the reducing process by H_2O_2 and the oxydizing process by $S_2O_8^{2-}$. Another possibility for oscillation is to study the high flow end of the bistable region ; very fast flow through reactors are presently built for this purpose in our laboratory.

The mechanisms are still not clearly understood even for the basic permanganate hydrogen peroxide redox reaction in spite of a wealth of literature (see references in [12]). There is no satisfactory explanation, specially for the *oscillatory* rate dependence of the reaction on initial H_2O_2 concentration. Formation of hydrogen peroxide dimer [22] or peroxide complexes with Mn(II) or Mn(III) [12] was proposed but there is neither evidence for hydrogen peroxide dimers nor for peroxide complexes of Mn(II). Moreover a Mn(III)-peroxide complex would have too short a lifetime if one considerers the reactivity of Mn(III) [23]. The very sharp drop off in the permanganate concentration after a relatively long induction period is also not an easy kinetic task to solve. It requires more than first order dependence in the autocatalytic species or the presence of more than one major autocatalytic species. This latter assumption is consistent with recent observations in stopped-flow experiments on the H_2O_2 - MnO_4^- reaction, which show that even with a 20 fold excess of Mn^{2+} over MnO_4^-, the reaction still exhibits an induction period [24]. Any mechanism for this reaction will also have to account for the quadratic rate dependence of the Guyard reaction [25] (oxidation of Mn^{2+} by MnO_4^-) on initial $[Mn^{2+}]$. MORROW and PERLMAN [26] results for this latter reaction suggest the formation of a binuclear intermediate such as $Mn_2O_4^+$ which may be thought of as an analog of Ag MnO_4 proposed in the Ag^+ catalyzed reduction of permanganate by hydrogen [27]. Preliminary kinetic modeling of the MnO_4^- - H_2O_2 shows that the quenching of such binuclear species by H_2O_2 can produce non-monotonous rate dependence of the reaction on initial $[H_2O_2]$.

References

1. M. Orbàn, C.E. Dateo, P. De Kepper, I.R. Epstein, J. Am. Chem. Soc. 104, 5911-5918 (1982)
   I.R. Epstein, K. Kustin, P. De Kepper, M. Orbàn, Scientific American, 248, n°3 96-108 (1983)
2. *Oscillations and traveling waves in chemical systems*, Ed. R.J. Field and M. Burger, Wiley (1984)
3. M. Burger, R.J. Field, Nature 307, 720-21 (1984)
4. M. Orban, I.R. Epstein, to be published
5. J.H. Jensen, J. Am. Chem. Soc. 105, 2639-2641 (1983)
   J.H. Jensen et al., this volume
6. J. Boissonade, P. De Kepper, J. Phys. Chem. 84, 501-506 (1980)
7. P. De Kepper, Springer Series in Synergetics, Eds C. Vidal, A. Pacault, 12, 192-196 (1981)
8. R.M. Noyes, Trans. New York Acad. Sci. 13, 314-316 (1951)
   S.J. Adler, R.M. Noyes, J. Am. Chem. Soc. 77, 2036-2042 (1955)
9. A.Y. Drummond, W.A. Waters, J. Chem. Soc., 2456 (1954)
10. K.C. Baley, G.I. Taylor, J. Chem. Soc. 994 (1937)
11. E. Abel, Monatsh 80, 455-462 (1949) ; Helv. Chim. Acta 33, 785 (1950) ; Monatsh 86, 952-957 (1955)
12. J. Casado, J. Lizado-Lamsfus, Anales Real. Soc. Esp. Fis. y Quim. 63, 739-48 (1967)
    S. Senent, J. Casado, J. Lizaso, Anales Real Soc. Esp. Fis. y Quim. 67, 1133-44 (1971)
13. J.D.H. Strickland, G. Spicer, Anal. Chim. Acta 3, 517-42 (1949)
14. G.H. Jones, J.C.S. Chem. Comm. 536-7 (1979)

15. I.R. Epstein, C.E. Dateo, P. De Kepper, K. Kustin, M. Orbàn, Springer Series in Synergetics, Eds C. Vidal, A. Pacault, 12, 188-91 (1981)
16. J.S. Reckly, K. Showalter, J. Am. Chem. Soc. 103, 7012 (1982)
17. C.E. Dateo, M. Orbàn, P. De Kepper, I.R. Epstein, J. Am. Chem. Soc. 104, 504-509 (1982)
M. Orban, P. De Kepper, I.R. Epstein, J. Am. Chem. Soc. 104, 2657 (1982)
18. W. Geiseler, J. Phys. Chem. 86, 4394 (1982)
19. P. De Kepper, J. Boissonade, chap. 7 in ref. [2]
20. V. Wannagat, E.M. Horn, G. Valk, F. Höfler, Z. Anorg. Allgem. Chem. 340, 181-8 (1965)
21. W.K. Wilmarth, A. Haim, pp. 175-225 in *Peroxide reaction mechanisms*, Ed. J.O. Edwards, Interscience, New-York (1962)
22. E.H. Riesenfeld, Z. Anorg. Chem., 218, 257 (1934)
23. G. Davies, L.J. Kirschenbaum, K. Kustin, Inorg. Chem. 7, 146-154 (1968)
24. P. De Kepper, K. Kustin, unpublished results
25. P. Gorgeau, Ann. Chim. Phys. 66, 153 (1962)
26. J.I. Morrow, S. Perlman, Inorg. Chem. 12, 2453-5 (1973)
27. A.H. Webster, J. Halpern, Trans. Faraday Soc. 53, 51-60 (1957)

Measurements and Modelling of Unstable Steady State, Separatrix, and Critical Point Behavior

Tammy Pifer, N. Ganapathisubramanian, and Kenneth Showalter
Department of Chemistry, West Virginia University, Morgantown, WV 26506-6045, USA

1. Introduction

The relaxation of a perturbed chemical system to its stable states provides a means for characterizing the essential dynamic features of the system. These features are contained in the phase portrait, a plot of the concentration of one variable species as a function of another variable species for a variety of initial conditions. A particular steady state may be characterized as a stable node or a stable focus according to the appearance of the relaxation trajectories around that state. In addition, basins of attraction in multistable systems may be determined as well as the separatrix partitioning these basins. The nature of the unstable steady state may be determined from relaxation experiments in systems that can be accurately described in terms of one or two variable species. This paper reports on measurements and modelling of the relaxation behavior of the bistable iodate-arsenous acid system. The stable and unstable steady states are characterized and the separatrix is located in the phase plane. In addition, the special case of relaxation to the critical point is investigated. A more detailed account of this study will be forthcoming [1,2].

2. Chemical System

We consider the iodate-arsenous acid system in a CSTR with arsenous acid in stoichiometric excess. A single stoichiometry describes the reaction at any time according to (I).

$$IO_3^- + 3H_3AsO_3 = I^- + 3H_3AsO_4 \qquad (I)$$

The reaction is autocatalytic in iodide with the rate of reaction governed by the empirical rate law α.

$$d[I^-]/dt = R_\alpha = (k_1 + k_2[I^-])[I^-][IO_3^-][H^+]^2 \qquad (\alpha)$$

Rate equations (R) provide a near quantitative description of the buffered reaction in a CSTR,

$$\begin{aligned} d[I^-]/dt &= R_\alpha + k_0([I^-]_0 - [I^-]) \\ d[IO_3^-]/dt &= -R_\alpha + k_0([IO_3^-]_0 - [IO_3^-]) \end{aligned} \qquad (R)$$

where k_0 is the reciprocal residence time and $[I^-]_0$ and $[IO_3^-]_0$ are the reactant concentrations in the combined feed stream.

Model R was developed to describe slowing down in the iodate-arsenous acid system [3] and a modified version has been used to describe mushroom and isola behavior [4]. A more detailed account of the model, which is a reduction of earlier, more elaborate models [5,6] may be found in [1-4].

3. One-Dimensional System

In solutions containing excess arsenous acid, iodide and iodate are the only stoichiometrically significant iodine containing species; therefore, $d[I^-]/dt + d[IO_3^-]/dt = 0$ and conservation relation (1) is obtained from (R).

$$[IO_3^-] + [I^-] = [IO_3^-]_0 + [I^-]_0 \tag{1}$$

Substitution of (1) into either rate equation (R) yields (R'), a one-variable model of the system.

$$d[I^-]/dt = (k_1 + k_2[I^-])[I^-]([IO_3^-]_0 + [I^-]_0 - [I^-])[H^+]^2 + k_0([I^-]_0 - [I^-]) \tag{R'}$$

3.1 The Unstable Steady State

Figure 1 shows the stable (solid circles) and unstable (open circle) steady states of the system for the composition and residence time indicated in the caption. The steady states lie on the composition line defined by (1); even away from the steady states the system is constrained to this line.

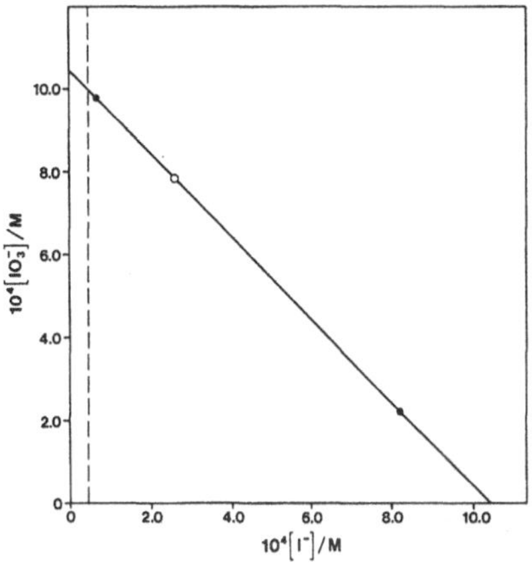

Fig. 1. Low iodide, unstable, and high iodide steady states on composition line. Combined reactant feed stream concentrations: $[IO_3^-]_0 = 1.00 \times 10^{-3}$ M, $[H_3AsO_3]_0 = 5.00 \times 10^{-3}$ M, $[I^-]_0 = 4.41 \times 10^{-5}$ M (indicated by vertical dashed line). Acidity maintained constant with sulfate/bisulfate buffer at pH = 2.122 ($[H^+] = 7.55 \times 10^{-3}$ M). Reciprocal residence time $k_0 = 8.20 \times 10^{-3}\ s^{-1}$. Temperature: 25.0 ± 0.2°C. From Ref. [1].

The location of the unstable steady state in Fig. 1 was determined by forcing the system to move along the composition line. Consider the system initially in the stable steady state with low iodide concentration. When the flow is temporarily stopped, iodide concentration increases and the system moves toward the unstable steady state along the composition line. Iodide concentration is monitored and at the desired concentration the flow is resumed at the original value of k_0. The system relaxes back to the low iodide steady state if the iodide concentration is below that of the unstable steady state in Fig. 1. When the iodide concentration exceeds that of the unstable steady state, a transition occurs to the high iodide steady state. Similar relaxation experiments can be carried out with the system initially in the high iodide steady state. Here, the flow rate is temporarily increased and iodide concentration decreases toward the unstable steady state. A repetitive sequence of experiments allows the location of the unstable steady state to be accurately determined.

Figure 2a shows a set of four experiments where the iodide concentrations, upon resuming the flow, were identical within the experimental resolution of three significant figures. The relaxation curves to each of the stable steady states are effectively superimposable when translated on the time axis; however, the relaxation half-times are 15.2, 31.0, 21.0 and 14.8 min for the upper and lower curves, respectively. Relaxation curves calculated [7] from (R') are shown in Fig. 2b. Here, the initial iodide concentrations were also the same within three

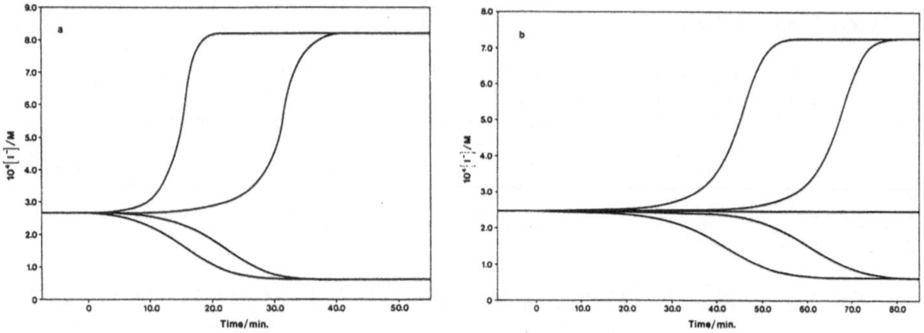

Fig. 2. Relaxation curves from unstable steady state of system in Fig. 1: (a) Experimental behavior; (b) calculated behavior according to (R') with $k_1 = 4.5 \times 10^3\ M^{-3}\ s^{-1}$, $k_2 = 4.5 \times 10^8\ M^{-4}\ s^{-1}$, and $k_o = 6.40 \times 10^{-3}\ s^{-1}$. Ref. [1].

significant figures. The initial iodide concentration for the middle curve, with a relaxation half-time of 2.6 hours, was the same within eight significant figures as that calculated for the unstable steady state. Figures 2a and 2b show that there is an extreme sensitivity to initial conditions around the unstable steady state. In addition, we see that the system may remain near the unstable steady state for extended periods of time when it is initially very close to that state.

4. Two-Dimensional System

Conservation relation (1) allows the system to be described in terms of one variable. This relation is no longer valid when the system is subjected to perturbations that introduce additional material to the reactor. Thus, model R is needed to describe experiments where the system is perturbed by sudden injections of reagent.

4.1 The Separatrix

Relaxation experiments were conducted to determine the location of the separatrix in the $[I^-]$-$[IO_3^-]$ phase plane. The method was similar to that used in the one-dimensional system. However, in these experiments the system was perturbed by injecting reagents containing KI, KIO_3, and $AgNO_3$. In all cases, the volume of the injected reagent (10-70 µl) was insignificant compared to the reactor volume (35.39 ml).

The solid curve in Fig. 3 shows the experimentally determined separatrix. Perturbations from the low iodide steady state were carried out using iodide and neutral iodide/iodate reagent. Perturbations from the high iodide steady state were carried out with silver reagent rapidly followed with iodate reagent. The unstable steady state determined by one-dimensional perturbations (open circle) compares well with the determination carried out by two-dimensional perturbations, indicated by the arrows on the composition line.

5. Critical Point Behavior

The critical point was found by locating the hysteresis limits for a series of reaction mixtures with increasing iodide concentration in the reactant stream. The critical iodide concentration, $[I^-]_0^c$, was iteratively determined to three significant figures.

One-dimensional relaxation experiments were carried out at the critical iodide concentration and different values of k_0. Each perturbation consisted of a

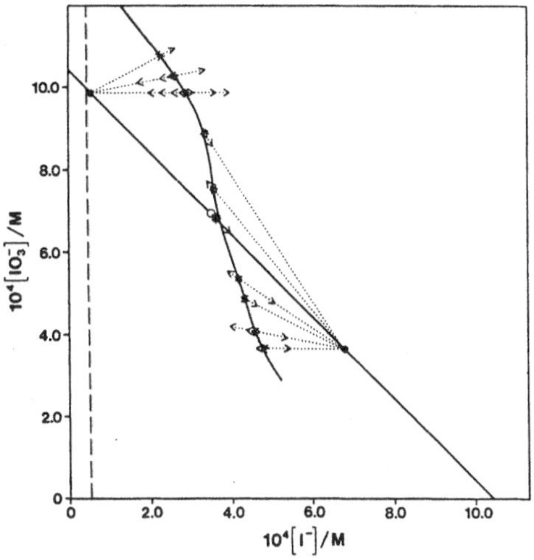

Fig. 3. Separatrix. Arrow heads show initial positions of perturbed states and their directions indicate to which steady state the system relaxes. Dotted lines indicate origin of perturbed states. Concentrations and conditions as in Fig. 1 except $k_0 = 9.82 \times 10^{-3}\ s^{-1}$. From Ref. [1].

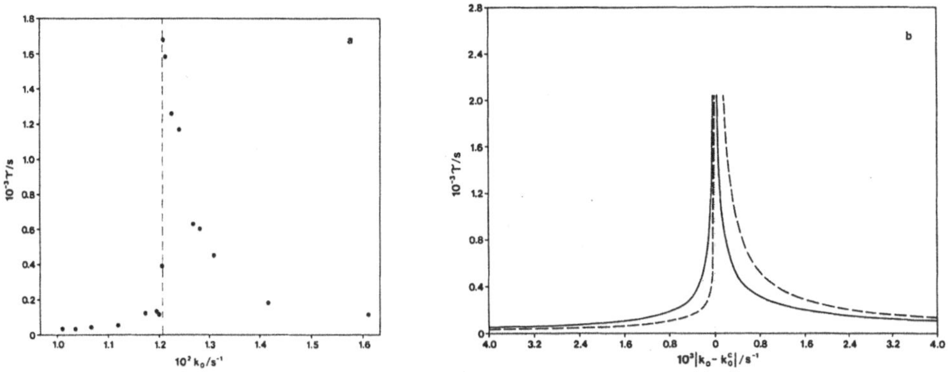

Fig. 4. Relaxation half-times τ as a function of k_0: (a) Experimental values. Concentrations and conditions as in Fig. 1 except $[I^-]_0^c = 1.47 \times 10^{-4}$ M and pH = 2.176 ($[H^+]_0 = 6.67 \times 10^{-3}$ M). (b) Half-times calculated from (R'). Concentrations as in (a) except $[I^-]_0^c = 1.20 \times 10^{-4}$ M. From Ref. [2].

2.0×10^{-4} M increase in iodide concentration above the steady state concentration. A dramatic slowing down is evident near the critical point in a plot of the relaxation half-times as a function of k_0, shown in Fig. 4a. However, the quantitative features of the slowing down are highly dependent on the size of the perturbation, the proximity to the critical point, and the particular definition of relaxation time [2].

Figure 4b shows relaxation half-times calculated [7] from (R'). Shown are half-times (----) for perturbations consisting of 2.0×10^{-4} M increases above the steady state iodide concentration and half-times (——) predicted from linear stability eigenvalues of (R'). Half-times calculated from (R') for perturbations consisting of increases less than ca. 1.0×10^{-7} M above the steady state fall on the linear approximation curve (——). Each curve can be approximately described by (2), where τ is the half-time, k_0^c is the critical reciprocal residence time, and z is the critical exponent.

$$\tau = |k_0 - k_0^c|^{-z} \qquad (2)$$

Experimental values of z in Fig. 4a of 0.50 ± 0.09 and 0.82 ± 0.13 compare with the values (----) in Fig. 4b of 0.58 and 0.85 before and after the critical point, respectively. Values for the linear approximation (——) are 0.77 and 0.64. For small perturbations very near the critical point [2], z ≈ 2/3, in accord with theoretical predictions [8].

Acknowledgment

This work was supported by the National Science Foundation (Grant CHE-8311360).

References

1. T. Pifer, N. Ganapathisubramanian, and K. Showalter, J. Chem. Phys., submitted.

2. N. Ganapathisubramanian, T. Pifer, and K. Showalter, J. Chem. Phys., submitted.

3. N. Ganapathisubramanian and K. Showalter, J. Phys. Chem. 87, 1098 (1983); 87, 4014 (1983).

4. N. Ganapathisubramanian and K. Showalter, J. Chem. Phys. 80, 4177 (1984).

5. G. A. Papsin, A. Hanna, and K. Showalter, J. Phys. Chem. 85, 2575 (1981).

6. P. De Kepper, I. R. Epstein, K. Kustin, J. Am. Chem. Soc. 103, 6121 (1981).

7. A. C. Hindmarsh, "LSODE", Lawrence Livermore Laboratory: Livermore, CA, 1980.

8. G. Dewel, P. Borckmans, and D. Walgraef, J. Phys. Chem., in press.

Isothermal Autocatalysis in Open Systems (CSTR): Simple Models and Complex Behaviour

Peter Gray and David Knapp
University of Leeds, Leeds LS2 9JT, United Kingdom
Stephen Scott
Macquarie University, N.S.W. 2113, Australia

Autocatalysis lies at the heart of all isothermal oscillatory systems. When the catalyst is not completely stable but undergoes decay [1-5], particularly varied behaviour is possible even in the simplest of open systems - the cstr operating isothermally. The kinetics may be represented by the scheme

quadratic	$A + B \rightarrow 2B$	rate $= k_1ab$	(*1a*)
cubic	$A + 2B \rightarrow 3B$	rate $= k_1ab^2$	(*1b*)
	$B \rightarrow C$	rate $= k_2b$	(*2*)

Cubic autocatalysis with catalyst decay can give rise to strange patterns of stationary states and to sustained oscillations. (This is also true of enlarged schemes [6] incorporating the reverse steps and the uncatalyzed reaction although the algebra is heavier.) Today we are mainly concerned with the birth, growth and extinction of stable oscillations and to the response of the system to continuous changes in residence time.

1 Exotic patterns of stationary-state dependence on residence time

Consider the simple, irreversible cubic autocatalysis (*1*) coupled with a first-order decay or poisoning reaction (*2*). The mass-balance equations can be readily written [2] in the following dimensionless form

$$\frac{d\alpha}{d\tau} = -\alpha\beta^2 + \frac{1}{\tau_{res}}(1-\alpha) \qquad (3a)$$

$$\frac{d\beta}{d\tau} = \alpha\beta^2 - \frac{\beta}{\tau_2} + \frac{1}{\tau_{res}}(\beta_0 - \beta) \qquad (3b)$$

where

$$\alpha = a/a_0 \qquad \beta = b/a_0 \qquad \beta_0 = b_0/a_0 \qquad (4)$$

$$\tau = k_1a_0^2t \qquad \tau_2 = k_1a_0^2/k_2 \qquad \tau_{res} = k_1a_0^2t_{res}.$$

Here a_0 and b_0 are the concentrations of A and B that would be established by the inflow in the absence of chemical reaction and t_{res} is the mean residence time. The stationary-state condition is

$$\alpha_{ss}(1 + \beta_0 - \alpha_{ss})^2 = \frac{1}{\tau_{res}}\{1 + \frac{\tau_{res}}{\tau_2}\}^2(1 - \alpha_{ss}). \qquad (5)$$

This is a cubic equation in α_{ss} and a quadratic in the residence time τ_{res}. In the special case of no B in the inflow ($\beta_0 = 0$), one solution of this equation is $(1 - \alpha_{ss}) = 0$. This factor may then be cancelled from each side to leave a quadratic in α_{ss} for the two remaining stationary-states

$$\alpha_{ss}(1 - \alpha_{ss}) = \frac{1}{\tau_{res}}(1 + \frac{\tau_{res}}{\tau_2})^2. \qquad (6)$$

Because the left-hand side has a maximum value of $\frac{1}{4}$ when $\alpha_{ss} = \frac{1}{2}$, and the right-

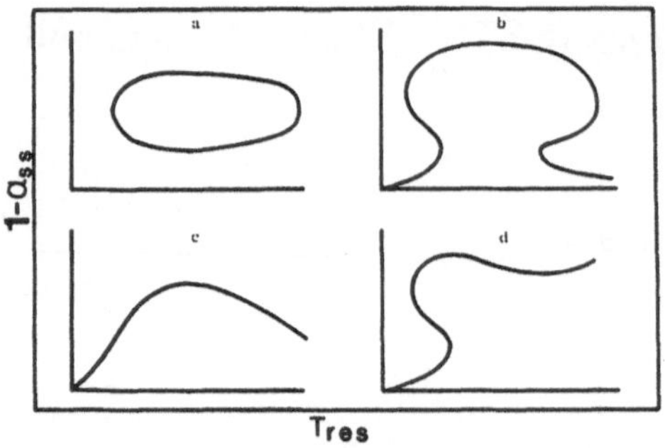

FIGURE 1 Different dependence of stationary-state extent of conversion found for cubic autocatalysis with catalyst decay

hand side a minimum of $4/\tau_2$ when $\tau_{res} = \tau_2$, this equation can only have real solutions if $\tau_2 \geqq 16$. With this condition satisfied, the resulting dependence of the stationary-state extent of conversion $1 - \alpha_{ss}$ on residence time portrays an isola pattern as shown in figure 1a.

A graphical route to evaluating the stationary-state patterns for the general case of non-zero catalyst inflow ($\beta_0 > 0$) has been given elsewhere [2,3]. As well as isolas, mushrooms and unique dependences of $1 - \alpha_{ss}$ on τ_{res} may be found (Figs. 1b and c). Figure 2 divides the $\beta_0 - \tau_2$ parameter plane into regions in which each of these different responses occur.

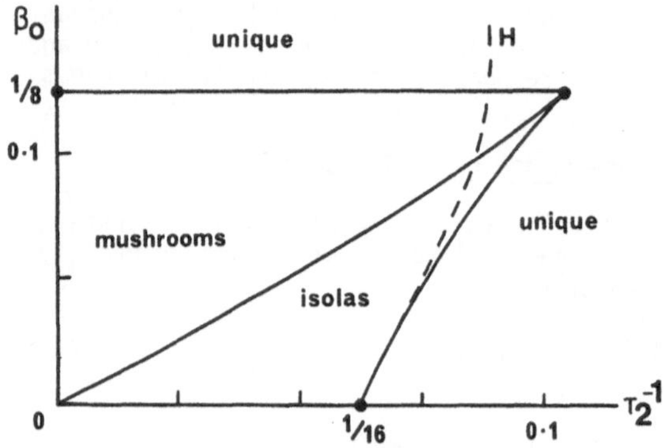

FIGURE 2 $\beta_0 - \tau_2$ regions for mushroom, isola and unique dependences of stationary-state on residence time. The line H divides systems which may show Hopf bifurcation (high τ_2) from those which may not (low τ_2)

2 Stability and character of stationary-states

The *character* and *local stability* of the stationary-state solutions along the different branches of the isola and mushroom *etc.* patterns are evaluated in the usual manner [7] *via* the eigenvalues of the Jacobian matrix.

In the case of no catalyst inflow, the lowest solution corresponding to no conversion is always a stable node (sn). The middle solution is always an unstable saddle point (sp). The nature of the highest extent of conversion varies with the residence time: it may be a stable node, stable focus (sf), unstable focus (uf) or unstable node (un). For non-zero catalyst inflow there is further complication as the character along the lowest branch, which now corresponds to non-zero extents of conversion, varies between stable and unstable node or focus. Some of the different sequences found are indicated on the mushrooms in Fig. 3.

3 Hopf bifurcation, limit cycles and sustained oscillations

Points at which a stable focus becomes unstable correspond to points of Hopf bifurcation [8]. These are the conditions at which limit cycles emerge in the α-β phase-plane. The line H in figure 2 divides the $\beta_0 - \tau_2$ parameter space into

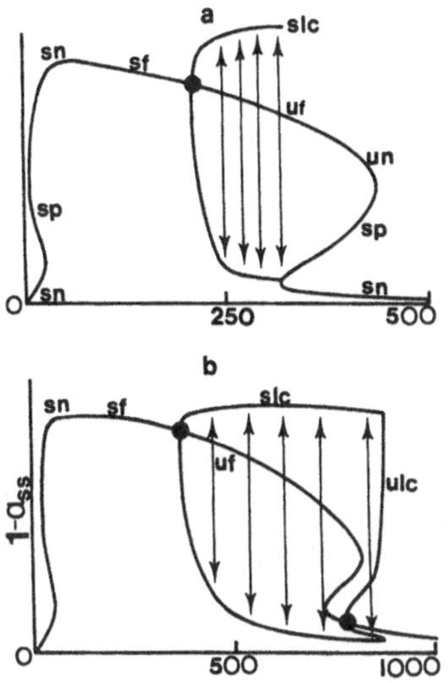

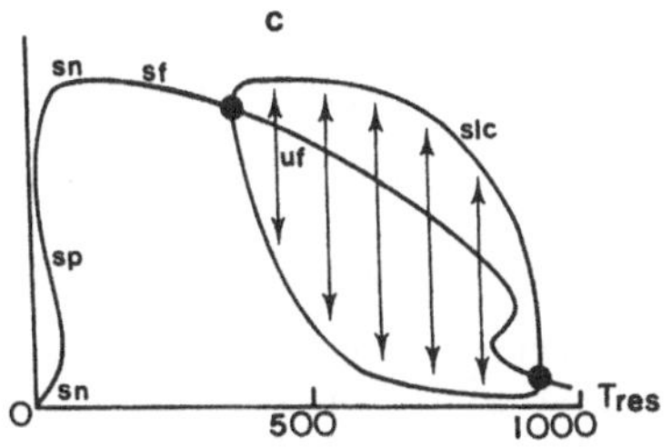

FIGURE 3 Mushroom dependences of $1 - \alpha_{ss}$ on τ_{res} showing different behaviour at long residence time: (a) $\tau_2 = 40$, $\beta_0 = {}^1/15$; (b) $\tau_2 = 50$, $\beta_0 = 0.1$; (c) $\tau_2 = 50$, $\beta_0 = 0.12$. Also shown are envelopes of limit cycles in $\alpha - \beta$ plane

two: systems lying to the left (at high τ_2) exhibit at least one Hopf bifurcation at some residence time, those lying to the right (at short τ_2) do not. For no catalyst inflow ($\beta_0 = 0$), limit cycle phenomena only occur if $\tau_2 > 16$.

Limit cycles may be stable or unstable. Only stable limit cycles lead to sustained oscillations in the reactant and catalyst concentrations. Unstable cycles cannot give rise to observable oscillations, but act in a similar way to the separatrices of a saddle point, as watersheds. Trajectories starting within an unstable limit cycle tend to the stable stationary-state lying inside: other trajectories move across the phase-plane to other stable states or cycles. The size of a limit cycle is thus important whether it is stable or unstable.

4 Experimental responses to variations in residence time: birth, growth and extinction of oscillatory behaviour

The three mushroom dependences of the stationary-state on residence time shown in Fig. 3 differ from each other, in the way in which oscillatory behaviour sets in and disappears. These in turn lead to different responses of the system as the residence time is varied. The observations which would be recorded by an experimenter slowly decreasing the flow-rate in each case are set out below.

Figure 3(a) For an experiment set up with an initial residence time $\tau_{res} = 100$, there is a unique stationary-state solution which has stable focal character. Small perturbations decay in a damped oscillatory fashion. As the residence time increases, the degree of damping becomes less until a point of Hopf bifurcation is encountered, in this case at $\tau^*_{res} = 200$. Beyond this point the system moves into a series of self-excited, sustained oscillations. These grow from a vanishingly small amplitude at τ^*_{res}: initially the amplitude is proportional to $(\tau_{res} - \tau^*_{res})^{\frac{1}{2}}$. The oscillations have a non-zero period at the point of Hopf bifurcation which is determined by the imaginary eigenvalues of the Jacobian matrix $\lambda = \pm i\omega_0$, $\rho = 2\pi/\omega_0$. There is no hysteresis at this "supercritical" bifurcation [8]; decreasing the residence time simply causes the amplitude of the oscillations to shrink back to zero at τ^*_{res}. As the residence time is increased further, the oscillations become larger and separated by longer periods. The envelope surrounding the upper branch of stationary-state in Fig. 3(a) shows the highest and lowest values of $1 - \alpha$ attained during the pulses as a function of residence time.

At τ_{res} = 312.5, a saddle point and a stable node appear in the phase-plane, corresponding to the two other stationary-states. Shortly after this, the limit cycle grows sufficiently large to touch the separatrix of the saddle point and form a homoclinic orbit. The period of motion around this cycle becomes infinite and beyond this point oscillations are not found. There is thus a sudden quenching of large amplitude oscillations. For all residence times greater than τ_{res} = 313 the system sits at the lowest, stable state. There is also hysteresis at this limit. As the residence time is decreased, we pass the point at τ_{res} = 313 at which the limit cycle burst on the upward journey, but do not now move from the lower branch. Only when there is a saddle-node bifurcation of the lower two stationary states at τ_{res} = 312.5 does the system jump back into large oscillations.

Figure 3(b) Again, if an experiment is started at τ_{res} = 150 there is a unique stable node. On increasing the residence time, a supercritical Hopf bifurcation of the above type is encountered at τ^*_{res} = 326 and approximately sinusoidal, small-amplitude oscillations set in. These become larger and have longer periods as τ_{res} increases. At τ_{res} = 782 there is a saddle-node bifurcation, but the two emerging stationary-states now lie within the limit cycle. The lowest extent of conversion is at first an unstable node and then becomes an unstable focus. For τ^*_{res} = 786 there is a point of Hopf bifurcation along the lower branch. Here an unstable limit cycle emerges to surround the stable focus, still within the original stable limit cycle. As the residence time is further increased, the upper two solutions merge and disappear. Finally, the unstable limit cycle becomes large enough to touch the stable cycle and oscillations are abruptly quenched. Beyond this point the system sits on the lowest branch.

Again there is hysteresis at this long-residence-time end of the range of oscillatory behaviour. If τ_{res} is now reduced, the system stays on the stable lower branch until the Hopf bifurcation point at τ^*_{res} = 786, whence it jumps back into large amplitude oscillations about all three stationary-states.

Figure 3(c) A further increase of the catalyst inlet concentration leads to the pattern in Fig. 3c. The behaviour at long residence-times is now quite different.

Once the oscillations have set in at the first point of Hopf bifurcation (τ^*_{res} = 349), there are two saddle-node bifurcations at τ_{res} = 921 and τ_{res} = 927 within the limit cycle. The first gives rise to the emergence of an unstable lower branch and the saddle point corresponding to the middle branch. At the second the middle branch merges with the highest extent of conversion. The latter occurs before the lowest stationary-state has regained its stability, i.e. before the second point of Hopf bifurcation. When the unstable focus finally becomes stable, at τ^*_{res} = 939, the limit cycle shrinks to a point and the oscillations vanish at this second supercritical bifurcation. There is now no hysteresis at either end of the oscillatory range.

4 Discussion and conclusions

The cubic rate-law model presented here affords a prototype for many systems of practical interest, including solution-phase kinetics and enzyme reactions. It cannot be stressed too strongly that steps 1(a) and (b) do not have to be regarded as elementary steps. They may be combinations of such steps. Thus the reaction between arsenite and iodate ions, which is autocatalytic [9,10] in the production of I^-, is well approximated at constant pH by the rate expression

$$\text{rate} \propto \{a + [I^-]\}[I^-][IO_3^-]$$

where $[I^-] + [IO_3^-]$ = constant, over some range of concentration.

The present model differs significantly from the so-called "pool chemical" approximation approach, which assumes that the concentration of reactant A would somehow be held constant, in two important ways. First, there is a finite maximum value which may be attained by the reaction rate, *viz.*

$$R_{max} = \frac{4}{27} k_1 a_0^3 ,$$

whereas the "pool chemical" rate shows hyperexponential growth to infinitely large values [11,12]. Secondly, multistability is a real feature of the cstr equations. Karmann and Hinze [13] have shown that the multiple solutions for the "pool chemical" approach are not stable to the inclusion of lower-order steps and their reverse reactions.

We may also note that the cubic dependence on concentration embodied in step (1b) is not a necessary feature for sustained oscillations. If the decay reaction for the catalyst has a less simple form, for example

$$B \rightarrow C \qquad \text{rate} = k_2 b/(1 + rb), \tag{7}$$

the quadratic non-linearity of step (1a) is sufficient [14].

A rate-law of the form (7) may arise in enzyme reactions, or in heterogeneous catalysis obeying a Langmuir-Hinshelwood law. The oxidation of carbon monoxide at low pressures displays sustained oscillations under isothermal conditions in an open system [15], and results indicate that an important role is played by the state of the reactor surface. This, and other features of the observed behaviour, are readily correlated qualitatively with this simple, two-step model.

Acknowledgments

We are grateful to Dr J.H. Merkin of Leeds and Dr C. Escher of Aachen for helpful discussion.

References

1 S.K. Scott: Ph.D. dissertation, University of Leeds (1982).
2 P. Gray and S.K. Scott: Chem. Engng. Sci. 38, 29-43 (1983): 39 (1984).
3 S.K. Scott: Chem. Engng. Sci. 38, 1701-1708 (1983).
4 P. Gray and S.K. Scott: J. phys. Chem. 87, 1835-38 (1983): J. Chem. Phys. 79 6421-23 (1983): Ber. Dtsch. Bunsenges 87, 379-382 (1983).
5 P. Gray and S.K. Scott: Chemical Instabilities (Eds. G. Nicolis and F. Baras); Reidel (1984).
6 B.F. Gray, P. Gray and S.K. Scott: J. Chem. Soc. Faraday Trans. 2 (1984) in the press.
7 A.A. Andronov, A.A. Vitt and S.E. Khakin: Theory of Oscillators Pergamon Press (1966).
8 B.D. Hassard, N.D. Kazarinoff and Y.-H. Wan: Theory and applications of Hopf bifurcation (Cambridge University Press 1981).
9 H.-G. Lintz and W. Weber: Chem. Engng. Sci. 35, 203-208 (1980): G.A. Papsin, A. Hanna and K. Showalter: J. phys. Chem. 85, 2575-82 (1981).
10 N. Ganapathisubramanian and K. Showalter: J. phys. Chem. 87, 1098-99 (1983): J. chem. Phys. 80, 4177 (1984).
11 G.A.M. King: J. chem. Soc. Faraday Trans. 1 79, 750-8 (1983).
12 R.M. Noyes: J. phys. Chem. (1984) in the press.
13 K.-P. Karmann and J. Hinze: J. chem. Phys. 72, 5476 (1980).
14 S.K. Scott: J. chem. Soc. Faraday Trans. 2 (1984) in the press: J.H. Merkin, D.J. Needham and S.K. Scott: Proc. R. Soc. Lond. A (1984) in the press.
15 P. Gray, J.F. Griffiths and S.K. Scott: Proc. R. Soc. Lond. A (1984): 20th Symp. (Int.) Combust. Ann Arbor (1984).

Composition Variables Needed to Model Complex Chemical Systems

Richard M. Noyes
Department of Chemistry, University of Oregon, Eugene, OR 97403, USA

1. Introduction

The state of a closed, uniform, thermostated chemical system is uniquely defined by specifying the temperature, the pressure, the number of moles of each species capable of undergoing any sort of reaction under the conditions of interest, and the numbers of moles of any species (such as inert solvent) which are incapable of undergoing any chemical change.

Such a system will ultimately decay to a state of chemical equilibrium calculable in principle by invoking the chemical potentials of all potentially reactive species in their standard states and the activity coefficients of species as functions of composition.

The trajectory of that decay may be very complex, including oscillations in concentrations of some species and even spontaneous nonuniformities of composition. However, if the homogeneous system does indeed remain uniform at all times, the trajectory can always be described in terms of a number of composition variables equal to the total number of species capable of reaction.

The change in composition of the system can also be described in terms of variables based on extents of advancement of various chemical processes. Because there are constraints such as conservation of the number of atoms of each element, the number of independent variables defined in terms of extents of reaction will always be less than the number of species undergoing reaction, and the state of even a very complex system can often be described in terms of only one or two such variables. Description of dynamic behavior will usually require more variables, but even this number will often be much less than the number of reactive species.

The objective of the present paper is to propose procedures for determining the number of independent composition variables necessary to describe the behavior of any specific system. The argument will invoke conservation of mass as expressed in chemical stoichiometry and will rely heavily upon a recent treatment by CORIO [1,2]. A further objective, not addressed in this manuscript, is to examine ways in which the directional constraints of thermodynamics can be applied to properly selected reaction variables.

2. Stoichiometrically Significant Change

Any net chemical change in a uniform system can be described by an equation like (T) involving ρ reactant species and π product species. Each of these species is distinguishable from all others because of differences in number of atoms of some element or because of some distinguishable characteristic such as isomerism, electronic charge, or electronic excitation.

$$\sum_{i=1}^{\rho} \alpha_{Ti} R_i + \sum_{j=1}^{\pi} \alpha_{Tj} P_j = 0 \qquad (T)$$

The α_i coefficients are all negative, and the α_j coefficients are all positive; they need not be integral, and their ratios need not even be rational. The R_i and P_j terms are formulas of chemical species. Let λ denote the total number of species represented by equation (T). Then

$$\lambda = \rho + \pi. \tag{1}$$

Conservation of mass dictates that it should in principle be possible to balance equation (T) to individual atoms of every element represented. However, neither analytical technology nor interest in the system could attain or justify such precision. It will be satisfactory if equation (T) neglects every species the magnitude of whose coefficient α is less than some arbitrary fraction (such as 0.01%) of the maximum α. An equation balanced to this accuracy is said to represent the stoichiometrically significant change in the system.

3. Multiplicity of Stoichiometrically Significant Change

Although λ species change concentration during the change in state of the system, these changes are not all independent because conservation of atoms of each element imposes a constraint on permissible behavior.

Let ε be the number of elements whose atoms are present in at least one reactant and one product species in equation (T). If the equation as written involves ions in addition to neutral molecules, conservation of electronic charge imposes an additional constraint which can be treated as a unit increase in ε. Finally, if isotopic composition of some element differs in different species, ε must be increased to represent, as a different kind of atom, each isotope whose fraction varies.

Although conservation of mass imposes ε constraints on the system, those constraints are not all necessarily independent. Let ν be the number of independent mass-conservation and charge-balance constraints imposed on equation (T). Of necessity $\varepsilon \geq \nu$. CORIO [1,2] has shown how ν can be evaluated objectively as the rank associated with the matrix obtained from the α coefficients in equation (T).

The stoichiometric multiplicity, ϕ, can be defined to be the number of independent variables needed to describe any arbitrarily selected change of state of the system. It is the number of potential composition variables less the number of constraints. Then

$$\phi = \lambda - \nu = \rho + \pi - \nu. \tag{2}$$

Of course ϕ is positive and integral for any real system.

4. Independent Component Processes

If consideration is directed to processes rather than to species, the multiplicity ϕ is the number of independent stoichiometric processes necessary to describe any arbitrary overall process. Then any possible combination of coefficients for which equation (T) is balanced can be generated by a linear combination of ϕ independent component processes each of the form of equation (C).

$$\sum_{i=1}^{\rho} \alpha_{mi} R_i + \sum_{j=1}^{\pi} \alpha_{mj} P_j = 0 \qquad m = 1,2,\ldots,\phi \tag{C}$$

The coefficients of that combination need not be rational but their ratios must be.

The number of independent component processes is uniquely defined by the procedures described above, but the specific processes may often be selected arbitrarily according to convenience. The situation is completely analogous to the selection of component species in an application of the Gibbs phase rule.

5. Extents of Advancement of Component Processes

Let ξ_m be the extent of advancement of the mth independent component process where ξ_m is defined to be zero for the initial state of the system and is positive in the equilibrium state.

Let c_n be the concentration of the nth of the λ species which are either reactants or products in equation (T). Then

$$dc_n = \sum_{m=1}^{\phi} \alpha_{mn} d\xi_m \qquad n = 1,2,\ldots,\lambda. \qquad (3)$$

The λ equations of the form of (3) permit a transformation of variables so that change in state of the system can be described in terms of the ϕ variables denoting extents of advancement of independent processes. As equation (2) shows, $\phi < \lambda$ for any conceivable system, and a smaller number of variables will be employed if changes of state are described in terms of advancement of independent processes rather than in terms of changes in concentrations of all species. In practice, the stoichiometrically significant change in even quite complex systems can often be described in terms of no more than two or three independent component processes. Thus, we have recently shown [3] how the stoichiometrically significant change in the Belousov-Zhabotinsky reaction can be described in terms of one or at most two component processes. However, even ϕ may become rather large for processes like the pyrolysis of some organic compounds.

6. Extension to All Species

The above argument was developed for treating stoichiometric changes in the total state of a system to the accuracy attainable experimentally. However, there may be species whose presence is essential to describing the dynamic behavior of the system even though they do not contribute significantly to changes in thermodynamic properties. Let there be ι of these intermediate species, each of which is either a very minor reactant or product or else is always formed and destroyed at such nearly identical rates that its absolute concentration never changes significantly. Then equation (T) can be rewritten as (T') which describes the exact stoichiometric change in the system.

$$\sum_{i=1}^{\rho} \alpha_{Ti} R_i + \sum_{j=1}^{\pi} \alpha_{Tj} P_j + \sum_{k=1}^{\iota} \alpha_{Tk} I_k = 0 \qquad (T')$$

The α_k coefficients may be positive, negative, or even exactly zero, but they are so small in magnitude that equations (T) and (T') yield insignificantly different results for any application to stoichiometric or thermodynamic considerations.

7. Multiplicity of Exact Stoichiometric Change

The multiplicity of the exact stoichiometric change can be determined by procedures analogous to those for the stoichiometrically significant change. The total number of composition variables is now κ given by

$$\kappa = \rho + \pi + \iota = \lambda + \iota. \qquad (4)$$

The number of independent constraints on equation (T') can be designated ν'.

The quantity ν' may be greater than ν because of homogeneous catalysis by an element which does not appear in equation (T) or because the additional species in (T') remove a degeneracy which had reduced the rank of the matrix of coefficients for equation (T). The multiplicity, θ, of the exact stoichiometric change is given by

$$\theta = \kappa - \nu'. \tag{5}$$

Here θ is the number of independent stoichiometric processes necessary to describe the exact evolution of the system. For many systems, there will be so many intermediates that θ will be much greater than ϕ.

8. Variables to Describe Dynamic Behavior

The reason for considering exact stoichiometric change is that the dynamic behavior of a system is often influenced by concentrations of species very much smaller than those necessary to define the state for purposes of evaluating thermodynamic properties. Dynamic modeling may need to be concerned with up to θ independent composition variables even though thermodynamic modeling need consider only ϕ such variables.

Fortunately, two additional types of constraint can reduce the number of independent composition variables needed to define the state for dynamic modeling. One such constraint exists if there are rapid reversible equilibria. Any such equilibrium requires that the average time for reaction of any species by the equilibrium process is very much less than the average lifetime before reaction of that species by a process considered in the dynamic modeling. The criterion for independence of equilibria will involve the rank of a tensor similar to that discussed in the evaluation of ν. Each independent equilibrium can reduce the value of θ by unity. If all species in the equilibrium are reactants or products, the value of ϕ will also be reduced.

A second powerful constraint arises because many complex mechanisms involve what CLARKE [4] calls flow-through intermediates. Such a species is always present at very low concentrations and must react either by the reverse of the elementary process by which it has just been formed or else by only one other process. Such a situation can be illustrated by the elementary processes (A1) and (A2).

$$R_1 + R_2 \rightleftarrows I_a + P_1 \tag{A1}$$

$$R_3 + I_a \rightleftarrows P_2 + P_3 \tag{A2}$$

If $c_{R1} \geq c_{R2}$ and $c_{P2} \geq c_{P3}$, then I_a is a flow-through intermediate if $k_{-1}c_{P1} + k_2c_{R3} >> k_1c_{R1} + k_{-2}c_{P2}$ and if I_a enters into no other reactions.

Let a particular system involve τ independent rapid equilibria, and let there be σ flow-through intermediates. Of course there must always be a degree of arbitrariness as to whether a specific equilibrium is established sufficiently rapidly compared to the time scale of importance for the modeling and whether a specific flow-through intermediate is present at sufficiently small concentration compared to its precursors and successors.

Let ω be the modeling multiplicity or number of independent composition variables needed to model the system to the desired precision. The argument as developed above has provided unambiguous ways to evaluate ω in terms of the relation

$$\omega = \theta - \sigma - \tau. \tag{6}$$

9. Concluding Comments

Some attentive readers may be concerned that the minimum number of composition variables necessary for dynamic modeling of the system has been defined in terms of the number of independent stoichiometric processes. The detailed mechanism of chemical change is ultimately the consequence of elementary processes each of which takes place in a single step. The number of elementary processes in a complex system may be greater than θ. However, those processes are not stoichiometrically independent. Different combinations of elementary processes then offer parallel paths to accomplish the same change of state. Dynamic modeling must recognize contributions from all of these parallel paths, but the number of truly independent composition variables needed to define the instantaneous state of the system will not exceed the value calculated by the method of this manuscript.

Acknowledgment. The development of these ideas was supported in part by a Grant from the United States National Science Foundation.

References

1. P. L. Corio: Trans. Kentucky Acad. Sci. 32, 51-56 (1970)

2. P. L. Corio: J. Phys. Chem. 88, 1825-1833 (1984)

3. R. M. Noyes: J. Chem. Phys. 80, 6071-6078 (1984)

4. B. L. Clarke: J. Chem. Phys. 64, 4168 (1976)

Modelling of Chemical Dynamics Towards a Methodological Approach

P. Hanusse

Centre de Recherche Paul Pascal, Domaine universitaire
F-33405 Talence Cédex, France

The modelling of chemical dynamics, in particular out of equilibrium, is a field where the variety and complexity of behaviour is at its greatest. It is, on the one hand, the subject of concern to theoriticians, chemists and mathematicians, who find in it remarkable examples of dynamical systems. They usually study formal simple models, using a low dimensional space of *pertinent* or *essential* state variables. They use rather sophisticated mathematical methods like bifurcation theory, interaction of singularities [1], graph theory etc... On the other hand the experimental chemist starts from a very different point of view. Given a set of reactants [to which he confers reaction properties using chemical knowledge which, for a given system, is far from being complete] he designs a model or reaction scheme, the complexity of which is not amenable to any systematic analysis.

It seems clear that the gap that exists between these two approaches is still so wide that it does not allow the experimentalist to obtain full benefit of the most recent or simply the most efficient relevant theoretical results. For this reason, among others, the modelling of the behaviour of an experimental chemical system is still very much of an art, even though progress has been achieved towards clarification and classification |3|. The gap will be reduced only through the development of a real computer-aided modelling methodology, providing the chemist with an effective, reliable and flexible tool through which, using his own language, he could have access to the ever-progressing theoretical knowledge of complex systems.

The usual approach of an experimental chemist can be considered as *deductive*, to the extent that, given a detailed chemical description of the system - the choice of reactants, of possible reactions between them, the choice of *physical* kinetic parameter values or *sensible* concentrations - he is essentially interested in finding under which conditions his model can produce the behaviours that he observes in his experiment. He will usually make a simulation of his model, which most of the time is just another way to repeat the experiment on the computer, or at least an idealized version of it. Yet he may learn something is this process. He may even be lucky enough to find a satisfactory agreement between the simulation and the experiment, between to two experiments we might say. But what does it mean as far as the intrinsic capabilities of the model to produce various dynamical behaviours are concerned ? All those who are familiar with this approach have quickly perceived its limitation and feel very frustrated by this random quest in a high dimensional parameter space, so high that it is very unlikely that the basic dynamical structure of the model could be detected.

Contrary to this deductive approach one could consider an *inductive* one, which starting from the minimal chemical description - choice of reactants and reaction network, with the minimum amount of restrictions - would a priori look for the structural properties of the model before any attempt to consider physical values of parameters and of course before any simulation. For example, graph theory techniques [2] are very much of this type.

For some time now we have been involved in such an attempt, trying to incorporate in a simulation system a reaction scheme analysis tool based on a *Reaction Scheme Translator* (RST). Starting from the description of the reaction scheme in chemical notation including experimental conditions, this program generates all the equations that are necessary for the analysis of the model. In a first step this program has been used to produce the system of differential equations as well as the jacobian of any complex system under various types of constraints - closed system, controlled concentrations, continuous stirred tank reactor etc... -. The generated program can be included in a simulation system which is described elsewhere |4,5|. It provides data-handling capabilities, various integration algorithms, file management, graphs, trajectories and profile display utilities and various analysis tools such as steady states calculation and stability analysis etc...

We shall focus on R.S.T. as a basis of a Reaction Scheme Analyser (RSA) which is presently being developed. First of all R.S.T. is not an *interpreter*. The reaction scheme is truly compiled and is submitted to a symbolic manipulation which allows the generated program to be optimized for the given model, leading to a very efficient code. Virtually any set of equations or algorithm can be generated in any programming language, thus allowing a computer-aided use of mathematical tools. Secondly this symbolic treatment guarantees the generation of error-free modules which is highly appreciable when dealing whith complex reaction schemes. Typically a 20 species model would lead to 20 differential equations containing 100 to 200 terms and a jacobian of 200 to 300 equations containing 300 to 500 terms. Nevertheless, designing an effective modelling system,which implements the inductive approach that we defined earlier, requires new theoretical methods that allow an explicit and systematic analysis of a complex system. We shall briefly present the basis of such a method. It relies on the fact that most dynamical behaviours that one observes in chemical systems have an effective low dimensionality. Indeed one to three dimensions are enough even to produce chaos. Starting from an n-variable model we would like to determine under which conditions, if they exist, the model is able to produce various given behaviours such as bistability, sub- or supercritical Hopf bifurcation, excitability, etc..

In the deductive approach, given rate-constants one looks for steady states. In general there is no way to know where they are and how many they are. In the inductive approach we shall give the value of a steady state and,since there could be several of them,we shall determine under which conditions they all merge, and from this point of maximum degeneracy we shall unfold the solutions to obtain the configuration of steady states for any given set of control parameters. For this procedure to work, it is necessary to free all rate constants of any a priori renormalization; also, the reaction scheme must be *generic* in a sense that we shall not make more precise here. Let us just mention that for any given model there exists an *equivalent* generic model, i.e. having the same bifurcation diagram.

Consider a general model with the following rate equations

$$\dot{X}_i = F_i(X,K)$$

where X is a vector of state variables and K a vector of control parameters. If there exists a stationary state X^s, for any generic model, it is possible to span the bifurcation space in such a way that the steady state is $X^s = 1$. This determines a set of relations between the K's : $0 = F_i(1,K)$. We can then work with reduced variables $x = X^s - 1$ in a restricted parameter space k. The dynamical equations are

$$\dot{x}_i = f_i(x,k)$$

with

$$0 = f_i(o,k)$$

Suppose now that one is looking for bifurcations with codimension 1. One variable only will be necessary to span the phase space. In other words, we shall first consider the interaction of steady states in phase space only in one dimension. In that case the solutions of the steady state equations will move along a line passing through the reference state $x_i = 0$, which can be projected out on any coordinate (in a typical situation). This one-dimensional variety can be parametrized by a Taylor expansion of the solutions around the zero state :

$$x_i = \sum_{\alpha=1}^{m} p_{i\alpha}\, x^{\alpha}$$

where x may be one of the x_i's. Inserting this expansion in the steady state equations, which are usually of polynomial type (in particular in chemistry due to mass action kinetic law), or can also be expanded around zero :

$$0 = J_{ij}\, x_j + K_{ijk}\, x_j\, x_k + \ldots,$$

where we use the summation convention, one obtains :

$$0 = J_{ij}\, p_{j\alpha}\, x^{\alpha} + K_{ijk}\, p_{j\alpha}\, p_{k\beta}\, x^{\alpha+\beta} + \ldots$$

We have n such equations. n-1 of them must be satisfied identically for all x. cially if one tries to avoid numerical calculations. Symbolic generation of the va-obtain :

$$0 = J_{ij}\, p_{j1} \quad , \quad i = 1,\ldots,n-1$$

$$0 = J_{ij}\, p_{i2} + K_{ijk}\, p_{j1}\, p_{k1} + \ldots,\ i = 1,\ldots,n-1$$

etc...

From these, the coefficients of the expansion, the p's, can be identified in terms of the kinetic parameters.

The last equation, for i = n, which has the same structure, once the expressions of the p's have been inserted, has the general form :

$$0 = A_1(k)\, x + A_2(k)\, x^2 + A_3(k)\, x^3 + \ldots$$

$$A_1(k) = J_{nj}\, p_{j1}$$

$$A_2(k) = J_{nj}\, p_{j2} + K_{njk}\, p_{j1}\, p_{k1} + \ldots$$

.....

The dependence on k expresses the fact that these terms are only function of the kinetic parameters. The above general equation describes the behaviour of the steady states along the one dimensional variety. By construction x = 0 is a solution. It is possible to have a double steady state if $A_1(k) = 0$. One sees that, together with the equation for the identification of p_{i1}, it says :

$$J_{ij}\, p_{i1} = 0 \quad \text{for all } i \text{ and } j,$$

that is, J has a zero eigen value, which is to be expected when two steady states coalesce. If $A_2(k) = 0$, there exists a triple steady state. If these conditions can be fullfilled, then the model can exhibit bistability. Clearly in one dimension the results of catastrophy theory [6] can be applied.

By this method it is possible to determine the conditions that the parameters should satisfy, to observe various types of bifurcations. It can be extended to higher dimensions to detect Hopf bifurcations, excitability, saddle-node bifurcations etc... For a given complex system this task is usually very difficult, especially if one tries to avoid numerical calculations. Symbolic generation of the various equations can be performed. This is the role of RSA, the Reaction Scheme Analyser. By this technique we think that it will be possible to analyze the dynamical structure of complex models, or even to classify elementary formal models.

References

1. J. Guckenheimer, P. Holmes, Nonlinear Oscillations, Dynamical Systems and Bifurcation of vector fields, Applied Mathematical Sciences, vol. 42, Springer (1984)
2. B. Clarke, Adv. Chem. Phys. vol. XLIII (1980)
3. J. Boissonade, P. De Kepper, J. Phys. Chem. 84, 501 (1980)
4. P. Richetti, Thesis in Chemical Physics, University of Bordeaux I (1982)
5. P. Hanusse, P. Richetti, Proceedings of the National Meeting of the "Société Française de Chimie", Nancy (1984)
6. R. Gilmore, Catastrophe theory for scientists and engineers, Wiley (1981)

Bistability and Oscillations in a Flow Reactor: The Systems H_2O_2-KI with and Without Iodate and $(NH_4)_2\ S_2O_8$-KI with and Without Oxalic Acid

Josette Chopin-Dumas and Marie-Noëlle Papel

Ecole Supérieure de Chimie de Marseille, Domaine Universitaire de St-Jérôme
F-13397 Marseille Cédex, France

We describe two bistable systems and the corresponding oscillators. The first system consists of H_2O_2 and KI in a dilute H_2SO_4 medium. An auxiliary flow of iodate IO_3^-, at elevated temperature only, causes oscillations to take place. The second system consists of $(NH_3)_2\ S_2O_8$ and KI in H_2SO_4 medium, is bistable at T $\gg$ 45°C, and it oscillates in the presence of a flow of oxalic acid $(COOH)_2$.

The redox potential of the solutions was monitored. Part of the reactor exit flux was passed through the cuvette of a spectrophotometer. It was thus possible to monitor the presence of I_3^- and I_2 and their concentration fluctuations. Since the electrode potential is governed by the rapid couples I_3^-/I^- and I_2/I^-, and since $[I_2]$ varies only a little, the potential indicates mainly pI = - log $[I^-]^2$.

1. The H_2O_2-KI system

The H_2O_2-KI reaction is well known. It forms I_2 and I_3^-. This can be observed also by injecting an iodide stream into a reactor fed with acidic hydrogen peroxide: the solution turns golden yellow. The evolution of the potential of this state γ is shown in figure 1 as a function of the KI concentration. An increase of the iodide concentration raises $[I_3^-]$ until the solution is saturated in I_2 and pink I_2 vapour and a black precipitate are formed.

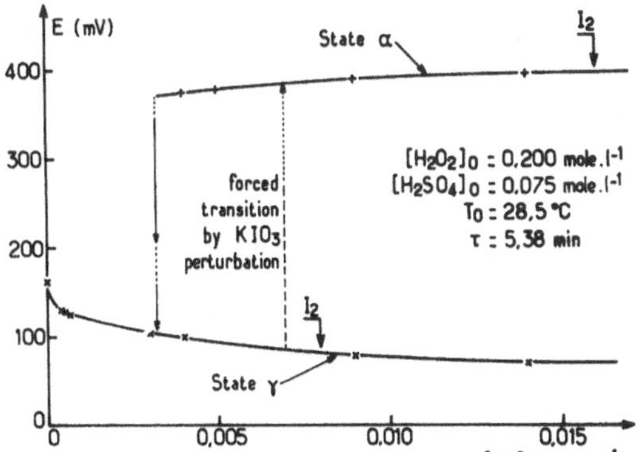

Fig. 1 H_2O_2-KI system: redox potential E (response) versus $[KI]_0$ (constraint)

Upon pertubation by KIO_3 injection, the potential grows rapidly (see dashed arrow in fig. 1), indicating that KIO_3 consumes I^- faster than does H_2O_2, and that I_3^- disappears causing a momentary increase on $[I_2]$. If the potential exceeds a sufficiently high value, it further increases suddenly (dotted arrow in fig. 1) while $[I_2]$ diminishes: the system has undergone a transition to a new state α where the iodine and iodide concentrations are lower than in the γ state, and $[I_3^-]$ is zero. The solution is orange.

When the system is in state α and $[KI]_0$ is increased, one observes, depending on the value of the other constraints, either the precipitation of iodine, at a $[KI]_0$

value higher than that of the γ state, or a transition to the γ state. On the other hand a decrease of $[KI]_0$ within the α state entails only a single response, i.e. the transition to γ. This spontaneous $\alpha \to \gamma$ transition is complex since the time-record of the potential shows two waves indicated by the successive arrows in fig.1. The iodine concentration was found to increase during the first wave, and $[I_3^-]$ reaches its normal level during the second wave. $\alpha \to \gamma$ transitions can also be induced by adding soda.

Hence we found that in a CSTR and in acidic medium, the H_2O_2-KI system can, only by adding iodate, be forced onto a stable state α which is different from the γ state in which the system usually finds itself. In this new state the iodide consumption is increased and the iodine formation is decreased, and KIO_3 is formed apparently [1]. By continuity between state diagrams it was possible to verify that the α and γ states in the H_2O_2-KI system are analogous to the α and γ states of the BRAY reaction [2]. The α state of the BRAY system is observed at high acidity or high iodate concentration, while the γ state corresponds to low acid and iodate concentrations. We suggest therefore that the iodide and iodine production is decreased in the BRAY system when it is fed strongly by KIO_3 or by acid and vice versa.

An increase of temperature and H_2O_2 concentration reduces the potential difference between the two states α and γ. Raising the temperature increases mainly the potential of the γ state, indicating an increased iodide consumption.

2. The H_2O_2-KI-KIO_3 system

At room temperature, addition of a KIO_3 flux to the bistable H_2O_2-KI system does not give rise to oscillations, regardless of concentrations. The H_2O_2-KI-KIO_3 system is also bistable; however a new aspect is that transitions between the stationary states can be reduced by variations of the constraints alone. We have studied this by adding each of the constituents to the two others: adding iodate to H_2O_2-KI, iodide to H_2O_2-KIO_3 and adding H_2O_2 to KIO_3-KI.

Addition of KIO_3 to H_2O_2-KI

This experiment confirms the findings from the transition addition of iodate to H_2O_2-KI. Figure 2 shows how the potential evolves as a function of iodate concentration. The initial H_2O_2-KI system is spontaneously in state γ, the potential is low, and I_3^- and I_2 are formed. A little iodate suffices to make $[I_3^-]$ drop to zero, $[I_2]$ to grow and the potential to rise; more iodide is therefore consumed. Following this, the potential grows slowly with iodate concentration while $[I_2]$ remains stationary: the system receives more iodate and consumes more iodide without producing more iodine. Finally the system transits spontaneously onto state α by drastically increasing its iodide consumption while the production of iodine diminishes.

In fig. 2 one sees that the potential and with it the free $[I^-]$ in state α are nearly independent of $[KIO_3]$ as well as $[I_2]$.

On decreasing $[KIO_3]$, the system way transits onto the γ-state or remains in the α-state, even at $[KIO_3]_0 = 0$. This alternative behaviour depends on the values of the other constraints.

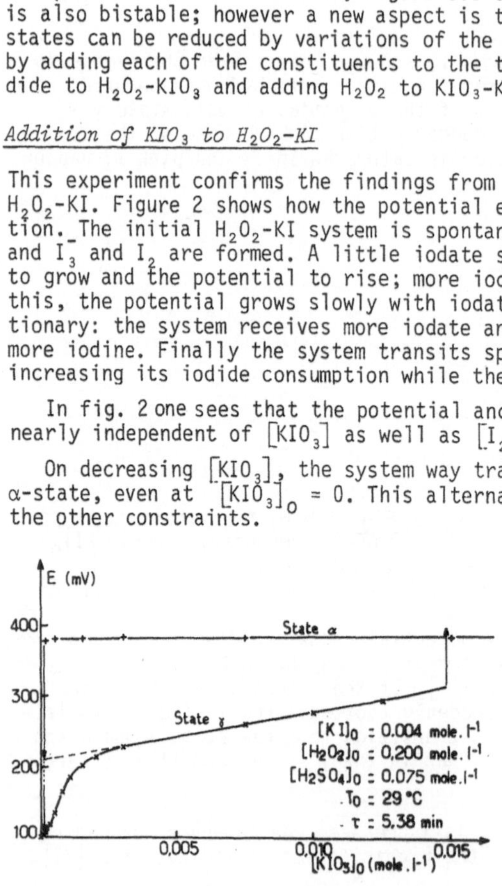

Fig. 2 H_2O_2- KI-KIO_3 system (T = 29°C); redox potential E (response) versus $[KIO_3]_0$ (constraint)

The temperature and $[H_2O_2]_0$ constraints have the same influence on this system as on the H_2O_2-KI system. Increasing their value causes the potential difference between states α and γ to diminish. Similarly when $[KI]_0$ is decreased, and if $[KI]_0$ is made to go to zero, the potentials of α and γ coalesce: it is thus this undifferentiated α-γ state in which the BRAY system H_2O_2-KIO_3 can be found at room temperature. As far as the KIO_3-KI system is concerned, which can be considered as the $[H_2O_2]_0 \rightarrow 0$ limit of the H_2O_2-KIO_3-KI, there exists a single stationary state γ.

Addition of KI to H_2O_2-KIO_3

Where the BRAY system H_2O_2-KIO_3 is in the undifferentiated α-γ state, its response to the addition of iodide is determined by its initial composition, and if the $[H_2O_2]_0$ is fixed it will be governed by $[KIO_3]_0$. At high iodate concentration a small addition of iodide suffices to deplace the system into state α. Transition to state γ may be forced by injecting; it returns however to α when the KI-feed diminishes. Conversely, the γ-state is privileged at low iodate concentrations, and the system behaves as shown in fig. 1. At intermediate KIO_3 concentrations, the bistability of the system is maintained, if we go to very low $[KI]_0 \geqslant 2.5 \times 10^{-5}$ mol.l^{-1}, as can be seen in the four graphes of fig. 3. The state of the system is therefore also determined by $[KIO_3]$. Likewise, in cases ⓐ and ⓑ where $[KIO_3] = 7.5$ and 6.25×10^{-3} mol.l^{-1}, state α is privileged, while γ is preferred in cases ⓒ and ⓓ where $[KIO_3]_0 = 5$ and 3×10^{-3} mol.l^{-1} respectively.

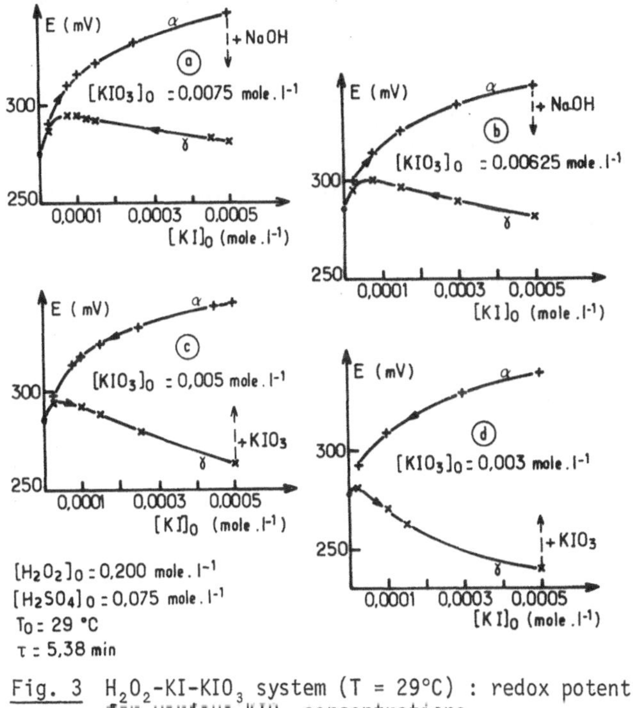

Fig. 3 H_2O_2-KI-KIO_3 system (T = 29°C) : redox potential E (response) versus $[KI]_0$ for various KIO_3 concentrations
ⓐ and ⓑ : state α is privileged ; ⓒ and ⓓ : state γ is privileged

Addition of H_2O_2 to KIO_3-KI

The initial KIO_3-KI system is in state γ. The potential is relatively high, and I_2 is formed without any trace of I_3^-. Addition of H_2O_2 slightly reduces $[I_2]$ and similarly raises the potential (fig. 4): more iodide is thus consumed and less is iodine produced. The γ→α transition corresponds to the same changes, only more accentuated and rapid.

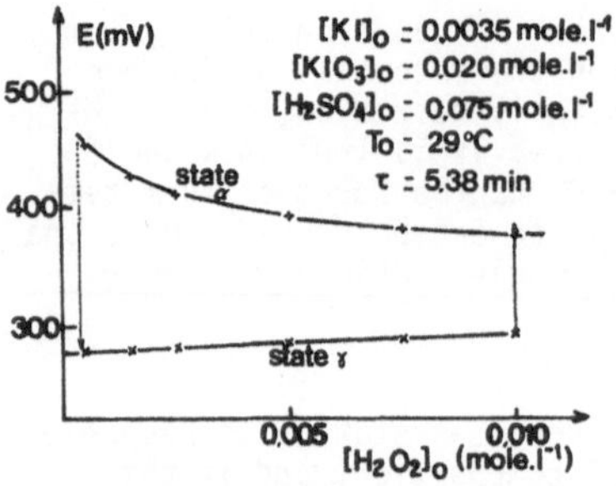

Fig. 4 H_2O_2-KI-KIO_3 system (T = 29°C): redox potential E versus $[H_2O_2]_0$ (constraint)

After the transition, the potential and the iodide consumption in state α increase upon decreasing $[H_2O_2]_0$. The α state way persists until very low $[H_2O_2]_0$ values. But the α - γ transition is inevitable when $[H_2O_2]_0 \to 0$.

At different values of $[KIO_3]_0$, the E = f$[H_2O_2]_0$ curves confirm the observations of fig. 2: the $[KIO_3]_0$ value does not affect the potential of the system in state α, while the potential of the γ state increases with $[KIO_3]_0$.

3. The $(NH_4)_2\ S_2O_8$-KI and $(NH_4)_2\ S_2O_8$-KI-$C_2O_4H_2$ systems

When iodide is added at room temperature to an acidic persulfate solution, the potential drops rapidly and then reaches a stable value. As the temperature is increased to 46.5°C, a shoulder develops on the E = f($[KI]_0$) curve (fig. 5). At and above 55°C the $(NH_4)_2\ S_2O_8$-KI system exhibits two stable states called α and γ, each of sigmoid form. They coexist over a narrow range of $[KI]_0$. Since the plot of E = f (flow) has the same shape as E = f($[KI]_0$), it is the α state which persists at zero flow, while the γ state is stabilized by flow and exists only at high flow rates.

The system remains bistable at 61°C at all persulfate concentrations studied, i.e. $0.01 < [S_2O_8^{=}]_0 < 0.4\ mol.l^{-1}$.

A supplementary flow of oxalic acid causes oscillations to take place. The state diagram of the $(NH_4)_2\ S_2O_8$-KI-$C_2O_4H_2$ system in the constraint space $[KI]_0$, $[C_2O_4H_2]_0$ exhibits a cross-shaped topology (fig. 6).

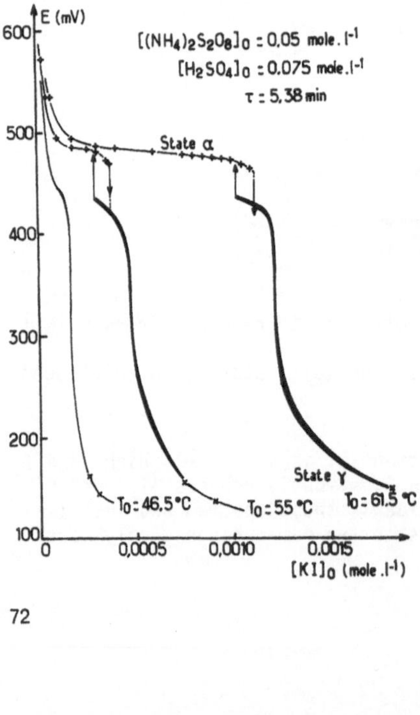

Fig. 5 $S_2O_8^{--}$ - I^- system: dependence upon temperature of the function E = f($[KI]_0$)

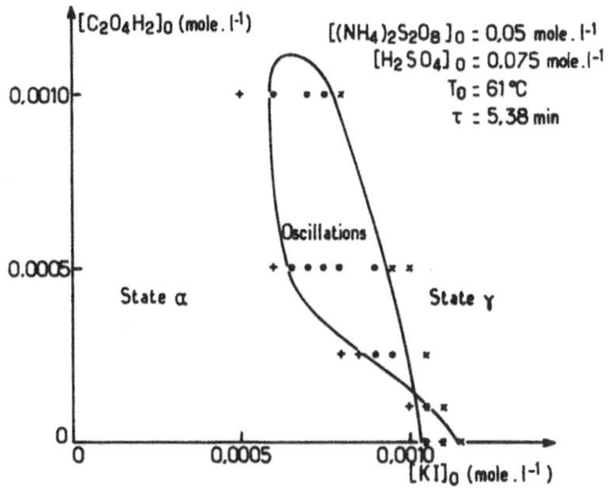

Fig. 6 Phase-state diagram of $S_2O_8^{--}$-KI-$C_2O_4H_2$ system in $[C_2O_4H_2]_0$, $[KI]_0$ space

References

1. W. Bray and H. Liebhafsky, J. Am. Chem. Soc., 53, 38 (1931)
2. J. Chopin-Dumas, C.R. Acad. Sci. t.287, 553 (1978)

Part III

Spatial Structures and Chemical Waves

Propagator-Controller Systems and Chemical Patterns

Paul C. Fife
Mathematics Department, University of Arizona, Tucson, AZ 85721, USA

1. Introduction

A (chemical) propagator-controller (PC) system is a reacting and diffusing system which supports chemical wave fronts in some chemical species (the propagator species), the velocity and wave forms of these fronts being modulated by the concentrations of other species (the controller species). The dynamics of the controller species may in turn be influenced by the concentrations of the propagator species.

This concept is best understood in the context of simple mathematical models in which the propagation and control processes can be closely examined. Fortunately, such simple models exist, and there is ample evidence of their relevance to real pattern-forming reagents. The idea of a modulated chemical front was present already in the work of Ostrovskii and Yahno (1975), as well as in the independent work of Ortoleva and Ross (1975) and Fife (1976).

In this talk, attention will be drawn to the ease by which spatio-temporal patterns observed in the Belousov-Zhabotinskii reagent can be understood by the use of PC models. My specific purposes are (1) to show that PC systems are present in model chemical networks which have been proposed for the BZ reagent; (2) to expound and analyze the simpler mathematical reaction-diffusion systems which are of PC type, and finally (3) to show that expanding ring and rotating spiral patterns can be understood within this context.

Most of this is a review and refashioning of previously presented material: for item (1), see (Tyson 1979), (Tyson and Fife 1980), and (Tyson 1984); for item (2) see (Fife 1976), (Zaikin and Kawczynski 1977), (Tyson and Fife 1980), (Fife 1984) and (Fife 1981b); and see (Tyson and Fife 1980) and (Tyson 1983) for the ring patterns in (3). What has remained up to now was to fit the phenomenon of spiral patterns into the PC framework. This turns out to be more difficult than the corresponding task for target patterns, and one approach to this is the principal new development being reported here. Other mathematical approaches to understanding these patterns, using less realistic models or else (in the case of spirals) excluding analyses within the core regions, have been pursued avidly in the past; I shall not report on them. Finally, I should mention that many other researchers, in many countries, have come a long way toward developing our intuitive understanding of rotors (spirals) and rings by combinations of careful experimentation and the fashioning of theories which are somewhat independent of the mechanistic details underlying their formation. For two of many examples of this, as well as good reviews, see (Winfree 1978) and (Kuramoto 1984).

2. Sigmoidal nonlinearities

The simplest mathematical realization of the PC concept is a system of two equations with suitable small parameter and with a sigmoidal nonlinearity. Specifically,

$$u_t = \varepsilon\Delta u + \frac{1}{\varepsilon}f(u,v), \qquad (1)$$

$$v_t = \varepsilon\Delta v + g(u,v), \qquad (2)$$

where ε is a small positive parameter, and the nullcline $f(u,v) = 0$ has the following S-shape in the u-v plane:

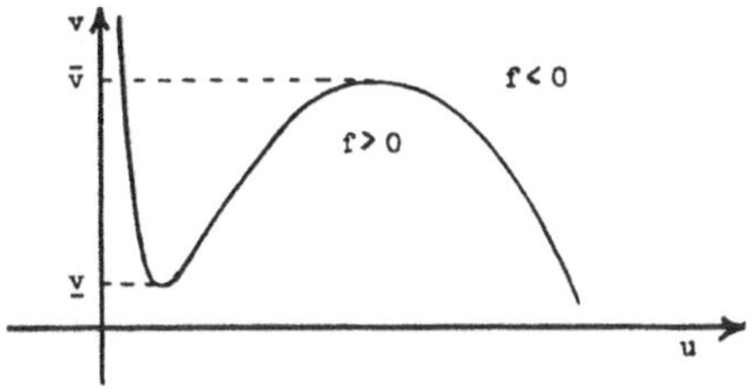

Figure 1

If the diffusion terms were absent from (1) and (2), the resulting system of ordinary differential equations might well be the familiar relaxation oscillator; the condition for this is that g be negative on the left branch in Figure 1, and positive on the right branch. So as a rule of thumb, PC systems are most simply modeled by adjoining diffusion to a kinetic system which has the same u-nullcline as a relaxation oscillator. On the other hand if the nullcline $g(u,v) = 0$ intersects either the left or the right branch, the intersection point may, under simple conditions, be a global attractor, so that no oscillations exist for the ordinary differential system. When diffusion is added, the spatio-temporal behavior of solutions may well be totally different. For example, as we shall see, there may exist stable oscillating solutions in the global attractor case.

A detailed description of the dynamics of the system (1), (2) will be given in Section 4, but the following can be said at this point in support of the claim that it has the properties of a PC system. The lowest order formal approximation to (1), (2) for small ε is the system

$$f(u,v) = 0, \qquad (3)$$

$$v_t = g(u,v). \qquad (4)$$

According to Figure 1, (3) can be solved in three ways for u as a function of v; these three solutions are represented by the three branches in the nullclines drawn there. The middle branch is unstable; the other two we denote by

$$u = h_{\pm}(v). \tag{5}$$

Solutions of (3), (4) may exist for which, at one instant of time, u is discontinuous in x; such discontinuities represent transition points where the relation between u and v changes from the $+$ relation in (5) to the $-$ one or vice versa. In this way, space is divided into regions in which one or the other of the signs $+$ or $-$ in (5) holds. Suppose the boundary between these two regions is smooth, with curvature not large. At a transition point on this boundary, if one reverts to the original equation (1) and stretches the space variable in the direction normal to the curve of discontinuity by a factor $\frac{1}{\varepsilon}$, then the resulting equation in u implies that in fact the transition curve must move with a normal velocity c which depends on v:

$$c = c(v), \tag{6}$$

as long as the value of v at the discontinuity lies strictly within the limits

$$\underline{v} < v < \overline{v}. \tag{7}$$

These moving curves of discontinuity in the variable u will represent very steep traveling fronts. u is thus the propagator variable. The fronts' velocities are modulated by the controller variable v through the relation (6).

3. Examples

Rather simple reaction schemes have been proposed as skeletons for the Belousov-Zhabotinskii and related reactions. See (Tyson 1983) for a critical review of them. The most famous is the Oregonator (Field and Noyes 1974), which tracks the changing concentrations of three species $HBrO_2$, Br^-, and Ce^{4+}. Although this model has been exceedingly successful in reproducing observed exotic phenomena of various types, alternate three-species skeletons have been proposed as being even more realistic. A recent one was the "explodator" (Noszticzius et al. 1984), followed by three revised versions of the Oregonator (Noyes 1984). Tyson (1979) and Tyson-Fife (1980) showed that a pseudo-steady-state relation legitimately reduces the original Oregonator to a system with the properties described in Section 2, provided that the rate constants are taken to have appropriate orders of magnitude. At the time of the appearance of the above four alternate models, Tyson (1984) further showed that if the explodator is supplemented by the first reaction in the Oregonator models, then all four of the new models have these same characteristics.

In all five cases, therefore, there is, for apparently reasonable ranges of the parameters, a pseudo-steady state reduction of the kinetic equations to a pair of equations supporting relaxation oscillations or excitable behavior. If diffusion is then added, a legitimate PC system of the type in Section 2 results. If reaction 1 is not added to the explodator, it turns out that a PC system is still obtained, but with the following feature which distinguishes it from the model system in Section 2: the left branch of the S-curve in Figure 1 now lies on the v-axis, and the lower knee on that curve now becomes the sharp corner formed by the intersection of another curve with the v-axis (Figure 2). The portion of the v-axis below the intersection point also lies in the null set $f(u,v) = 0$, but is unstable, so should be considered an extension of the unstable middle branch.

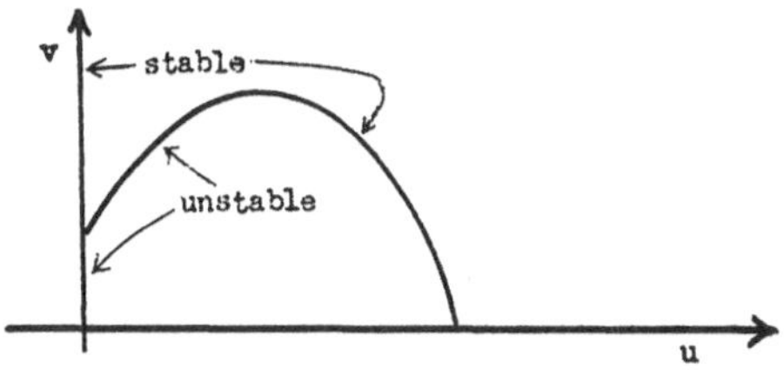

Figure 2

For completeness, it may be instructive to indicate the pseudo-steady-state reduction process in the case of the explodator with reaction 1 added ($\alpha = \beta = 1$ in their notation) and Noyes' third variant of his revised Oregonator. These two models turn out to be formally almost the same, although their authors assign different chemical identities to one of the variable species. The following reproduces the analysis in (Tyson 1984). In the network below, A denotes a species whose concentration is held fixed, and 0 denotes an inert product:

(1) $A + Y \to X + Z$

(2) $X + Y \to 2Z,$

(3) $A + X \to 2X,$

(4) $2X \to A + Z,$

(5) $Z \to Y,$

(6) $Y \to 0.$

The rate constants will be denoted by k_{1-6}, and the concentrations of the various species by the same identifying symbols used in the above network. Then the kinetic equations are as follows:

$$\dot{X} = k_1AY - k_2XY + k_3AX - 2k_4X^2, \quad (8)$$

$$\dot{Y} = -k_1AY - k_2XY + k_5Z - k_6Y, \quad (9)$$

$$\dot{Z} = k_1AY + 2k_2XY + k_4X^2 - k_5Z. \quad (10)$$

Use the scaled variables

$$u = (2k_4/k_3)X, \quad y = (k_2/k_3A)Y, \quad v = (2k_4k_5/k_3^2A^2)Z, \quad \tau = k_5t,$$

and the parameters

$$\varepsilon \equiv k_5/k_3A, \quad \delta \equiv 2k_4k_5/k_2k_3A, \quad q = 2k_1k_4/k_2k_3, \quad \gamma \equiv 2k_4k_6/k_2k_3A.$$

Then using τ as the time variable, one obtains

$$\varepsilon u' = u(1 - u) - y(u - q), \tag{11}$$

$$\delta y' = v - y(u + q) - \gamma y, \tag{12}$$

$$v' = (2u + q)y + u^2/2 - v. \tag{13}$$

It is assumed (Tyson presents evidence in favor of this) that

$$\delta << \varepsilon << 1.$$

the pseudo-steady-state aproximation which follows from (12) is

$$y = \frac{v}{u+q+\gamma}.$$

When substituted into (11) and (13), this results in a system like (1) and (2), except for the diffusion terms, but with the u nullcline having the general features shown in Figure 1. If diffusion is now added, a PC system is thereby obtained.

4. Limiting evolution problem: trigger and phase fronts

Here it will be shown that the lowest order formal approximation to the evolution problem (1), (2) is a simpler evolution problem which can be solved by trivial methods. For more details and graphical examples, see (Fife 1984). As mentioned in Section 2, excluding points on the fronts themselves, space-time can be divided into 2 regions: the "+" region consisting of all points where the relation (5) holds with the + sign, and the "-" region similarly defined. Let $s(x,t)$ denote the function yielding the state of the system at the point (x,t): s then takes two possible values, + or -. Substituting (5) into (4), we find that

$$v_t = g_\pm(v) \equiv g(h_\pm(v),v) \quad \text{when} \quad s = \pm. \tag{14}$$

Now let a spatial location x be fixed, let s be a given sign, and consider the evolution of v according to (14). There are two ways in which a switch in the value of s can come about:

1) At some future time, a front as described above, moving with velocity c given by (6), may pass by the position x, while the value of v at that point lies in the range (7).

2) At some future time the variable v, evolving according to (14), may reach the upper limit $\overline{v}$ while $s = +$, or the lower limit $\underline{v}$ while $s = -$. The former case may happen, for example, if $g_+(v) > 0$, so that v steadily increases, and no front of the first kind reaches x before v reaches $\overline{v}$.

Moving fronts triggering switches of type 1) have been called trigger fronts. They are very stable structures. If switches of the second kind occur, their time of occurrence will generally depend on x. This dependence generates a switching surface traveling through x-space with the property that $v = \overline{v}$ or $\underline{v}$ at every point of it. These moving discontinuities are called phase fronts. Of course in either case, any sort of switch at the location x causes the dynamics of v (14), in particular the value of v_t, to change abruptly.

One phenomenon important in the following is the conversion of a phase front to a trigger front. Both kinds of fronts have more or less the same profile, so a phase front profile may well evolve into a trigger front, the latter being very stable. This will happen precisely when the velocity of the trigger front to which it would evolve (and this velocity can readily be determined) is at least as large as the instantaneous velocity of the phase front. In other words, if the trigger front can outrun the phase front, it will immediately form and do so. On the other hand if the velocity of the phase front is high, no trigger front could be formed: the faster front wins out. Graphical examples were given in (Fife 1984).

The evolution of v according to (14), together with the above rules governing the changes of s, constitute the limit evolution problem. Given initial values $v(x,0)$ and $s(x,0)$ of these two functions, their subsequent evolution may be determined by elementary, for example graphical, methods, as will be explained in Section 5.

5. Expanding ring patterns

This outlines the basic construction of expanding target patterns given in (Tyson and Fife 1980) and made graphical in (Fife 1984). I shall work within the context of the revised Oregonator model given above. Thus, the starting point is a system of the form (1), (2), with f as in Figure 1. It will now be important to know the relative position of the other nullcline, $g(u,v) = 0$. Notice that in the present case, the functions f and g in (1) and (2) depend on two parameters, q and γ. Assume, as does Tyson (1984), that $q \ll 1$, and let the parameter $\rho \equiv \gamma/q$ be an $O(1)$ quantity. It is a matter of direct verification that

1) this nullcline is monotone increasing in the u-v plane,

2) $g > 0$ below the curve, $g < 0$ above it;

3) there is a number ρ_0 close to 3 such that for $\rho < \rho_0$, the g nullcline intersects the f one at only one point, to the left of the lower knee in Figure 1; for $\rho > \rho_0$, the intersection is to the right of it; and in the latter case for ρ not too large, the intersection is on the middle increasing branch in Figure 1. These cases are shown in Figure 3.

Suppose now that the parameter ρ depends on x; specifically, it depends on $r \equiv |x|$ only, and satisfies $\rho < \rho_0$ for $r > 1/2$, and the opposite for $r < 1/2$. Also suppose that $\rho \equiv \rho_1 < \rho_0$ for $r > 1$. In

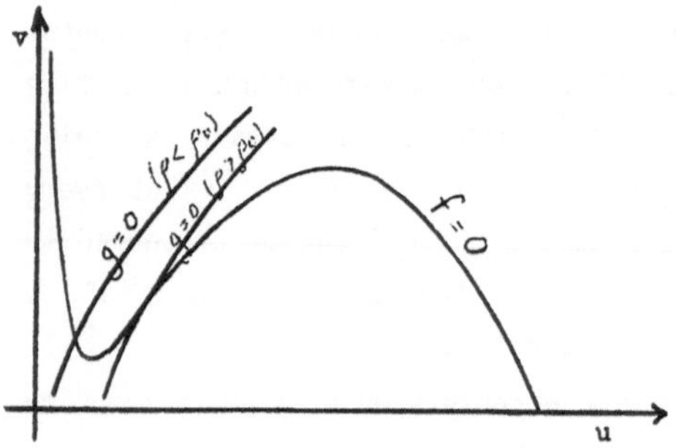

Figure 3

(Tyson and Fife 1980) we took this to represent the situation that there is a catalyst particle at one position (the origin) which affects the chemistry by changing the rate constants. There is evidence that such particles may act as leading centers to cause the propagation of ring structures into the surrounding medium, which may be an excitable one.

In the present case, according to the intersection properties mentioned above, there exists a stable rest state (u_0, v_0) on the left branch for $r > 1$. Let us take the uniform initial condition $v(x,0) \equiv v_0$, $s(x,0) \equiv -$, and solve the limit problem for $t \geq 0$ according to the technique alluded to above. First, integrate (14), for each fixed x, forward in time. For $r < 1$, the value of $\rho(r)$ is such that $g < 0$, so that v decreases. For $r < 1/2$, it approaches and eventually attains the value $\underline{v}$, at which time s switches to "+". This means there is an eventual phase front which might be encountered for r in this range. It is shown as a dotted line in Figure 4. However, at some value of r in this range, the velocity of this phase front matches the velocity of the trigger front associated with that value of $v = \underline{v}$ (although the trigger velocity was only defined for v in the range $\underline{v} < v < \overline{v}$, we now define it when $v = \underline{v}$ to be the limiting value of $c(v)$ as $v \downarrow \underline{v}$). It is at this value of r that the phase front P_1 is replaced by a trigger front T_1, which propagates into the excitable portion of the medium $(r > 1)$ as shown by the solid line in Figure 4.

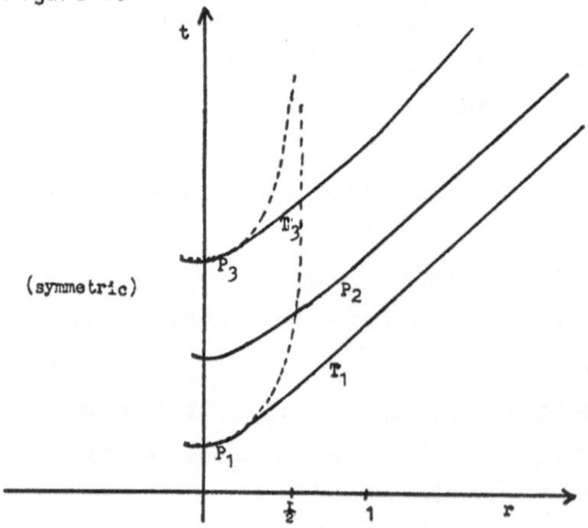

Figure 4

In the region $r > 1$, $v = v_0$ and the velocity of the trigger front is constant at $c(v_0)$. Thus the first trigger front in Figure 4 is eventually a straight line. Immediately above it, the medium is in state +, where g remains strictly positive. This means that v, by (14), will increase, for each x, until its upper limit $\bar{v}$ is reached. This forms the phase front P_2 as shown in the figure. It is roughly parallel to T_1, because v takes roughly the same amount of time to reach $\bar{v}$ no matter where on the front T_1 it starts from. The question now arises as to whether or not a trigger front can develop spontaneously from P_2. For this to happen, P_2's velocity would have to drop to the value of $c(\bar{v})$. It turns out that when the nullcline is asymmetric as it is in Figure 1, then this will not happen, so that a trigger front cannot be born from P_2. This asymmetry will in fact hold in the case of the revised Oregonator example being considered. The third front, like the first, begins as a phase front, but is necessarily changed into trigger for the same reasons. Continuing in this manner yields a succession of expanding fronts, hence an expanding target pattern, as shown in Figure 4.

6. Spirals

These rotating structures are born when a ring is broken. This could be done by mechanically mixing the reagent at some small location. The ends of the broken ring then begin to curl up, and eventually develop into a steadily rotating spiral. The initial curling-up phenomenon is exhibited clearly by the PC model; in fact, it is evident from a simple thought experiment, as described in (Fife 1984). I shall not repeat the argument here, but rather go on to the question of modeling and describing the fully developed spiral. When a ring, originally expanding in a regular fashion, is broken and begins to curl, the curling branches become somewhat narrower and more closely packed than were the original expanding rings. The period of the oscillation also decreases. In the context of the PC model, it turns out that spirals can be described by using tighter space and time scales than those of the ring structure, and in this approach the new space and time scales are related to the old by multiplicative factors

$$\delta^2 \equiv \varepsilon^{2/3} \quad \text{and} \quad \delta \equiv \varepsilon^{1/3},$$

respectively. Also, because the spatial structure is finer and the function v has no sharp gradients, it may be suspected that the function v has small total variation within a spatial wave length.

To avoid irrelevant complications, I shall make two simplifying assumptions about the functions f and g in (1), (2). The first is a symmetry assumption on f: $f(-u,-v) = -f(u,v)$ (Figure 5). One effect of this is to shift the u-v coordinate system so the origin lies half-way up the middle ascending branch of the u-nullcline. The actual nullcline in the Oregonator models is asymmetric, and the asymmetry was important in

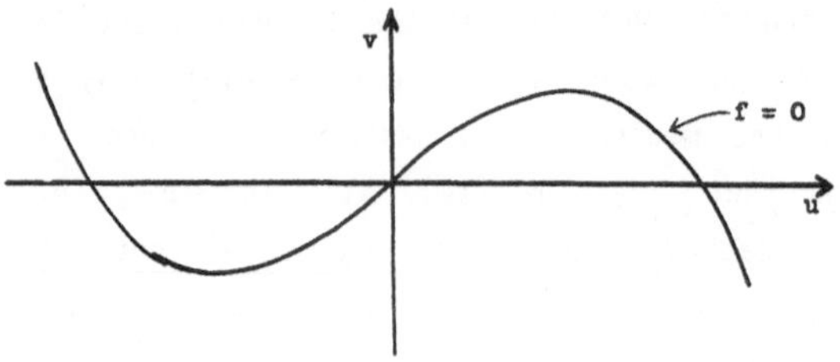

Figure 5

describing the rings, but it is of no consequence in the study of spirals. The second assumption is that

$$g(h_\pm(v),v) = \pm 1, \tag{15}$$

at least for v near zero. This certainly gives the right sign for g in the chemical models mentioned before, and the assumption is made only for convenience.

Let u^* be the value of u on the right branch, so that

$$f(\pm u^*,0) = 0. \tag{16}$$

In accordance with the scale changes mentioned above, set

$$\xi \equiv \delta^{-2}x, \quad \tilde{v} \equiv \delta^{-1}v. \tag{17}$$

In terms of the polar coordinates $(r,\tilde{\theta})$ in the ξ-plane and some scaled angular velocity ω, the spiral solution will be represented by functions

$$u = u(r,\tilde{\theta} - \delta^{-1}\omega t), \quad \tilde{v} = \tilde{v}(\text{same}). \tag{18}$$

It then follows from (1), (2) that

$$f(u,\delta\tilde{v}) + \delta^2(\Delta_\xi u + \omega u_{\tilde{\theta}}) = 0, \tag{19}$$

$$\Delta\tilde{v} + \omega\tilde{v}_{\tilde{\theta}} + g(\cdots) = 0. \tag{20}$$

From this point on, the tildes will be dropped from the symbols $\tilde{v}$.

The problems (19), (20) can be approximated, to lowest order in the small parameter δ, by a free boundary problem. In fact, to lowest order, (19) says that $f = 0$, so that (u,v) lies on either the right or left branch in Figure 5, and (5) holds. In view of (15), we have that $g = \pm 1$ in (20), which becomes

$$\Delta v + \omega v_{\tilde{\theta}} \pm 1 = 0. \tag{21}$$

In this lowest order approximation, then, the plane will be split into two portions, which we call $\Omega_\pm$, in which the corresponding sign in (21) holds. One seeks for solutions such that the interface Γ between Ω_+ and Ω_- is spiral shaped, and is given in two parts: Γ_+, described by $\tilde{\theta} = h(r)$ for some function h, and $\Gamma_-: \tilde{\theta} = h(r) + \pi$. As $r \to \infty$, we require that

$$h = \gamma r + O(1) \tag{22}$$

for some γ; this will guarantee that the spiral is Archimedian, which appears usually to be the case.

New polar coordinates, with angle referenced to Γ, will be useful:

$$\theta = \tilde{\theta} - h(r),$$

so that Γ_+ is given by $\theta = 0$, and Γ_- by $\theta = \pi$. If one also defines

$$H(r) = \frac{h'(r)}{r}, \qquad (23)$$

then the Laplacian may be written

$$\Delta_\xi u \equiv \frac{1}{r}(ru_r)_r + \frac{1}{r^2}u_{\theta\theta} - \frac{1}{r}(r^2H)'u_\theta - 2rHu_{r\theta} + r^2H^2u_{\theta\theta}. \qquad (24)$$

An interface condition on the free boundary Γ is needed to complete the formulation of the problem. It can be obtained by a fine structure analysis in the vicinity of Γ. Stretch variables near Γ by defining

$$\Psi \equiv \frac{\theta}{\delta}; \quad U(r,\Psi) \equiv u(r,\theta); \quad V(r,\Psi) \equiv v(r,\theta).$$

In this way, (1) becomes

$$\frac{\delta^2}{r}(rU_r)_r + \frac{1}{r^2}U_{\Psi\Psi} - \frac{\delta}{r}(r^2H)'U_\Psi - 2rH\delta U_{r\Psi}$$

$$+ r^2H^2U_{\Psi\Psi} + \delta\omega U_\Psi + f(U,\delta V) = 0. \qquad (25)$$

Originally, the interface Γ was specified as the location of the boundary between the regions Ω_+ and Ω_- where the outer solution is in states $+$ and $-$ respectively. Now, however, along with the stretch of the variable θ into the variable Ψ, with its concomitant smoothing out of the interface, we must specify more clearly the definition of the interface $\theta = h(r)$. We simply define it to be the location where $u = 0$, 0 being a convenient value between $-u^*$ and u^*:

$$u = 0 \quad \text{for} \quad \theta = \Psi = 0. \qquad (26)$$

The next step is to formally expand

$$U = U^0 + \delta U^1 + \delta^2U^2 + \ldots; \qquad V = V^0 + \ldots .$$

Then to lowest order, (24) becomes

$$\left(\frac{1}{r^2} + r^2H^2\right)U^0_{\Psi\Psi} + f(U^0,0) = 0, \qquad (27)$$

and (26) becomes

$$U^0(r,0) \equiv 0. \qquad (28)$$

The function $f(U,0)$ appearing in (27) is of "bistable type": as U increases from $-u^*$ to u^*, f is successively zero, negative, zero, positive, and zero again. Further, by symmetry,

$$\int_{-u^*}^{u^*} f(u,0)du = 0.$$

It is well known (see (Fife 1979), for example), that under these conditions, the ordinary differential equation $U_{\Psi\Psi} + f(U,0) = 0$ has a unique solution $U = \chi(\Psi)$ satisfying $\chi(-\infty) = -u^*$, $\chi(\infty) = u^*$, $\chi(0) = 0$. Therefore (27), (28) has the transition solution

$$U^0(r,\Psi) = \chi(p(r)\Psi), \tag{29}$$

where

$$p(r) \equiv (\frac{1}{r^2} + r^2H(r)^2)^{-1/2}. \tag{30}$$

The first order terms in (25) yield the following:

$$\begin{aligned} &p^{-2}U^1_{\Psi\Psi} + f_u(U^0,0)U^1 + f_v(U^0,0)V^0 - \frac{1}{r}(r^2H)'U^0_\Psi \\ &\quad - 2rHU^0_{r\Psi} + \omega U^0_\Psi = 0. \end{aligned} \tag{31}$$

We also have that $U^0_{\Psi\Psi} = 0$, so that $U^0 = V^*(r)$ for some function V^*. This order is sufficient to provide the desired interface condition.

Considered, for each fixed value of r, as a second order operator on $L_2(-\infty,\infty)$, the operator L defined by $LU \equiv p(r)^{-2}U_{\Psi\Psi} + f_u(U^0(r,\Psi),0)U$ has a single nullvector $U^0_\Psi(r,\Psi)$. Therefore the solvability condition for (31), written as $LU^1 = F$, is that F be orthogonal in L_2 to that nullvector. This means that

$$\int_{-\infty}^{\infty} U^0_\Psi[f_v(U^0,0)V^*(r) - \frac{1}{r}(r^2H)'U^0_\Psi - 2rHU^0_{r\Psi} + \omega U^0_\Psi]d\Psi = 0. \tag{32}$$

We let $\nu \equiv \int_{-\infty}^{\infty} f_v(U^0,0)U^0_\Psi d\Psi$, and assume that $\nu > 0$; then the above orthogonality condition may be solved for $V^*(r)$. Specifically, to do so we note that $U^0_r = p'(r)\chi'(p\Psi)$, etc. It follows that

$$V^*(r) = \nu^{-1}[(\frac{1}{r}(r^2H)' - \omega)p + rHp']N, \tag{33}$$

where $N \equiv \int(\chi'(\tau))^2d\tau$.

Equation (33) is now to be considered an interface condition which must be imposed on Γ. Note that it relates the value of v on the interface to ω, to the inclination of the interface, and to its curvature (the latter two quantities being expressed in terms of H and H'). This condition replaces the usual relation (6) between the normal velocity of a front and the value of v there, which is valid when the curvature is not too great and/or v is not too small. Symmetry requires v to be odd.

To summarize, the problem for the fully developed spiral is, to lowest order in δ, the following <u>free boundary problem</u>: Find a function $v(r,\theta)$, a function $H(r)$, and constants ω and γ, satisfying

$$\Delta_H v + \omega v_\theta \pm 1 = 0$$

(where Δ_H is given by the expression (24));

$$v(0,r) = N\nu^{-1}[(\frac{1}{r}(r^2H)' - \omega)p(r) + rHp']$$

where $p(r)$ is given by (30);

$$v \text{ is } 2\pi\text{-periodic in } \theta, \quad v(r,\theta + \pi) = -v(r,\theta);$$

and

$$H(r) = \frac{\gamma}{r} + O(r^{-2}) \quad \text{as} \quad r \to \infty.$$

The expected solution will have an interface of the following form, drawn in the original ξ-plane:

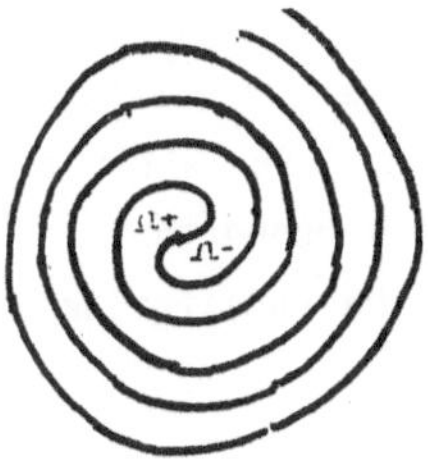

Figure 6.

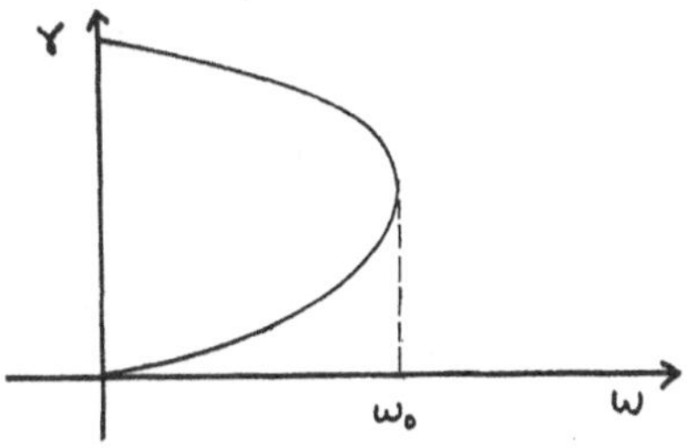

Figure 7.

A complete solution would have to be obtained numerically; but approximate solutions for large r and for small r can be obtained by the usual expansion methods. I'll not give the details; just the results.

For large r, asymptotic solutions exist for each ω in some finite range $0 < \omega < \omega_0$. Given such an ω, there are two possible choices for the solution triple (v, H, γ). The possible values of γ are depicted in Figure 7. I conjecture that the spiral selected by the system through the curling-up process is the one with the maximum value $\omega = \omega_0$ (then there is a unique solution). The reason is that at the "curling center", in the initial stages of formation, the system tries to twist as much as possible; this twisting is only tempered by the diffusion of v. Therefore it is reasonable to expect that the maximum value of ω is selected.

For small r, the function $v(r,\theta)$ may be expanded in a Fourier series in θ, the coefficients being power series in r. Likewise, the function H is expandable in even powers of r. Certain symmetry relations simplify the computations. Formal solutions of this sort are available for arbitrary ω. for example, it turns out that $H(0) = -\omega/6$. For the function v, there is some arbitrariness in choosing the coefficients in the expansion, even though ω is prescribed.

It remains to be seen:

1) which of these small-r expansions represent real solutions which can be continued to Archimedean spirals for all r;

2) which ω is actually selected by spirals forming in the fashion indicated; and

3) which small-r solutions connect with the corresponding large-r solutions for the same value of ω.

More on the mathematical side, an existence theory should be developed for the free boundary problem, not to speak of computational techniques for handling it.

It is important to mention the limitations of this (and any other) approach to the modeling of spirals in the BZ reagent. In Oregonator-type models, the parameter ε in (1), (2) is given explicitly in terms of rate constants. These constants are not really known, but there is argument in favor of ε being small (Tyson 1984). Of course this does not imply that $\varepsilon^{1/3} = \delta$ is also small, as I have assumed. That is a more stringent condition. Experimental evidence does not suggest a wide discrepancy in space-time scales between spirals and targets, although there is some tightening and acceleration in passing from the latter to the former (Winfree, personal communication). Nevertheless, it is hoped that the asymptotic treatment here, based on the smallness of δ, will be qualitatively correct at least, in describing the properties of spirals.

Research supported by N.S.F Grant DMS-8202056.

References

R. J. Field and R. M. Noyes 1974 J. Chem. Phys. 60, 1877-1884.

P. C. Fife 1976 J. Chem. Phys. 64, 854-864.

--1979 Mathematical Aspects of Reacting and Diffusing Systems, Lecture Notes in Biomath. 28, Springer-Verlag.

--1981a in: Analytical and Numerical Approaches to Asymptotic problems in Analysis, O. Axelsson, L. S. Frank, A. van der Sluis, eds., Math. Studies 47, North-Holland, 45-56.

--1981b in: Application of Nonlinear Analysis in the Physical Sciences, H. Amann, N. Bazley, K. Kirchgassner, eds., Pitman, 206-228.

--1984 in: Proc. Conf. on Nonequilibrium Phenomena in Physics and Related Fields, M. G. Velarde, ed., Plenum Press, to appear.

Y. Kuramoto 1984 Chemical Oscillations, Waves and Turbulence, Springer Series in Synergetics #19.

Z. Noszticzius, H. Farkas, and A. Z. Schelly 1984 J. Chem. Phys. 80, 6062-6070.

R. M. Noyes 1984 J. Chem. Phys. 80, 6071-6078.

P. Ortoleva and J. Ross 1975 J. Chem. Phys. 63, 3398-3408.

L. A. Ostrovskii and V. G. Yahno 1975 Biofizika 20, 489-493.

J. Tyson 1979 Ann. N.Y. Acad. Sci. 36, 279-295.

--1983 in: Oscillations and Travelling Waves in Chemical Systems, R. J. Field, M. Burger, eds., Wiley, New York.

--1984 J. Chem. Phys. 80, 6079-6982.

J. Tyson and P. C. Fife 1980 J. Chem. Phys. 73, 2224-2237.

A. T. Winfree 1978 Theor. Chem., Vol. 4, Academic Press, 1-51.

A. N. Zaikin and A. L. Kawczynski 1977 J. Non-Equilib. Thermodyn. 2, 39-48.

A. N. Zaikin and A. M. Zhabotinsky 1970 Nature 225, 535-537.

The Speed of Propagation of Oxidizing and Reducing Wave Fronts in the Belousov-Zhabotinskii Reaction

John J. Tyson* and V.S. Manoranjan

Centre for Mathematical Biology, Oxford University, Oxford OX1 3LB, United Kingdom

1. Introduction

Periodic expanding target patterns of chemical activity are observed in thin unstirred layers of solution containing bromate, malonic acid and ferroin in dilute sulfuric acid [1,2]. Commonly these patterns appear as thin blue rings propagating outward from a central point into red bulk medium. Since the indicator, ferroin (Iron 1,10-phenanthroline), is blue in the oxidized state (Fe^{3+}) and red in the reduced state (Fe^{2+}), the thin blue ring is a zone of oxidation in a predominantly reduced background. The zone of oxidation is delimited by two wave fronts: a wave front of oxidation carrying the medium from the reduced state to the oxidized state, and a wave front (or, we might say, wave "back") of reduction restoring the medium to the original reduced state. These two waves travel at the same speed, which depends on the exact chemical constitution of the reactive solution but is practically independent of the temporal frequency and spatial wavelength of the target pattern.

Under special conditions the opposite pattern can be observed: red waves of reduction propagating through an oxidized bulk medium [3]. In this case the wave front of reduction travels slowly away from the center, and after a considerable time delay a wave back of oxidation sets off from the center at about twice the speed of the wave front. As the wave back catches up with the wave front, the red ring of reduction shrinks dramatically in width and finally disappears altogether.

2. Qualitative Description of Target Patterns

TYSON and FIFE [4] have presented a theory of target pattern formation in the BZ reaction, based on the assumption that at the center of each pattern there is a heterogeneity which periodically triggers waves of excitation (either oxidation or reduction) which then propagate away from the center at speeds determined by the chemical composition of the medium at the wave front. They describe the chemistry of the reaction in terms of the highly successful Oregonator model [5,6]. In suitably scaled and reduced form the Oregonator equations are

$$\varepsilon\, dx/dt = x(1-x) - fz(x-q)/(x+q) \tag{1a}$$

$$dz/dt = x - z \tag{1b}$$

where x and z are proportional to the concentrations of $HBrO_2$ and of Fe^{3+}, respectively, and ε, f, q are parameters ($\varepsilon \ll 1$, $f \approx 1$, $q \approx$

*Permanent address: Department of Biology, Virginia Polytechnic Institute and State University, Blacksburg, VA 24061, USA.

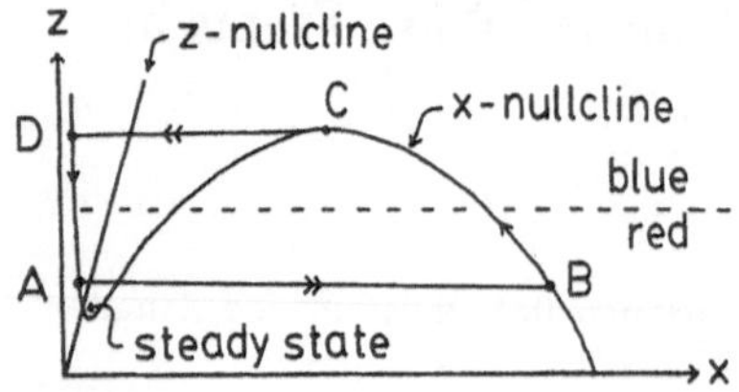

Fig.1 Phase plane for the Oregonator. The x- and z-nullclines intersect at a stable but excitable steady state in which the indicator-catalyst is reduced. The horizontal lines AB and CD indicate, respectively, a wave front of $HBrO_2$ production and a wave back of $HBrO_2$ destruction.

10^{-3}). The nullclines of system (1) are illustrated in Fig. 1, for the case of an asymptotically stable, excitable, reduced steady state.

According to TYSON and FIFE target patterns in the BZ reaction can be constructed locally by singular perturbation theory from wave fronts and wave backs suitably pieced together. For example, the jumps in Fig. 1 illustrate the situation far from the center of a standard pattern [1,2] of thin oxidation waves in a reduced bulk medium. The wave front is a wave of $HBrO_2$ production connecting state A (low $HBrO_2$) with state B (high $HBrO_2$) at fixed $[Fe^{3+}]$ slightly larger than the steady state level. The speed of propagation of this wave depends on the precise concentration of Fe^{3+} in the wave front, as well as $[BrO_3^-]$ and $[H^+]$ in the bulk medium. The visual color change lags behind the actual wave front by a small fixed distance determined by the time necessary to increase z from state B to the color-change-level (indicated by a dashed line in Fig. 1). The wave back is a wave of $HBrO_2$ destruction connecting state C with state D at fixed $[Fe^{3+}]$ at the local maximum of the x-nullcline. Strictly speaking the wave back is a phase wave traveling at the same speed as the wave front, at a fixed distance behind the front determined by the time necessary to increase z from B to C. The color change from blue to red lags behind the actual wave back. Since the time needed to reduce Fe^{3+} from D to A at low $[HBrO_2]$ is considerably longer than the time needed to oxidize Fe^{2+} from B to C at high $[HBrO_2]$, the blue annulus of oxidation is narrow relative to the intervening red regions of reduction and the red-to-blue color change is sharp compared to the blue-to-red color change.

The situation for waves of reduction in an oxidized medium is illustrated in Fig. 2. In this case the wave front (CD) is a "trigger" wave of low speed, determined by the precise $[Fe^{3+}]$ in the front (i.e., the value of z along the line CD). The wave back (AB) is also a trigger wave, traveling at a speed (determined by the value of z along line AB) which is generally larger than the speed of the wave front. As the wave back catches up with the wave front (i.e., as the red annulus moves farther away from the organizing center), the level of the wave-back-jump (AB) must increase and the speed of propagation of the wave back must decrease. Far enough from the center, one of two things must happen: either the levels AB and CD stabilize and an annulus of reduction of fixed width proceeds off to infinity (the annulus may or may not be visible depending on whether level AB is below or above the level of color change), or levels AB and CD coalesce and the annulus of reduction actually disappears at some finite distance from the center.

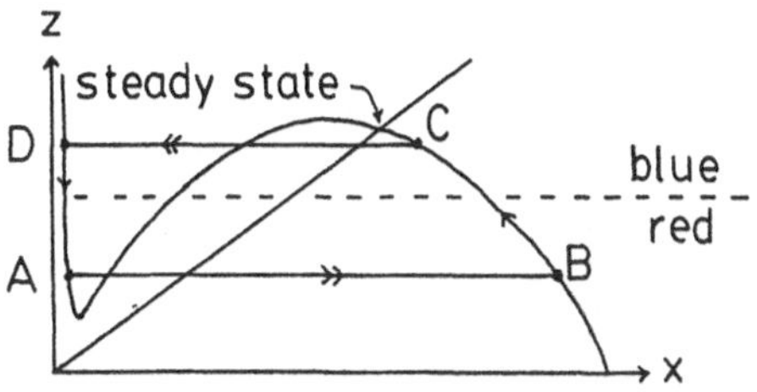

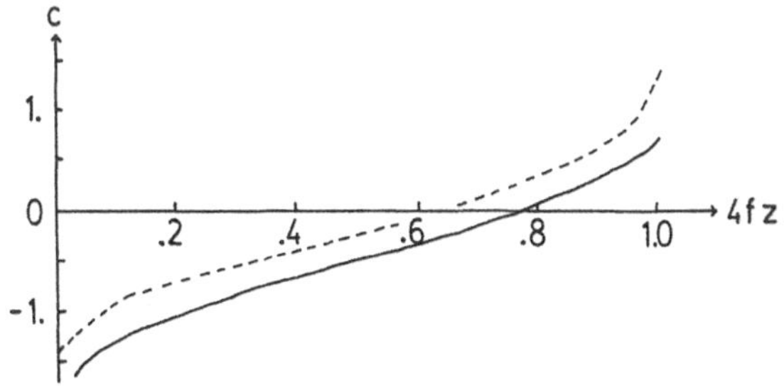

Fig.2 Phase plane for the Oregonator, as in Fig. 1 except that the stable excitable steady state corresponds to an oxidized state of the indicator-catalyst.

Fig.3 The dependence of wave speed c on indicator-catalyst concentration z in the wave front.

All of these conclusions agree qualitatively with experimental observations. To test the theory quantitatively we must have actual numerical values for the wave speed as a function of z in the wave front.

3. Wave Speeds

To calculate the dependence of wave speed, c, on Fe^{3+} concentration, z, TYSON and FIFE replaced the nonlinear Oregonator kinetics (1) by a piecewise linear approximation. They found that c=c(z) is given implicitly by the transcendental equation

$$\frac{1}{4fz} = 1 + \frac{\exp[-(c/2\omega)\arcsin(\omega)]}{c + (2+c^2)^{1/2}}, \qquad (2)$$

where $2\omega = (2-c^2)^{1/2}$ and $3\pi/4 < \arcsin(\omega) < \pi$. The wave speed as a function of 4fz is plotted as a dashed line in Fig. 3.* As c varies from $-\sqrt{2}$ to $+\sqrt{2}$, 4fz varies from 0 to 1. The wave front is stationary (c=0) at $4fz = 2-\sqrt{2} \approx 0.586$.

We have computed c(z) for the fully nonlinear problem

$$x'' + cx' + x(1-x) - fz(x-q)/(x+q) = 0, \qquad (3)$$

$$x(-\infty) = x_- \;,\; x(+\infty) = x_+,$$

where $x_- \le x_o \le x_+$ are the three real roots of $x(1-x) - fz(x-q)/(x+q) = 0$. The numerical calculations were done by the Galerkin finite element method, as described by MANORANJAN and MITCHELL [7]. (The numerical method was tested by computing wave speeds for the piecewise linear model; the numerical results agreed excellently with the analytical values of c.) For $q=10^{-3}$ in (3), the curve c(z) is plotted as a solid line in Fig. 3. Notice that, in agreement with the comparison theorem for wave speeds of TYSON and FIFE [4], the trigger wave speed of $HBrO_2$-production fronts developing from the local minimum of the x-nullcline is larger (in absolute value) than the trigger wave speed of $HBrO_2$-destruction fronts developing from the local maximum of the x-nullcline.

*Figure 8 in Ref. [4] is incorrect because $\arcsin(\omega)$ was computed in the range $(0, \pi/4)$ instead of the range $(3\pi/4, \pi)$.

Recall now the situation in SMOES' [3] target patterns (Fig. 2). The wave front is a reducing wave (blue-to-red transition following on the heels of an $HBrO_2$-destruction front) traveling at a (dimensionless) speed of about 0.6. The wave back is an oxidizing wave (red-to-blue transition following an $HBrO_2$-production wave) which, close to the organizing center, starts off from the local minimum of the x-nullcline,

$$4fz \approx 4(1+\sqrt{2})^2 q \approx 0.023, \quad x \approx (1+\sqrt{2})q \approx 0.0024, \tag{4}$$

at a speed of about 1.6. As the wave back moves away from the center, the value of z in the wave increases and the speed of propagation decreases (in absolute value). The value of 4fz must increase twenty-fold (to approximately 0.5) before the wave back slows down to a speed of 0.6. This is probably sufficient to bring $[Fe^{3+}]$ above the color-change-level, so that the annulus of reduction is no longer visible.

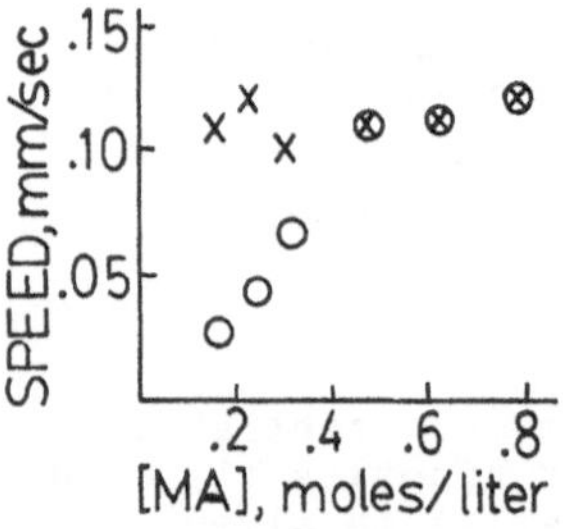

Fig.4 The dependence of wave speeds of oxidizing waves (x) and reducing waves (o) on initial malonic acid concentration, from SMOES [3]. At low [MA] oxidizing waves slow down as they propagate away from the center: the maximum speed of oxidizing waves is marked by an x, the minimum speed is close to the speed of the reducing wave.

In Fig. 4 we reproduce SMOES' measurements of wave speeds as a function of initial malonic acid concentration in the reaction mixture. At low values of [MA] one observes slow reducing wave fronts and rapid oxidizing wave backs. As [MA] increases, the speed of the reducing wave fronts increases (presumably because the intersection of the x- and z-nullclines in Fig. 2 is moving up closer to the local maximum of the x-nullcline) but the maximum speed of the oxidizing waves remains roughly constant (because they all take off from the local minimum of the x-nullcline, at $4fz = 4(1+\sqrt{2})^2q$, which is independent of [MA]). Moreover, the oxidizing waves travel about twice as fast as the reducing waves, in quantitative agreement with the solid line in Fig. 3.

At high values of [MA] one observes the standard BZ target patterns: thin blue annuli in a red background, with wave fronts (oxidizing waves) and wave backs (reducing waves) traveling at the same speed (see Fig. 4). The wave backs are phase waves, as discussed by TYSON and FIFE [4], and the wave fronts are trigger waves traveling at (dimensionless) speed of roughly 1.5 (compare Fig. 3 with 4fz slightly larger than $4(1+\sqrt{2})^2q$).

All that remains is to calculate the dimensioned speed (in, say, cm/sec) of oxidizing wave fronts. Undoing the scaling used by TYSON and FIFE to obtain (3), we find that

$$\text{wave speed} = (k[H^+][BrO_3^-]D)^{1/2}\, c \tag{5}$$

where D is the diffusion coefficient for $HBrO_2$ and k is the third-order rate constant for the autocatalytic production of $HBrO_2$,

$$H^+ + BrO_3^- + HBrO_2 \xrightarrow{k} Br_2O_4 + H_2O$$

$$Br_2O_4 + 2Fe^{2+} + 2H^+ \xrightarrow{fast} 2HBrO_2 + 2Fe^{3+}.$$

ROVINSKII and ZHABOTINSKII [8] have presented kinetic evidence that k $\approx 10\ M^{-2}s^{-1}$. Using this value of k and $D = 2x10^{-5}\ cm^2s^{-1}$ and c = 1.5 in (5), we obtain,

$$\text{wave speed} = 0.02\ cm\ s^{-1}\ M^{-1}\ \sqrt{[H^+][BrO_3^-]}, \qquad (6)$$

which agrees within a factor of 2 to the experimental observations of FIELD and NOYES [9] and SHOWALTER [10]. For $[H^+] = 0.25$ M and $[BrO_3^-] = 0.3$ M, as in the experiments of SMOES in Fig. 4, (6) predicts wave speed = 0.06 mm/sec which is also only two-fold smaller than the observed wave speed. (The discrepancy of a factor of 2 is not significant given the uncertainty in the experimental value of k and in the activity coefficient of H^+ in sulfuric acid solution).

4. Conclusion

We have shown that the Oregonator can describe in quantitative detail the behavior of propagating waves of oxidation and reduction in thin, unstirred layers of BZ reagent. In WINFREE's reagent [2] wave fronts are trigger waves of $HBrO_2$ production and wave backs are phase waves of $HBrO_2$ destruction, which travel necessarily at the same speed as the preceeding wave front. In SMOES' reagent [3] wave fronts are slow trigger waves of $HBrO_2$ destruction and wave backs are fast trigger waves of $HBrO_2$ production. The wave backs start off at approximately twice the speed of the wave fronts but slow down as they catch up with the preceeding wave front.

The dimensioned value of the wave speed agrees well with experimental measurements if we accept the ROVINSKII - ZHABOTINSKII [8] value for the rate constant describing the autocatalytic production of $HBrO_2$. This observation is important in light of the considerable disagreement among chemical kineticists about the values of three of the rate constants involved in the five-step Oregonator model [6].

This work was supported by the Science and Engineering Research Council (Great Britain) Grant GR/C/53595 to the Centre for Mathematical Biology and by the National Science Foundation (USA) Grant MCS-8301104 to JJT.

1. A. N. Zaikin, A. M. Zhabotinskii: Nature (London) 225, 535 (1970)
2. A. T. Winfree: Science 175, 634 (1972); Sci. Am. 230 (6), 82 (1974)
3. M. L. Smoes: in Dynamics of Synergetic Systems, edit. by H. Haken (Springer, Berlin, 1980) pp. 80-96
4. J. J. Tyson, P. C. Fife: J. Chem. Phys. 73, 2224 (1980)
5. R. J. Field, R. M. Noyes: J. Chem. Phys. 60, 1877 (1974)
6. J. J. Tyson: in Oscillations and Traveling Waves in Chemical Systems, edit. by R. J. Field, M. Burger (Wiley, New York, in press), Chapter 3
7. V. S. Manoranjan, A. R. Mitchell: J. Math. Biol. 16, 251 (1983)
8. A. B. Rovinskii, A. M. Zhabotinskii: Theor. Exp. Chem. 15, 17 (1979)
9. R. J. Field, R. M. Noyes: J. Am. Chem. Soc. 96, 2001 (1974)
10. K. Showalter: J. Phys. Chem. 85, 440 (1981)

Macroscopic Self Organization at Geological and Other First Order Phase Transitions

Peter J. Ortoleva

Department of Chemistry, Indiana University, Bloomington, IN 47405, USA

I. Instability and Nonlinear Restabilization During First Order Phase Transitions

A central prerequisite for the establishment of macroscopic self organization is that a system be sufficiently far from equilibrium. During the transformation from one *thermodynamic phase* to another, some systems can be so arranged that they can exhibit macroscopic self organization phenomena. One interesting feature of such self organization phenomena is that the spatial patterns that result from them can persist for very long periods after the original pattern was generated and the thermodynamic driving force for the processes involved has essentially gone to zero. Thus many systems undergoing first order phase transitions can not only generate patterns but can also "freeze in" patterns.

The freezing-in property has interesting consequences for biological systems because it implies that after the original pattern has been created only very little expense of free energy is needed to sustain the structure. Since such a mechanism of maintaining order exists it is hard to believe that the evolutionary pressures have not selected many organisms which use it as a method of self organization in embryonic development or lim and other regenerative processes.

Geological systems are the most striking manifestation of freezing in phenomena. In this case the patterns can persist for millions of years. These rock patterns can exist in many geometries - bands, concentric rings, spot patterns and spirals - and are reviewed in Refs. 1-5. The patterns may result from interdiffusion of coprecipitates [6], flows of reactive waters through rocks to form precipitate banding [7] and dissolution fingering [8], growth of solid solution crystals to form periodic zoning [9], pressure solution to form periodic layering [10] and dissolution seams [11] and many other phenomena [1-5]. The realization that the interplay of first order phase transitions with transport could lead to macroscopic order dates back to the last century [6,12], and seems to stem mainly from the work of Liesegang.

Here we discuss a few systems selected to show the range of the possible. It is clear that this area is rich with phenomena and worthy of more attention in the future. These systems show a complete spectrum of instability and nonlinear restabilization phenomena familiar from now "traditional" reaction-diffusion theory. However they also show new phenomena not contained in the latter theory.

II. Competitive Particle Growth Theory

When a uniform sol of PbI_2 in agar is allowed to age it is observed to form mottled patterns of precipitate and precipitate-free domains. The driving force for this instability of the uniform sol has been conjectured to be due to the competitive growth of the particles due to the particle radius dependence of the equilibrium constant as a result of surface tension [13]. The simplest mathematical model of this mechanism is

$$\frac{\partial R}{\partial t} = k[c - c^{eq}(R)] \qquad (1)$$

$$\frac{\partial c}{\partial t} = D\nabla^2 c - n\rho\frac{\partial}{\partial t}\left(\frac{4}{3}\pi R^3\right). \tag{2}$$

Equation (1) is a growth law for particles of radius $R(\vec{r},t)$ in the vicinity of spatial point $\vec{r}$ and time t; k is a rate constant, c is a monomer concentration and $c^{eq}(R)$ is the R-dependent equilibrium concentration. As shown in Ref. 13, the fact that c^{eq} decreases with R for supracritical particles implies that the uniform sol is unstable to pattern-forming perturbations.

Recently the above CPG equations were simulated numerically in two spatial dimensions. A number of interesting spatial patterns resulted. Cases of concentric rings (Fig. 1), speckles (Fig. 2) and spirals (Fig. 3) were shown to evolve from very minimal deviations from nonuniformity. See also Ref. 14.

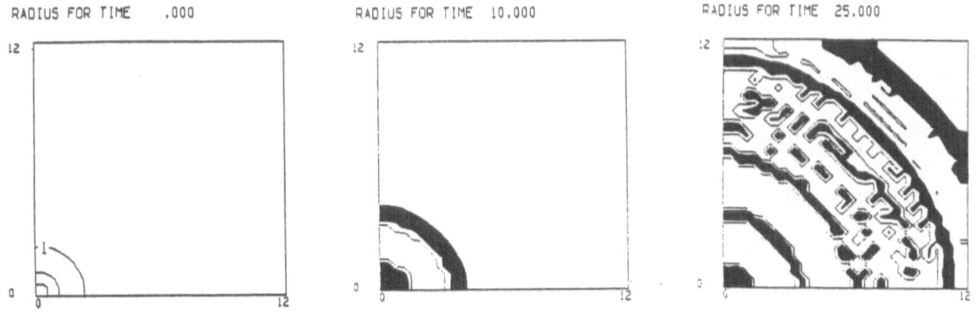

Fig. 1

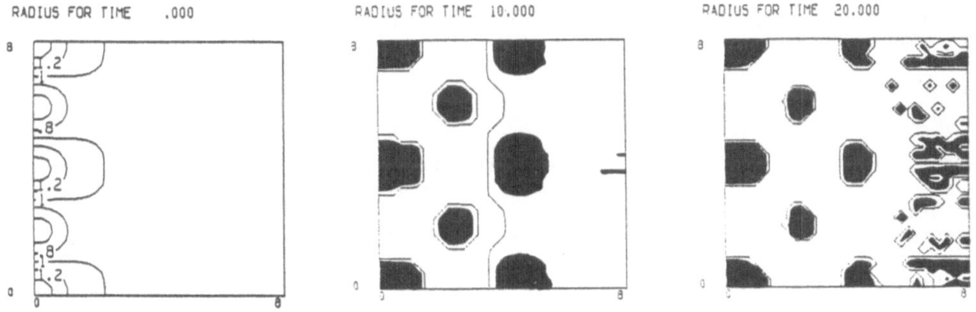

Fig. 2

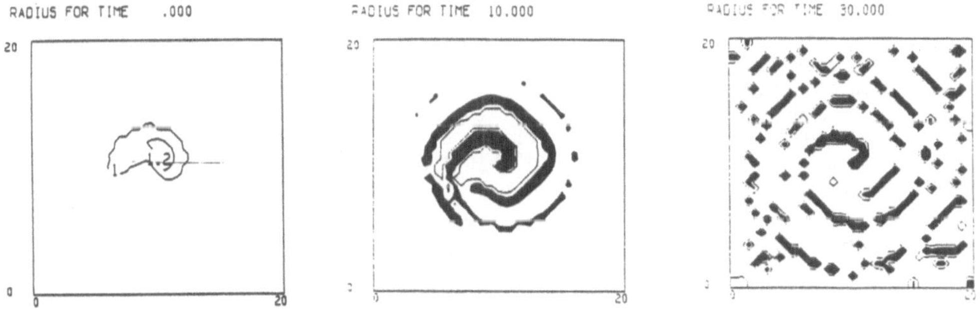

Fig. 3

The experiments of Ref. 13 show that between the patches of high precipitate content, a few "giant" particles survive. The monodisperse theory of (1) and (2) cannot describe this phenomenon. A particle size distribution formalism studied numerically in Ref. 15 and analytically and numerically in Ref. 16 shows that the modes of patterning wherein the tail and maximum of the particle size distribution are out of phase in space can explain these "greedy giants" in the clear areas of the PbI_2 experiments.

III. Ostwald-Liesegang Cycle in a Pyrite-Hermitite Percolation System

Consider a flow of oxygen (X) containing water through an aquifer containing the mineral pyrite (Py). Crudely speaking this leads to the following sequence:

$$X + Py \rightarrow F + \ldots \tag{3}$$

$$X + F \rightleftarrows Hm \tag{4}$$

where Hm denotes the iron oxide mineral hermitite. Let the rate of oxidative dissolution of Py be given by

$$\frac{\partial Py}{\partial t} = k(Py)^{2/3}X \tag{5}$$

where k is a rate coefficient and the 2/3 power is due to the surface area dependence of the process. Here Py represents the moles of Py per rock volume. The rate of Hm growth is more interesting:

$$\frac{\partial \Delta}{\partial t} = q(\Delta, XF)[XF-Q] \tag{6}$$

Here Q is the Hm equilibrium constant. Since Hm often forms as a thin coating on the other grains of the rock, we keep track of Hm content through the thickness Δ of the coating. The factor q accounts for nucleation: q vanishes if $\Delta=0$ unless XF exceeds a nucleation threshold ηQ, $\eta \geq 1$. In that case and for $\Delta>0$ q is a constant q_o. The concentration of X satisfies

$$\frac{\partial X}{\partial t} = D_X \frac{\partial^2 X}{\partial r^2} - v\frac{\partial X}{\partial r} + \frac{\partial Py}{\partial t} - \frac{\partial Hm}{\partial t}$$

and similarly for F. Here r is the spatial coordinate along which a percolation flow of speed $v>o$ takes place. This model is in the spirit of the original supersaturation - nucleation - depletion picture of Ostwald to describe Liesegang banding (except for the flow term) - see Ref. 17 for more details.

Unlike earlier models of the Ostwald type, this model yields bands of finite width. Most interestingly it appears to be the first well-posed pde model capable of analysis by techniques of bifurcation and linear stability analysis. Fig. 4 shows the results of analytical calculations showing that a steady Hm pulse solution exists in the shaded region of the η-q_o plane. These steady solutions exist for $1<\eta<\eta_{max}$ where η_{max} was calculated in terms of the boundary data-X and F at the flow inlet. Fig. 5 shows a sequence of simulations showing how the pulse becomes undulatory and then breaks up into discrete bands as q_o is increased beyond the domain of steady pulse propagation. Although the system has some features in common with a Hopf bifurcation, the fact that no steady solution exists for large q_o is the underlying reason for interesting differences under study at present. Finally we note that the occurrence of this type of banded hemitite is common in rocks.

IV. Reactive Percolation Morphological Instabilities

Consider the flow of water through a two-mineral rock such that one of the minerals is dissolved out. Because of the coupling of flow and dissolution through Darcy's law (that states that the flow speed is proportional to the pressure gradient) one

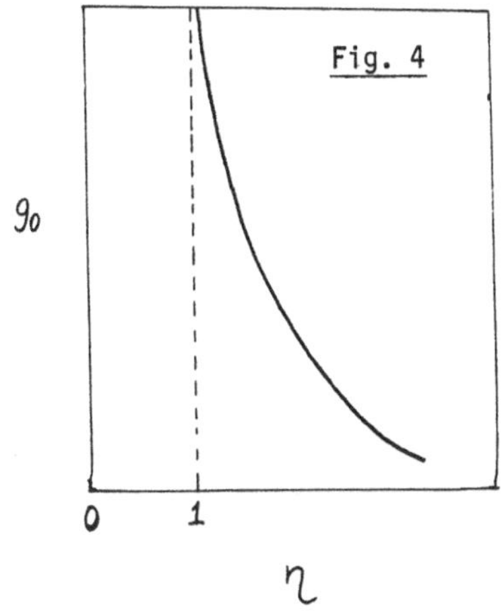

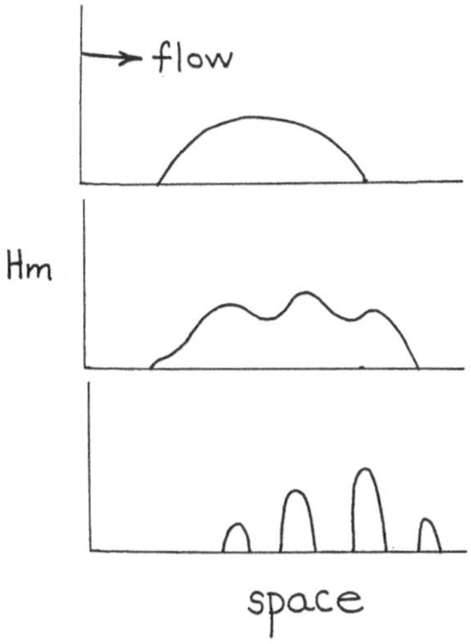

Fig. 5

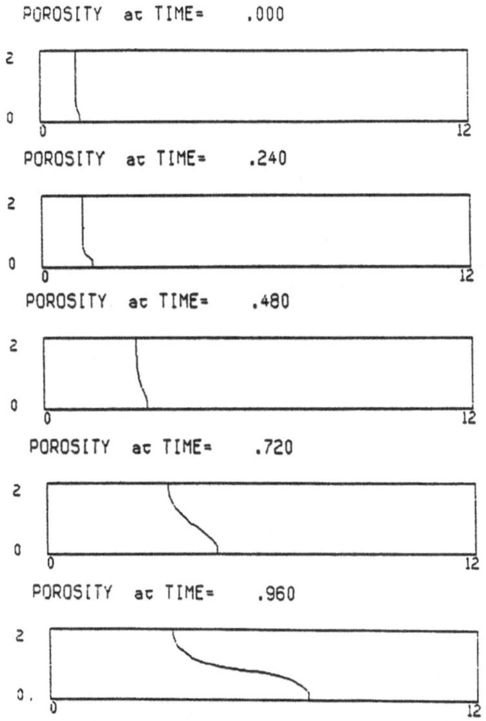

Fig. 6

can argue that the planar dissolution front is unstable to the formation of fingers. Numerical simulations of the phenomenon are seen in Fig. 6. Linear stability, bifurcation and matched asymptotic methods have been applied to this problem [18,19].

V. Self Organization in Geological Systems

It is hoped that the above phenomena and the cited references indicate the wealth of instability, nonlinear restabilization and other self-organization effects due to the interaction of transport and first order phase transitions. These problems are both interesting in themselves and for the new nonlinear problems they pose. In a number of cases they also have economically interesting applications in the oil and mineral industries.

This work was supported by the U.S. Department of Energy (Office of Basic Energy Sciences, Engineering Research Program) and the U.S. National Science Foundation.

1. P. Ortoleva: "The Multi-faceted Family of the Nonlinear: Waves and Fields, Center Dynamics, Catastrophes, Rock Bands and Precipitation Patterns," 1978 Bordeau Conference on Far From Equilibrium Phenomena (Springer-Verlag, 1978)
2. A.R. McBirney and R.M. Noyes: "Crystallization and Layering of the Skaergaard Intrusion," J. Petrol., v. 20, p. 487-554 (1979).
3. E. Merino: "Survey of Geochemical Self-patterning Phenomena." G. Nicolis and J.S. Turner, editors, in Proceedings of the NATO Conference on Instabilities in Chemistry, Geology, and Materials Science, Austin, Texas (1983)(in press).
4. P. Ortoleva, J. Chadam, M. El-Badewi, R. Feeney, D. Feinn, S. Haase, R. Larter, E. Merino, P. Strickholm and S. Schmidt: "Mechanisms of Bio- and Geo-Pattern Formation and Chemical Signal Propagation," in the Proceedings of the Conference on Instabilities, Bifurcations, and Fluctuations, Austin, Texas, University of Texas Press, pp. 125-195 (1980).
5. E. Merino and P. Ortoleva: The Self-Organization of Geo-Chemical Periodicity. (book manuscript in preparation).
6. E.S. Hedges and J.E. Meyers: The Problem of Physico-Chemical Periodocity, New York, Longmans, Green and Co., 95 p. (1926).
7. E. Merino, C. Moore, E. Ripley, G. Auchmuty, J. Chadam, J. Hettmer and P. Ortoleva: "Kinetic Modeling of Redox Roll Fronts and Their Instabilities," Geochimica et Cosmochimica Acta (submitted for publication).
8. J. Chadam, E. Merino, P. Ortoleva, and A. Sen: "Reactive Percolation Porosity Instabilities in Roll-front Propagation," (in preparation).
9. J. Chadam, C.S. Haase, D. Feinn, and P. Ortoleva: "Oscillatory Zoning in Plagioclase Feldspar," Science 209, 272.
10. E. Merino, P. Ortoleva and P. Strickholm: "Kinetics of Metamorphic Layering in Anisotropically Stressed Rocks," Amer. Jour. Science 282, 617 (1982).
11. E. Merino, P. Ortoleva, and P. Strickholm: "A Kinetic Theory of Spontaneous Stylolite Generation and Spacing," Contrib. Mineral. Petrology 82, 360.
12. R.E. Liesegang: Geologische Diffusionen, (Dresden, Steinkopff, 1913).
13. P. Ortoleva: "The Self Organization of Liesegang Bands and Other Precipitate Patterns," (an invited review for the Proceedings of the NATO Conference on "Instabilities and Fluctuations in Chemical, Engineering and Geological Systems," Austin, Texas, March, 1983).
14. P. Strickholm, J. Hettmer and P. Ortoleva: "Halo, Spiral, Speckled and Branched Band Precipitate Patterns: CPG Theory in Two Dimensions," (in preparation).
15. P. Ortoleva and P. Strickholm: "Particle Size Distribution Effects in the CPG Model of Liesegang Banding," J. Stat. Phys. (submitted for publication).
16. F. de Pasquale, P. Ortoleva, P. Strickholm and P. Tartaglia: "Multiple Scaling Approach to Particle Size Distribution: Analysis of the CPG-Precipitate Patterns," (in preparation).
17. P. Ortoleva, P. Tartaglia, F. de Pasquale: "Unstable Nucleation Fronts: Ostwald-Liesegang Banding," (in preparation).
18. J. Chadam, J. Goncalves, S. Hagstrom, J. Hettmer, D. Hoff, J. Johnson, R. Larter, E. Merino, P. Ortoleva and A. Sen: "Instability at Reactive Percolation," (submitted for publication).
19. J. Chadam, P. Ortoleva and A. Sen: "Bifurcation of Reaction-Percolation Fingers," (in preparation).

Temporal and Spatial Structures in Chemical Systems Far from Equilibrium

John Ross

Department of Chemistry, Stanford University, Stanford, CA 94305, USA

1 Introduction

In this lecture we present an overview of some recent results on temporal and spatial structures in chemical systems far from equilibrium. We discuss experiments on multiple stationary states (bi- and tristabilities), hysteresis of unstable stationary states, limit cycles, generation of limit cycles by periodic external perturbations, regular and Hopf bifurcations, chemical fronts, time-independent spatial structures and periodic precipitation processes.

2 Photo-illuminated Reactions

We begin with the photo-illuminated gas phase reaction, $S_2O_6F_2 \rightleftarrows 2\ SO_3F$. SO_3 absorbs light of a given frequency which is turned into heat; the temperature of the gas rises which increases the concentration of SO_3F and this in turn leads to an increase in light absorption. Hence the system has a positive feedback loop which leads to the possibility of multiple stationary states. (For representative articles on this subject see [1-3]). When the power incident on the gaseous mixture is slowly increased and then decreased over a range of light power a hysteresis loop is traced out in the plot of steady state absorption vs. incident power. Sharp transitions are observed between the stable branches at the marginal stability points. Two stationary states can coexist, one on the low absorption branch and one on the high absorption branch. The experimental results [4] confirm the theoretical predictions [5].

A simple modification of this experimental system leads to a number of other interesting results [6]. Light transmitted by the sample of sulphur compounds can be detected and amplified, and the signal can be delayed prior to transmission to a modulator which controls the light input into the system. By appropriate choice of the degree of amplification and the extent of delay it is possible to stabilize with the external feedback loop the unstable stationary state of the system without external feedback. The location in phase space of the stable and unstable states of the system without feedback are not altered by the external feedback. Furthermore the system with external feedback shows limit-cycle oscillations and inverted Hopf bifurcations, and is predicted to have chaos.

3 Chemical Fronts and Waves

We turn next to chemical fronts and waves (for some representative experimental and theoretical articles on this subject see [7-10]). We have developed a technique for the quantitative measurement of propagating chemical profiles in reaction systems far from equilibrium [10]. The measurements are made on circular waves in a thin layer of quiescent but excitable solution of the Belousov-Zhabotinsky reaction by means of light absorption of ferroin. Wave initiation is achieved by the application of a voltage pulse to wire electrodes dipped into the solution. The transmission profile of the wave is measured with a linear photo-diode array to a resolution of 50 microns. The chemical front is about 300 to 500 microns in width and we confirm experimentally the remarkable property of the constancy of the profile in space and time. The relaxation of the wave profile behind the front is characterized by two distinct time constants which reflect rate-limiting processes corresponding to the reduction of ferroin during the regeneration of bromide ion. Further we report on front velocities at various initial reactant concentrations, temperatures, and depths of solution. At given initial conditions the front velocity is constant and the temperature dependence of the velocity has an apparent activation energy of 34.0 kJ/mole. In experiments in which two fronts are initiated and made to collide with each other we measure the process of annihilation of the front profiles. For a variety of initial conditions we observe, after the passage of the front, the formation of a stable stationary spatial structure, the onset of a mosaic pattern. This constitutes a transition from an initially homogeneous solution to an inhomogeneous one.

4 External Perturbations

In regard to the subject of temporal structures we discuss briefly the generation of multiple attractors by means of appropriate external perturbations in oscillatory chemical reactions [12], resonance effects [13], and the possibility of the control of distribution of dissipation in such systems [14].

5 Periodic Precipitation

The periodic precipitation process known as Liesegang band formation has been investigated for many years [15]. A theory based on a chemical instability has been proposed for the formation of periodic precipitation processes [16-18]. The essential point of the theory is the hypothesis that band formation is a post-nucleation process and is due to an instability in the autocatalytic growth of colloidal particles coupled with diffusion. We discuss a series of experiments [19, 20] which substantiate this hypothesis including the determination of the temporal and spatial evolution of band formation as measured by light scattering, by light deflection and visual observation. Furthermore, in the case of PbI_2 precipitate we measure the total lead concentration as a function of time and position in a one-dimensional band formation experiment, the total iodide concentration, and the free iodide ion concentration; at the same time we take photographs of microscope observations of particle formation and distribution. There is no evidence of flocculation in the lead iodide system. Nucleation is continuous in space,contrary to the Ostwald theory and its many variants, and band formation occurs within a region of continuously nucleated material by a self-focussing, instability mechanism.

Acknowledgements

I wish to thank my coworkers, M, LeVan, M. Schell, P. Wood, and C. Zimmermann for their participation.

References

1. R. Schmitz, in Chemical Reaction Engineering Reviews, ed. H. M. Hulburt, p. 165, Am. Chemical Society, Washington, D.C. 1975.
2. J. Ross: Berichte der Bunsen-Gesellschaft für physikalische Chemie, Bd. 80, Nr. 11 (1976).
3. I.R. Epstein: J.Phys.Chem. 88, 187 (1984).
4. E.C. Zimmermann and J. Ross: J. Chem. Phys. 80, 720 (1984).
5. A. Nitzan and J. Ross: J. Chem. Phys. 59, 241 (1973).
6. E.C. Zimmermann, M. Schell and J. Ross, J. Chem. Phys. 81, 1327 (1984).
7. A.T. Winfree, Science 175, 634 (1972); ibid, 181, 937 (1973); Theor. Chem. 4, 1 (1978).
8. M. Collins and J. Ross, J.Chem.Phys. 68, 4468 (1978); J. Ross and P.Ortoleva, ibid,63, 3398 (1975); P. Ortoleva and J. Ross, ibid, 60, 5090 (1974); P. Ortoleva and J. Ross, ibid, 58, 5673 (1973); R. Feeney, S.L. Schmidt and P. Ortoleva, Physica 2D, 536 (1981).
9. J.J. Tyson and P.C. Fife: J.Chem.Phys. 73, 2224 (1980).

10. A. Hanna, A. Saul, and K. Showalter, J. Amer. Chem. Soc. 104, 3838 (1982); R.J. Field and R.M. Noyes, J. Amer. Chem. Soc. 96, 2001 (1974); K. Showalter, R.M. Noyes and H. Turner. J. Amer. Chem. Soc. 101, 7463 (1979); K. Showalter, J.Chem.Phys. 73, 3735 (1980); A. Hanna, A. Saul, and K. Showalter, J.Amer. Chem. Soc. 104, 3838 (1982).
11. P.M. Wood and J. Ross: "A quantitative Study of Chemical Waves in the Belousov Zhabotinsky Reaction," submitted to J.Chem.Phys.
12. P. Rehmus, W. Vance and J. Ross, J. Chem. Phys. 80, 3373 (1984).
13. Y. Termonia and J. Ross: PNAS, USA, 78, 2952 (1981); ibid, 78, 3563 (1981); ibid, 79, 2878 (1982).
14. P. Rehmus, E.C. Zimmermann, H.L. Frisch and J. Ross: J. Chem. Phys. 78, 7241 (1983).
15. K. W. Stern, Chem. Rev. 54, 79 (1954).
16. M. Flicker and J. Ross, J. Chem. Phys. 60, 3458 (1974); R. Lovett, P.Ortoleva and J. Ross, J. Chem. Phys. 69, 947 (1978).
17. D. Feinn, P. Ortoleva, W. Scalf, S. Schmidt and M. Wolff, J. Chem. Phys. 69, 27 (1978).
18. G. Venzl and J. Ross, J. Chem. Phys. 77, 1302 (1982); ibid, 77, 1308 (1982).
19. S. Kai, S.C. Müller and J. Ross: J.Chem.Phys. 76, 1392 (1982); S.C. Müller, S. Kai and J. Ross, SCI 216, 635 (1982); S.C. Müller, S. Kai and J. Ross, J. Phys.Chem. 86, 4078 (1982); ibid, 86, 4294 (1982); S. Kai, S. Müller and J. Ross, J. Phys. Chem. 87, 806 (1983).
20. M. LeVan and J. Ross, to be published.

Experimental Study of Target Patterns Exhibited by the B.Z. Reaction

J.M. Bodet, C. Vidal, A. Pacault, and F. Argoul

Centre de Recherche Paul Pascal, Domaine universitaire
F-33405 Talence Cêdex, France

1. Introduction

After BELOUSOV [1] who discovered in 1958 an oscillating chemical reaction (involving a dicarbonic acid - e.g. malonic acid - potassium bromate, and an inorganic redox couple - e.g. $Fe(phen)_3^{3+}/Fe(phen)_3^{2+}$ in a strongly acidic medium) ZHABOTINSKY [2] showed its periodicity in space as well [3,7]. This spatio-temporal self-organization can take different forms depending on the geometry of the experimental device (1D, 2D or 3D-imensions). We are interested in the bidimensional waves, usually appearing as expanding concentric rings (disks or targets) wich are called target patterns. How to apprehend such a spontaneous spatial phenomenon occurring far from thermodynamic equilibrium? We are faced with two drastically different interpretations. On the one hand one can adopt a deterministic point of view and look at the heterogeneities as the single reason [5,6,7,8] for the centre birth ; on the other hand one can agree with the Brussels school [4] which assumes that the existence of internal fluctuations is at the origin of these spatial patterns. It is an experimental challenge to try to solve this controversy. In this paper we describe an equipment which has been especially designed to investigate the formation of the target patterns and which is adapted to a statistical study of the number and emission frequencies of centres, and of the characteristics of the waves too.

2. Experimental device and conditions

Our apparatus is so constituted that both the temporal and spatial evolution can be investigated abreast. The main parts are :

- a cylindrical stirred tank reactor, where we study the temporal evolution of stirred mixtures. A platinum thread and a reference electrode are used to record this evolution which is thus monitored through the redox potential of the bulk ;

- a plexiglass cell, which sandwiches a disk of reacting liquid at rest between two plates (Ø = 100 mm ; e = 1 mm). The 2D behaviour is recorded on a video tape at a rate of 25 images per second given by a standard video camera.

Both parts of this device are closed, that is to say there is no permanent input of reagent. At time t = 0 the tank reactor (or the cell) is fed and we then let the reaction evolve towards thermodynamic equilibrium. Accordingly we only observe transient regimes. In this spirit, our attention has been mainly focused on the early stages of the evolution, while the system is far away from equilibrium, which we will call later on the *initial state*. Let us summarize the following technical details :

- we set the whole in a thermostated room : 22°C ± 1°C
- the reactant's solutions are filtered through 0.22 µm millipore filters
- in a premixing tank we use a filtered nitrogen bubbling to mix the reactants. The purpose of this gazeous agitation is to homogenize the mixture. Then we pour the solution either into the tank reactor for a bulk investigation, or into the cell for the thin layer investigation.

3. Experimental observations

3.1 "Initial state" diagram

We have studied the behaviour of the B.Z. reaction scanning a set of initial concentrations in sulphuric acid, malonic acid and potassium bromate. We report in Fig. 1a - 1b the salient features of both the stirred and thin layer observations.

Stirred behaviour

The domain where the reaction oscillates is bounded by a zone of monotonic evolution (which, in some sense, would correspond to a stationary state for an open system). Inside this domain the way into which the (pseudo-) period and (pseudo-) amplitude of an oscillation change, while the system evolves towards equilibrium, depends on the initial composition (see below §3.2 and Fig. 2).

Thin layer behaviour

Two phenomena are worth noting :

- a pseudo-wave arising from spontaneous phase gradients [3]
- the so-called target patterns : disks and targets.

First of all, the domain where the reaction actually oscillates in the cell is included in the area of stirred oscillations. However the overlap is not complete : there are several ranges of initial composition leading to oscillations in the reactor whereas one observes a monotonic evolution in a thin layer geometry. Moreover, the contrary seems to occur in a small domain of the plane $[BrO_3^-]_0$ = 1 M (according to only two points in fig. 1.b). The experimental conditions (stirring, gazeous, exchanges with the surrounding, etc..) are so different that many reasons can account for this difference. But perhaps the most striking result is certainly the fact that, even when the layer does oscillate, there is still a significant composition domain where the system never exhibits any centre. This experimental fact calls for a relevant theoretical explanation and deserves a more complete and systematic investigation.

3.2 Evolution of the oscillations

Wanting to probe experimentally the statistical predictions made by WALGRAEF et al. [4] on the basis of their stochastic approach, we have first to select compositions of the B.Z. system exhibiting a limit cycle type behaviour, at least for a while. In this respect figure 2 shows two typical examples encountered during this study. The redox potential of certain mixtures (Fig. 2a) actually keeps an almost constant amplitude and period for some time : such mixtures therefore seem appropriate to the statistical analysis we intend to carry out. On the contrary, some others are definitively inadequate and must be rejected, as can be seen in figure 2b where the amplitude continuously decreases even though the period still exhibits a plateau.

3.3 Limit cycle behaviour and birth of centres

In spite of the link which might be established between the limit cycle type behaviour and the occurrence of target patterns, it is tempting to search very quickly for some qualitative correlation between the two. One can hopefully imagine, for instance, that a difference in the properties of the limit cycle has something to do with the above-mentioned fact that oscillations in thin layer do not necessarily give rise to centres.

In order to clarify this assumption we have normalized the representation of a redox oscillation (in the "limit cycle stage") to the unit square : i.e. the amplitude variation and the period are always set equal to 1 (in arbitrary units) whatever their absolute values. We are thus able to compare the "shape" of two oscillations.

The general relaxation character of all the oscillations is obvious at a glance at figure 3. The essential difference between two of them only lies in the relative

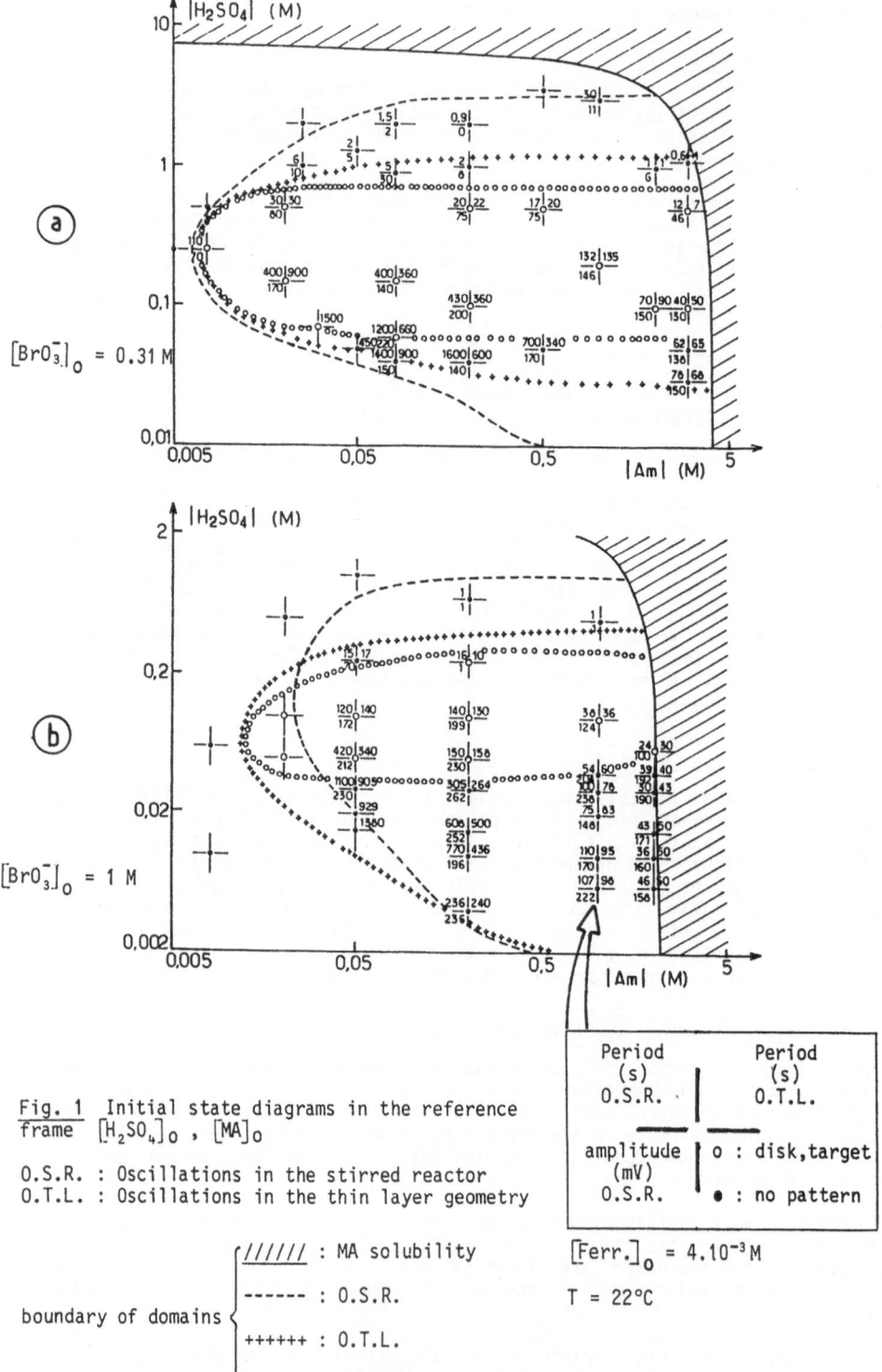

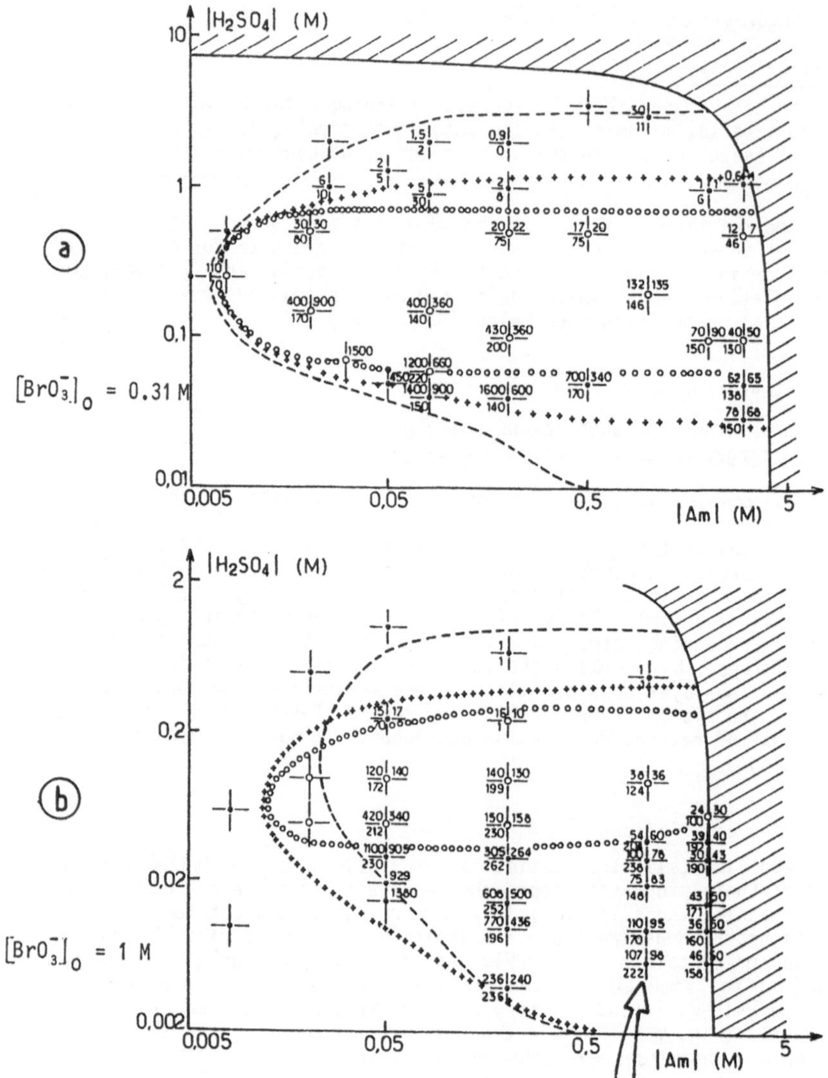

Period (s) O.S.R. | Period (s) O.T.L.

amplitude (mV) O.S.R. | o : disk, target
• : no pattern

Fig. 1 Initial state diagrams in the reference frame $[H_2SO_4]_0$, $[MA]_0$

O.S.R. : Oscillations in the stirred reactor
O.T.L. : Oscillations in the thin layer geometry

$[Ferr.]_0 = 4.10^{-3}$ M

T = 22°C

boundary of domains:
- ////// : MA solubility
- ------ : O.S.R.
- ++++++ : O.T.L.
- oooooo : patterns

(a) $[MA]_0 = 0.08$ M, $[H_2SO_4]_0 = 0.6$M

(b) $[MA]_0 = 3$ M, $[H_2SO_4]_0 = 0.1$ M

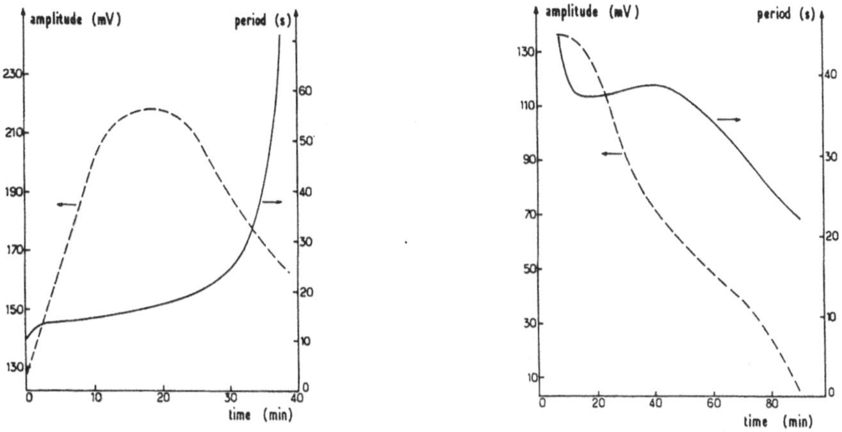

Fig. 2 Two kinds of transient evolution towards equilibrium (temporal variation of amplitude and period of (pseudo-) oscillations in the stirred reactor) $[BrO_3^-]_0 = 0.31$ M, $[Ferr]_0 = 4.10^{-3}$M, T = 22°C

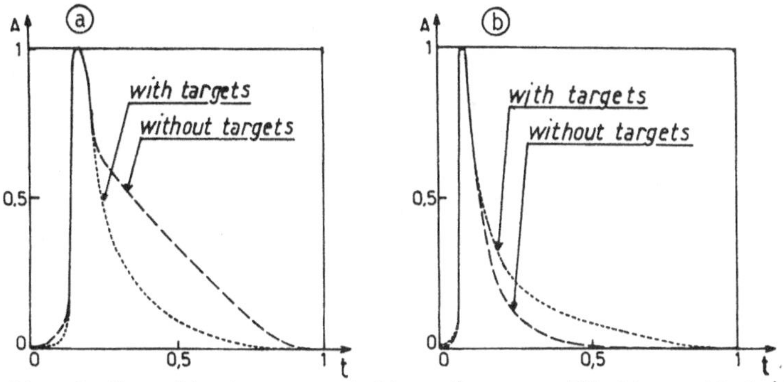

Fig. 3 Normalized representation of one oscillating period ($[Ferr.]_0 = 4.10^{-3}$ M)

(a) $[BrO_3^-]_0 = 0.31$ M ; $[MA]_0 = 0.5$ M { — — — $[H_2SO_4]_0 = 0.05$ M ; ----- $[H_2SO_4]_0 = 0.5$ M

(b) $[BrO_3^-]_0 = 1$ M ; $[MA]_0 = 1$ M { — — — $[H_2SO_4]_0 = 0.01$ M ; ----- $[H_2SO_4]_0 = 1$ M

reduction rate. Figure 3a, drawn for $[BrO_3^-]_0 = 0.31$ M, shows how the deformation occurs when $[H_2SO_4]_0$ is increased, for several $[MA]_0$. It is seen that the reduction takes place faster and faster as $[H_2SO_4]_0$ grows, centres appearing when the reducing phase becomes short enough (less than about 1/5 of the whole period). Nevertheless, this is no more the case in figure 3b, which corresponds to $[BrO_3^-]_0 = 1$M. There, on the contrary, the reduction rate slows down when $[H_2SO_4]_0$ increases, and centres are only observed if the reducing phase exceeds some time threshold. Hence, at this preliminary stage, looking for a very simple and qualitative correlation between the oscillating properties of the reaction and the birth of centre obviously fails. However, taking into account more relevant variables, especially the pH [9], we are now trying to improve the analysis of these results as will be seen in another paper.

4. Conclusion

Having collected these preliminary results, we now intend to carry out a complete statistical study of both the number and emission frequencies of centres and of the wavelength of waves within the initial concentrations range of reactants investigated in the present paper.

This analysis, which requires an automatic image-processing device and a fast computer, is in progress. When completed, we then hope to be in a position to discriminate among the different available theoretical interpretations.

References

1. B.P. Belousov, Sb. Ref. Radiat. Med. Za. 1958 (Russian), Medzig, Moscou (1959)
2. A.N. Zaïkin, A.M. Zhabotinsky, Nature, 225, 535 (1970)
3. A. Pacault, C. Vidal, J. Chim. Phys., 79, 691 (1982)
4. D. Walgraef, G. Dewell, P. Borckmans, J. Chem. Phys. 78, 3043 (1983)
5. P. Ortoleva, J. Ross, J. Chem. Phys. 60, 5090 (1974)
6. N. Kopell, L.N. Howard, Studies in Appl. Math. 52, 291 (1973)
7. Y. Kuramoto, J. Yamada, Progress Theor. Phys. 56, 724 (1976)
8. J.J. Tyson, P.C. Fife, J. Chem. Phys. 73, 2224 (1980)
9. A.B. Rovinsky, J. Phys. Chem. 88, 4 (1984)

Kinematic Analysis of Wave Pattern Formation in Excitable Media

John Rinzel* and Kenjiro Maginu

Mathematical Research Branch, NIADDK, National Institutes of Health
Bethesda, MD 20205, USA

1. Introduction

A homogeneous excitable medium supports the steady propagation of impulse waves of excitation in response to a spatially localized stimulus. In a one-dimensional medium, for which the nerve axon is a prototype, a periodic stimulus produces a train of impulse waves which propagates away from the stimulus site. Successive impulses in the train interact with each other during propagation and consequently form a spatio-temporal wave pattern. When an equally spaced impulse train, whose temporal period equals the period of the stimulus, develops as a consequence of the interaction, we say the medium is entrained to the input stimulus. In some cases however, the impulses exhibit bunching during propagation to form an unequally spaced impulse train. Here we investigate these phenomena by a kinematic analysis based upon an approximation of the wave interaction. We also illustrate that, in certain parameter ranges, the one-dimensional medium can support a center wave pattern in the absence of a maintained stimulus. This wave pattern is a one-dimensional analog of the target wave pattern which is observed in some two-dimensional excitable media such as the Belousov-Zhabotinskii reagent in a shallow dish.

An impulse wave of excitation typically consists of a rapid upstroke, a plateau (the excited state), and a fast downstroke followed by a slowly decaying tail. The temporal profile of the tail, observed at a fixed location, represents the time course of recovery of the medium from the excitation. The propagation speed of the impulse is determined almost completely by the recovery level of the medium (from the excitation of the predecessor impulse) just ahead of the upstroke. Moreover, in some media (in which the diffusion of an "activator" is dominant over the diffusion of the other reacting components), the recovery level of the tail can be regarded as a function of time from the upstroke of the impulse; in other words, the temporal profile of the tail is not affected by the influence of predecessor impulses. (Keener [2] shows this in the case of relaxation kinetics.) Under these two approximations, the instantaneous speed of an impulse, when it passes through a location of the medium, is regarded as a function of time since the upstroke of the predecessor impulse passes through the same location. In such a medium, the interaction of successive impulses in the course of propagation can be described by a simple kinematic equation.

Figure 1-A shows the temporal profile (observed at a fixed location) of a solitary impulse wave propagating in the following FHN (FitzHugh-Nagumo) medium:

$$v_t = v_{xx} - f(v) - w + I, \qquad w_t = \sigma^2 w_{xx} + b(v - \gamma w), \tag{1}$$

where $f(v) = v(v-0.14)(v-1.0)$, $b = 0.007$, $\gamma = 2.5$, $I = 0.0322$ and $\sigma^2 = 0$. This equation, when $\sigma^2 = 0$, is a simplified qualitative model of nerve impulse conduction, in which v corresponds to membrane potential and w to a slow recovery current; see [1] for references on the FHN equation. The upstroke of the impulse is taken as the

* J. R. performed a part of this work while a visitor at the Center for Mathematical Biology, University of Oxford (supported by SERC grant GR/C/6359.5)

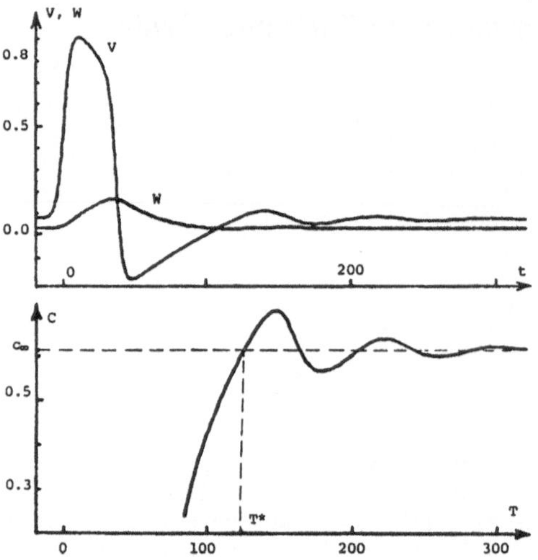

Fig.1-A. Temporal profile of a solitary impulse of (1) when $I = 0.0322$ and $\sigma^2 = 0$. The upstroke of the impulse is taken as the origin of time T.

Fig.1-B. Dispersion relation (speed c(T) versus temporal period T) for periodic wave trains in the FHN medium (1).

(Solutions for these figures are computed numerically by using a two-point boundary value problem solver PASVA 3. See [1] for details of such computations.)

origin of t in Fig.1-A. This impulse has a damped-oscillatory tail. So the level of recovery from excitation is a decaying oscillatory function of time. The propagation speed c(T) of a periodic impulse train with the temporal period T is computed numerically by using a two-point boundary value solver. Figure 1-B shows the speed c(T) as a function of T, which is called the dispersion relation for the impulse trains. When the above approximations are applicable to this medium (which will be verified by numerical solutions of the next section), we can apply c(T) to describe the speed of an individual impulse in a wave train. (The shape of c(T) coincides almost completely with the shape of the tail of the solitary impulse. This means that the speed of an impulse can be regarded as a function of the recovery level from the predecessor impulse.)

2. Kinematic Description of a Propagating Impulse Train

A spatially localized periodic stimulus generates a train of impulses which propagates into the medium away from the stimulus site. Let $x_k(t)$, $k = 1,2,\cdots$, be the position of the upstroke of the kth impulse. We consider its inverse function $t^k(x)$, $0 \leq x \leq L$, which represents the trajectory of the impulse in the x-t plane. If the above approximations are applied, the speed $dx_k(t)/dt$ of the impulse when it passes the location x is given by $c(T^k(x))$, where $T^k(x) \equiv t^k(x) - t^{k-1}(x)$ is the time since the predecessor impulse passed the same location. In this way, by using $dt^k(x)/dx = dt/dx_k(t)$, we obtain the following kinematic equation for the trajectories $t^k(x)$ [1]:

$$(d/dx)t^k(x) = \beta(T^k(x)), \qquad \beta(T) \equiv 1/c(T), \qquad T^k(x) = t^k(x)-t^{k-1}(x). \tag{2}$$

The function $t^0(x)$, $0 \leq x \leq L$, must be specified; if the medium is initially in the rest state then $t^0(x) = -\infty$. The "initial data" for $t^k(x)$ at the stimulus site $x = 0$ are given by $t^k(0) = (k-1)T_0$ if T_0 is the period of the stimulus.

The dotted curves in Fig.2-A show the wave trajectories $t^k(x)$ in the FHN medium ($I = 0.0322$) when the stimulus has the period $T_1 = 100$. (The input is a current injection at the terminal $x = 0$, i.e. the boundary condition $(\partial/\partial x)u(0,t) = \psi(t)$ is assumed, where $\psi(t)$ is a T_1-periodic square wave that is alternately zero or positive. The zero flux condition is assumed at the other boundary. The Crank-Nicholson method with spatial and temporal mesh $\Delta x = 0.75$ and $\Delta t = 1.0$ is used to solve (1).) The dashed curves, which almost perfectly coincide with the dotted curves, are solutions of the kinematic equation (2) (subject to the initial conditions $t^k(0) = (k-1)T_1$) based upon the dispersion relation of Fig.1-B. The curves in Fig.2-B show the speeds $c^k(x) \equiv dx/dt^k(x)$ of the impulses in the x-c plane.

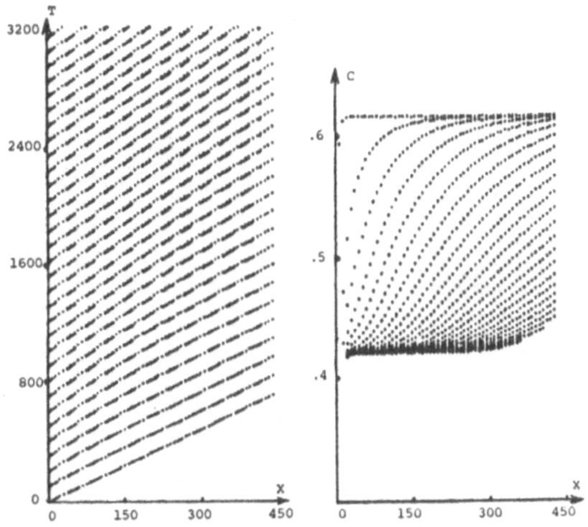

Fig.2-A. Trajectories $t^k(x)$ of the impulse waves in the FHN medium. The medium is entrained by the stimulus of period T_1 = 100 which is injected at x = 0. The dashed curves are solutions of the kinematic equation (2).

Fig.2-B. The dotted curves are the instantaneous speeds $c^k(x)$ of the impulses ($1 \leq k \leq 32$) when they pass the location x.

The first impulse propagates into the resting medium with a constant speed c_∞ = 0.614. The successive impulses propagate away from the stimulus site x = 0 with the slower initial speed $c(T^k(0)) = c(T_1) = 0.423$. These impulses accelerate during propagation (as a consequence of the increase of $T^k(x)$) until they are "phase-locked" to the first impulse, i.e. until the speed $c^k(x)$ and the interpulse time interval $T^k(x)$ increase to attain c_∞ and T^*, respectively, with the increase of x, where T^* = 126 is the smallest T for which $c(T) = c_\infty$ (see Fig.1-B). The kth impulse travels with the speed $c(T_1)$ (maintaining the initial time lag T_1) longer than the (k-1)st impulse and phase-locks to the first wave later than the (k-1)th impulse. To demonstrate this behavior analytically we first approximate $\beta(T)$ by a linear function on the range $T_1 \leq T \leq T^*$. Then (2) can be solved explicitly to yield

$$T^k(x) = T^* - (T^*-T_1)\exp(-\alpha x)\sum_{n=0}^{k-2}(\alpha x)^n/n!, \quad \alpha \equiv \{\beta(T^*)-\beta(T_1)\}/(T_1-T^*) = 0.0283. \quad (3)$$

From this solution we can deduce that at a location x far away from the stimulus site (a) the medium (which is initially in the rest state) begins to fire repetitively at first with period T^* (b) then, as k increases, the period $T^k(x)$ of the firing increases and exceeds $(T_1+T^*)/2$ at about $(\alpha x+2)$nd firing and (c) the period $T^k(x)$ converges to T_1 with the increase of k, i.e. the medium is eventually entrained to the T_1-periodic input stimulus.

The condition $c'(T_0) < 0$ implies the instability of equally spaced T_0-periodic wave trajectories of (2). Hence, in contrast to the above example, entrainment to the periodic stimulus does not occur if the period T_0 of the stimulus satisfies $c'(T_0) < 0$. Figures 3 and 4-A show the wave trajectories when the stimulus has periods T_3 = 165 and T_2 = 152.5, respectively. (The dashed curves are solutions of (2). The smooth curves in Fig.3 are from numerical solutions of (1) with coarse mesh Δx = 1.5 and Δt = 2.0. The disparity between the two types of curves disappears almost completely when a finer mesh is used to solve (1). The smooth trajectories of Fig.4-A are obtained from numerical solutions of (1) with the finer mesh Δx = 0.75 and Δt = 1.0.) The periodic impulse train, which propagates away from the stimulus site x = 0, reorganizes during propagation to form an unequally spaced impulse train. When the stimulus has the period 165, the train forms a "period 2" wave pattern in which two temporal periods 125 and 205 appear alternately. When the stimulus has the period 152.5, the train eventually forms a "period 2" pattern in which periods 120 and 185 appear alternately. This pattern does not develop so soon, or so close to x = 0, as that in Fig.3; presumably, it is less stable. The process of the organization of these impulse trains can be very sensitive to noise. Figure 4-B shows the trajectories for the T_2-periodic stimulus when a coarser mesh, Δx = 1.5 and Δt = 2.0, is used in the numerical solution. A completely different wave pattern is formed due to the influence of small numerical errors.

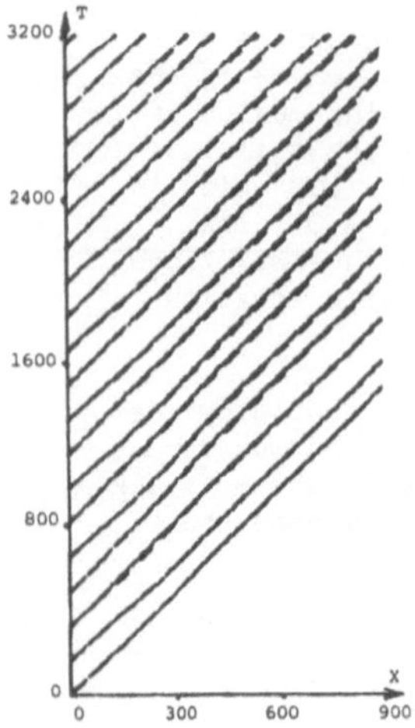

Fig.3. An unequally spaced impulse train forms during propagation when the period of stimulus is $T_3 = 165$.

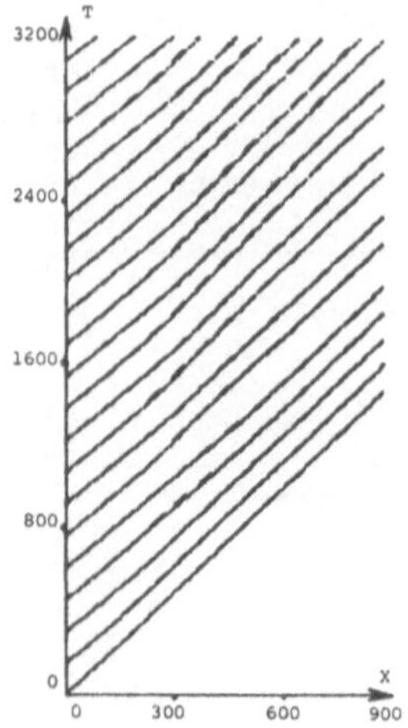

Fig.4-A. Wave trajectories when the period of the stimulus is $T_2 = 152.5$.

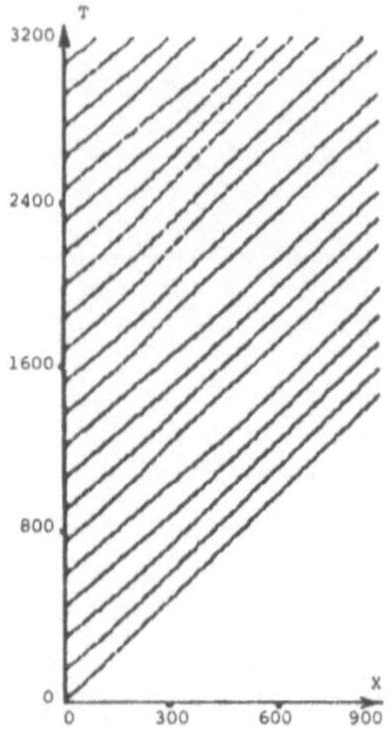

Fig.4-B. Trajectories when a coarser mesh is used to solve (1) for $T_2 = 152.5$.

The mechanism for formation of the unequally spaced patterns can be understood by examining the dispersion relation of the medium. When c(T) is as in Fig.5 (which is an enlargement of Fig.1-B), its inverse function T(c) is a multi-valued function of c, i.e. the curve c = c(T) is decomposed into the branches A,B,··· on each of which the single valued inverse functions $T_A(c), T_B(c), \cdots$, respectively, are defined. In other words, for a specified speed c, the medium has different (equally spaced) impulse trains with periods $T_A(c), T_B(c), \cdots$. By concatenating the spacings of the "A" and "B" trains (with speed c) alternately, we can formally compose an unequally spaced train "AB" with speed c, which corresponds to a phase-locked solution of the kinematic equation. The curve AB in Fig.5 shows the "exact" dispersion relation for the "AB" travelling wave solution of (1) obtained by using a two-point boundary value problem solver. The mean period $T_{AB}(c)$ of this train coincides almost perfectly, as the above formal composition predicts, with $(T_A(c)+T_B(c))/2$. The "period 3" train "AAB" is composed by concatenating two "A"s and one "B" alternately. The mean period $T_{AAB}(c)$ of this train coincides almost perfectly with $(2T_A(c)+T_B(c))/3$. The other impulse trains "AC", "AAAB", "ACC", etc., whose dispersion relations are shown in Fig.5, are composed in the same manner.

The equally spaced trains "B" and "D" are unstable since their dispersion relations satisfy $c'(T) < 0$. The unequally spaced trains "AB", "BC", "AD", etc., which contain the unstable spacings "B" and "D", are also unstable. (Stability of these impulse trains has been analysed based upon the kinematic description [Maginu & Rinzel, in

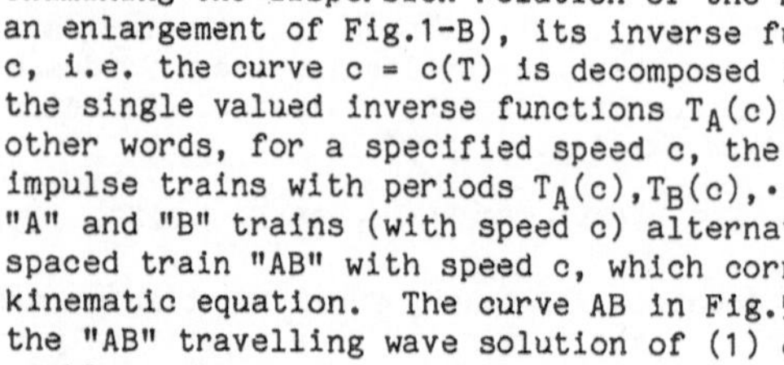

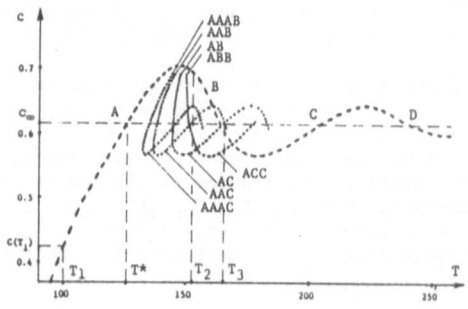

Fig.5. The dispersion relation (branches A, B,C,D) for equally spaced impulse trains in the FHN medium (I = 0.0322) in the T-c plane. The curves AB, AC, etc., are the dispersion relations (mean temporal period versus speed) of the unequally spaced impulse trains.

preparation].) On the other hand, the trains "AC", "AAC", "ACC", etc., which are composed of stable spacings only, are stable. For a given mean period T there may be many stable wave patterns corresponding to the various branches of the equally and unequally spaced impulse trains. The particular time course of the periodic stimulus and the initial condition will determine which pattern is eventually realized in the medium. We anticipate, in such cases of multi-stability, hysteresis phenomena and the capability of a brief perturbation to switch the response from one pattern to another. The simplest stable wave pattern "AC" is realized in response to the T_2-periodic stimulus in the numerical solution of Fig.4-A. However, as shown in Fig. 4-B, small numerical error can switch the response to another complicated pattern.

When the dispersion relation takes the form as in Fig.6-B, the medium has at most two different equally spaced impulse trains with the same propagation speed c: the stable "A" train and the unstable "B" train. Any compound trains, which are composed by the concatenations of "A" and "B", must be unstable since they contain the unstable spacing "B". Hence this medium has no stable periodic impulse trains whose mean temporal period is larger than T_{max}. When the stimulus period T_o is larger than T_{max}, no periodic wave patterns are realized during propagation. Instead, impulses group together to form successively larger (and less unstable) bunches. This is illustrated in Fig.6-A, which shows the wave trajectories $t^k(x)$ for the kinematic description (2) when the dispersion relation c(T) is given by Fig.6-B and the period T_o of the stimulus is larger than T_{max}. (The FHN medium (1) has such a dispersion relation c(T) with a single hump when $-0.01 \leq I \leq 0.012$.)

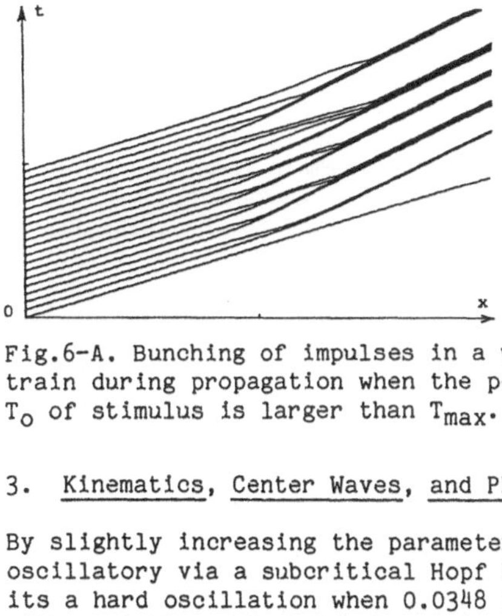

Fig.6-A. Bunching of impulses in a wave train during propagation when the period T_o of stimulus is larger than T_{max}.

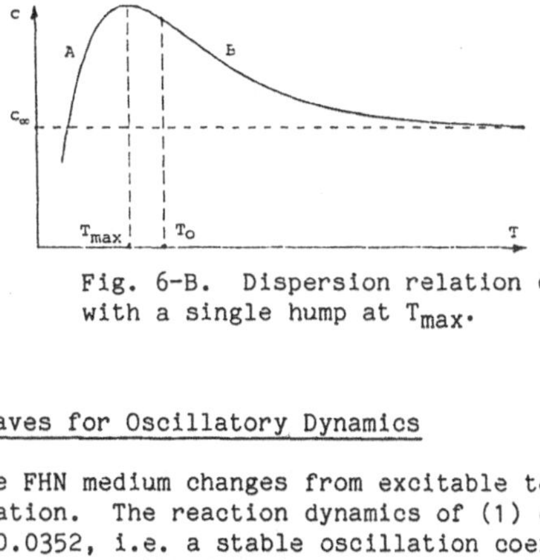

Fig. 6-B. Dispersion relation c(T) with a single hump at T_{max}.

3. Kinematics, Center Waves, and Phase Waves for Oscillatory Dynamics

By slightly increasing the parameter I the FHN medium changes from excitable to oscillatory via a subcritical Hopf bifurcation. The reaction dynamics of (1) exhibits a hard oscillation when $0.0348 \leq I \leq 0.0352$, i.e. a stable oscillation coexists with a stable rest state when I is in this range. The dynamics has a stable large amplitude oscillation, and the rest state is barely unstable, for I just above the hysteresis zone. In the simulations described below we imagine that the excitable medium is at rest when the parameter I is changed suddenly (but slightly) to exceed $I_c = 0.0352$ (from 0.035 to 0.036 say) and the medium is locally perturbed near x = 0 (by adding initial disturbance for v near x = 0). Under such situation we shall use σ^2 as a parameter and study different wave propagation phenomena which lead to phase-resetting or target pattern formation.

We observe that the medium tends to the homogeneous bulk oscillation when σ^2 is small. The smooth curves in Fig.7-A are the wave trajectories for the numerical solution of (1) in the case I = 0.036 and $\sigma^2 = 0.2$. (The bulk reaction dynamics of (1) has a stable periodic oscillation with period $T_p = 159$ when I = 0.036.) The local disturbance around x = 0 initiates the bulk oscillation in this region which transiently acts as a pacemaker. The leading wave propagates with nearly constant velocity $c_\infty = 0.57$ as it advances into the medium which is lingering for a long time near the (barely) unstable rest state. The next several succeeding waves initially travel fast and then slow down as they phase-lock to their predecessors with the temporal spacing T* (see Fig.7-B). Later waves, however, penetrate farther into the medium with their high velocity and with the pacemaker period T_p. These high velocities correspond to "phase waves" which eventually bring the medium to the homogeneous bulk oscillation. This behavior can be understood from the kinematic description based upon the dispersion relation (Fig.7-B) of this medium. The speed c(T) in Fig.7-B tends to $+\infty$ when $T \to T_p$, which means that the periodic wave train tends to the spatially homogeneous oscillation in this limit; the high velocity waves in this limit are sometimes called phase waves. By applying this dispersion relation quantitatively via the kinematics of equation (2) we obtain the dashed trajectories of Fig.7-A which coincide with the smooth curves with excellent precision. In this example the wave velocities during propagation follow a monotone branch of the dispersion curve with $c'(T) > 0$ so that the medium eventually entrains to the pacemaker period T_p. The wave trajectories become horizontal (because $c(T_p) = +\infty$) corresponding to the uniform bulk oscillation. This convergence can be shown analytically as was done for (3). We note that kinematic analysis with this type of dispersion relation also predicts how the uniform oscillation is recovered (how the spatio-temporal phase-resetting takes place) when the bulk-oscillating medium suffers a spatially localized brief perturbation.

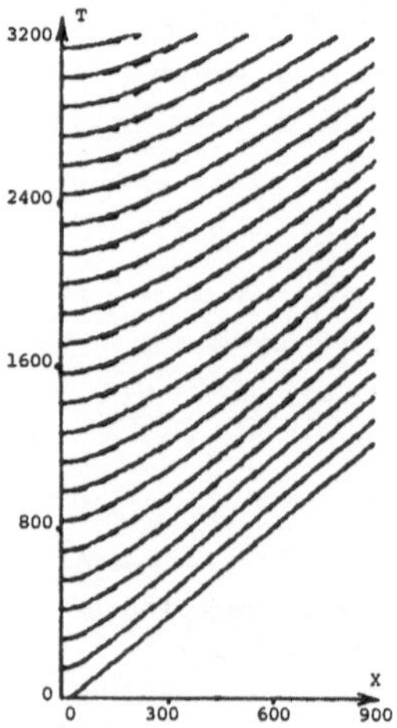

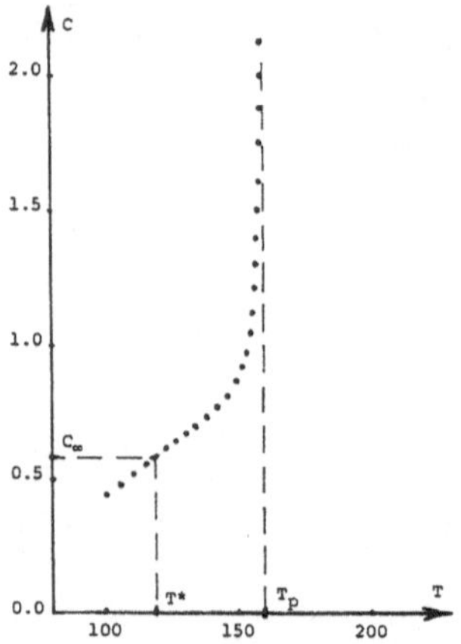

Fig.7-A. A T_p-periodic pacemaker is created near x = 0 by a localized initial perturbation when $\sigma^2 = 0.2$. This oscillation does not remain localized as a pacemaker; the medium tends to the spatially uniform bulk oscillation.

Fig.7-B. The dispersion relation c(T) for the periodic wave trains in the FHN medium when I = 0.036, $\sigma^2 = 0.2$.

We observe that, when diffusivity σ^2 of the "inhibitor" w is adequately increased, the pacemaker region near x = 0 remains spatially localized and initiates finite velocity waves; it forms the center of a target pattern. The wave trajectories of Fig.8-A illustrate this for the FHN medium (1) with I = 0.036, $\sigma^2 = 0.8$ and initial conditions as in the above example. The behavior is best understood by considering the dispersion relation (Fig.8-B) of this medium. We show two components of the curve c(T): the upper branch "H", which satisfies $c(T) \to +\infty$ as $T \to T_p$, and the lower branch "L". Speeds of waves initiated near the center follow the branch "L" rather than "H". Intuitively one would not expect "H" to be followed since a wave initiated time T_p after its predecessor must travel very fast (if it follows the branch "H") and thereby decreases the time interval below T_p where "H" is not defined. Because $c'(T_p) < 0$ on "L", the waves reorganize during propagation to form an unequally spaced wave pattern with mean temporal period approximately equal to the pacemaker period T_p. The kinematic description (with speed c(T) taken from the branch "L") similarly generates a target pattern (Fig.8-C) which develops unequal spacing during propagation. The kinematic pattern differs quantitatively, but not qualitatively, from that in Fig.8-A likely because the assumption for the applicability of the kinematic approximation may be less valid when the "inhibitor" has high diffusion rate. We further note that the particular unequal spacings of Fig.8-A show sensitivity to numerical truncation errors; parameters may be near a stability boundary for this particular pattern. For slightly different parameters, for which $c'(T_p) > 0$, we would expect a target pattern with equally spaced impulses far from the center.

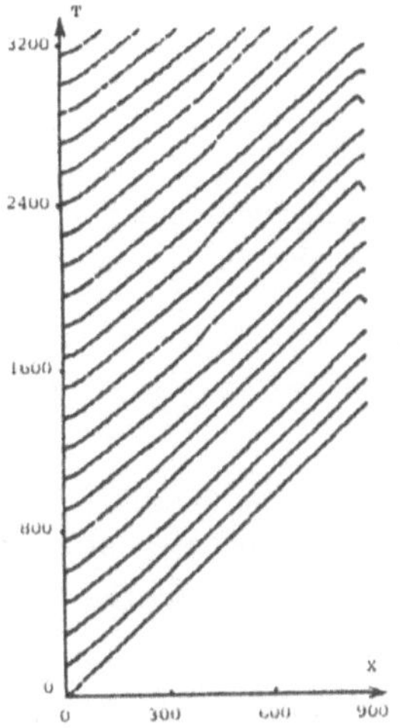

Fig.8-A. The oscillation of the pacemaker generates a center wave when I = 0.036 and $\sigma^2 = 0.8$.

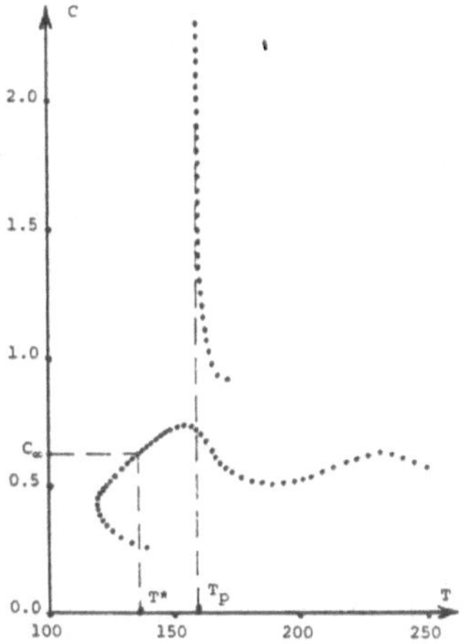

Fig.8-B. The dispersion relation c(T) for perodic wave trains in the FHN medium when $\sigma^2 = 0.8$.

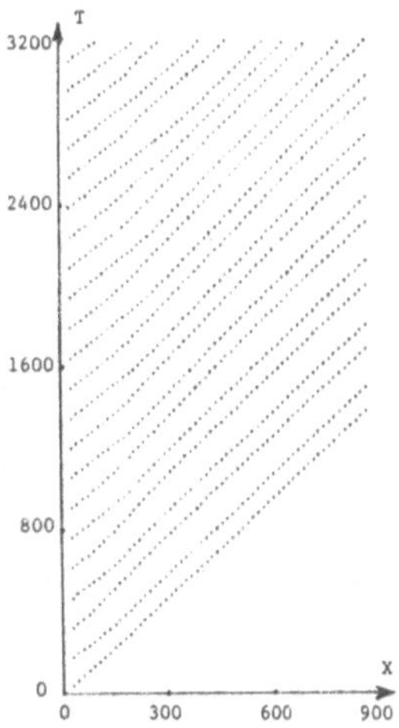

Fig.8-C. The wave trajectories obtained as solutions to (2) based upon the dispersion relation of Fig.8-B.

It is reasonable to expect a target pattern for some region of parameters in the hysteresis zone of the hard oscillation. In such a case a target could be initiated in a stably resting medium by a brief, spatially localized disturbance; the medium ahead of the center's leading wave would remain at the stable rest state. Our attempts to illustrate such a center wave for (1) in this parameter regime have so far been unsuccessful (the disturbed region near x = 0 eventually stops oscillating when σ^2 is large) probably because the hysteresis zone for the FHN medium is too narrow and the oscillation in this zone is not sufficiently stable.

Various features of the FHN reaction dynamics (e.g., excitability, self-sustained oscillations) are shared by many other systems including models for the Belousov-Zhabotinskii reagent. Hence some propagation phenomena in those media might be interpreted kinematically in terms of the dispersion relation as we have done here. For example, the one-dimensional center waves described above may be the analog of certain circular and spherical B-Z target patterns observed in two and three dimensional geometries. Such a mechanism might best be exposed by considering parameter conditions near to the hysteresis zone of a hard oscillation. There is some indirect evidence for a hard oscillation in the B-Z system: the induction phase of the bulk medium is terminated by a sudden jump to large amplitude oscillations. Thus one expects that a spatially localized disturbance, just prior to the jump while the medium is still in the (slowly changing) stable rest state, can initiate a target pattern. Other mechanisms for target pattern formation have been recently reviewed in [3].

1. R. N. Miller and J. Rinzel: "The Dependence of Impulse Propagation Speed on Firing Frequency, Dispersion, for the Hodgkin-Huxley Model", Biophys. J. 34, 227-259, 1981.
2. J. P. Keener: "Waves in Excitable Media", SIAM J. Appl. Math. 39, 528-548, 1980.
3. J. J. Tyson: "A Quantitative Account of Oscillations, Bistability and Travelling Waves in the Belousov-Zhabotinskii Reaction", in Oscillations and Travelling Waves in Chemical Systems (R. J. Field and M. Burger, eds.), Wiley, New York, 1984.

Pattern Formation in Chemical Systems: The Effect of Convection

Daniel Walgraef

Chercheur Qualifié au Fonds National de la Recherche Scientifique, Service de Chimie Physique II, Université Libre de Bruxelles, Campus Plaine, CP 231 B-1050 Bruxelles, Belgium

The spontaneous nucleation of spatial patterns far from thermal equilibrium has long been a puzzling phenomenon both from experimental and theoretical point of views. Numerous questions related to pattern formation and pattern selection in physical, chemical and biological systems remain still unanswered. However, great progress in the understanding of these phenomena have been achieved in the framework of reaction-diffusion equations which are believed to accurately describe many nonequilibrium systems (1). It has also been suggested that, like in equilibrium phase transitions and hydrodynamic instabilities, when chemical spatial or temporal structures appear through the breaking of a continuous symmetry(e.g. translational or rotational symmetry for spatial structures, phase or gauge symmetry for temporal oscillations) they are particularly sensitive to small external fields or to internal fluctuations (2). Effectively, in this case, long range fluctuations may spontaneously develop, leading to topological defects in the structure (e.g. dislocations in hydrodynamical patterbs (3), chemical waves in oscillating or excitable media (4)). A stochastic analysis leads to the evaluation of the probability of such fluctuations and consequently to their statistics. It may then be shown that in the case of chemical oscillators concentric waves or target patterns should be distributed according to their wavelength and to the characteristics of the oscillation. On the other hand spiral waves should only appear in clusters of zero total vorticity as isolated spiral waves are unlikely to appear spontaneously in nearly two-dimensional systems (5).

When the chemical reactions take place in fluid phases, concentration fluctuations may also result from the coupling with convective motion induced for example by local temperature gradients, surface effects like evaporative cooling and Marangoni effects or even by stirring memory. Due to the extreme sensitivity of the phase dynamics to fluctuations, this coupling is very likely to affect the overall behaviour of the system and should be incorporated in its description.

If we restrict ourself to the case of passive convective effects, the nonlinear chemical kinetics may be written as:

$$\dot{X}_i = F_i((X_j),b) + D_i \nabla^2 X_i + \underline{v} \cdot \underline{\nabla} X_i \qquad (1)$$

where X_i are the concentrations of the reacting species, F_i are nonlinear functionals, b the bifurcation parameter, D_i is the Fick diffusion constant and $\underline{v}$ the velocity field of the solvent.

In the case of a Hopf bifurcation the normal form corresponding to the dynamics of the unstable modes in the vicinity of the instability may then be written as:

$$\dot{A} = r_o A - u|A|^2 A + c \nabla^2 A + \underline{v} . \underline{\nabla} A \qquad (2)$$

where r_o, u and c are complex and depend on the various kinetic constants of the chemical network. The real part of r_o is proportional to $b-b_c$, b_c being the critical value of the bifurcation parameter.

In this analysis the homogeneous limit cycle appearing for $b > b_c$ is not affected by the convective term, which, however, modifies the inhomogeneous phase dynamics. Effectively, the adiabatic elimination of the amplitude of the oscillations ($A = R \exp iW$) leads to the following kinetic equation for the phase W:

$$\dot{W} = w + c' \nabla^2 W + c''(\underline{\nabla} W)^2 + \underline{v} . \underline{\nabla} W \qquad (3)$$

(where $w = \mathrm{Im}(r_o - uR^2)$) and the autowave propagation should be affected by the convective motion within the solution.

Let us for example consider the simple case of a layer of solution of thickness d where convective Bénard rolls develop along the Ox direction and where a chemical spiral scroll with vertical axis is nucleated (cf. fig.1) .
The first order correction ϕ to the wave shape W_N (W_N=N arctg(y/x) +Max($wt - k(x^2+y^2)^{1/2}$,0) is inferred from equation (3) and is given,

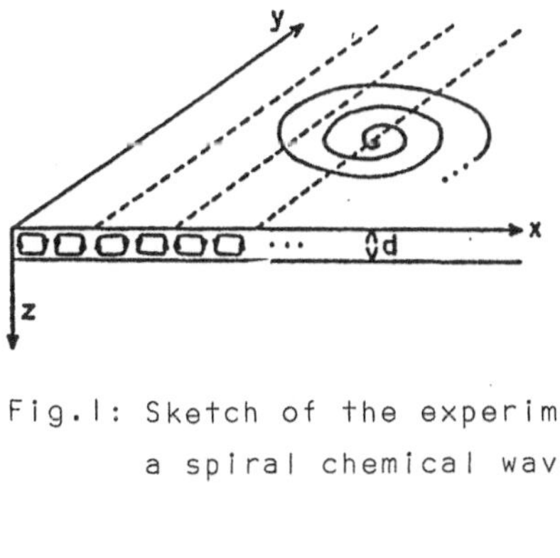

Fig.1: Sketch of the experimental situation described in the text: a spiral chemical wave is initiated in a convective layer.

far from the scroll axis, by:

$$\dot{\phi} = c' \nabla^2\phi + v_x \nabla_x W_N = c' \nabla^2\phi + 2v_x kx/(x^2+y^2)^{1/2} \qquad (4)$$

$$(wt-kr > 0)$$

In the case of free boundary conditions v_x behaves as $v_x = v_o.\sin(\Pi z/d).\sin q_o(x-a)$ and ϕ may be written as $\phi = \sin(\ z/d)\ \psi\ (x,y)$ leading to:

$$\dot{\psi}\ (x,y) = c'((\nabla_x^{\ 2} + \nabla_y^{\ 2}) - (\Pi/d)^2)\psi\ (x,y) + \frac{2kv_o x}{(x^2+y^2)^{1/2}}.\sin q_o(x-a) \qquad (5)$$

It turns out that the propagation of the spiral wave is practically unaffected near the surface or in the y direction while it is strogly affected in the center of the layer and in the x direction and the isoconcentration lines are given in good approximation by:

$$N\mathrm{arctg}(y/x) - k(x^2+y^2)^{1/2} - \frac{2kv_0}{3c'q_0^2} \cdot \frac{x}{(x^2+y^2)^{1/2}} \cdot \sin q_0(x-a) = cst \qquad (6)$$

leading to an irregular modulation of the wavefronts which increases with the thickness of the layer. The irregularity of this modulation may also be shown to be more important in the case of polygonal convective structures du to periodicities in the x and y directions. These effects are in qualitative agreement with the experimental observations of Krinsky et al.(6) and illustrated in fig.II.

When one of the reacting species shows a vertical concentration gradient like for example in the case of evaporative cooling or adsorption effects at the top surface, the phase dynamics becomes:

$$\dot{W} = w + v_z \nabla_z g.\sin W + c' \nabla^2 W + c''(\underline{\nabla}\ W)^2 + \underline{v}.\underline{\nabla}\ W \qquad (7)$$

where g is a functional of the concentration gradients and a decreasing function of the oscillation amplitude R. If the solution of this

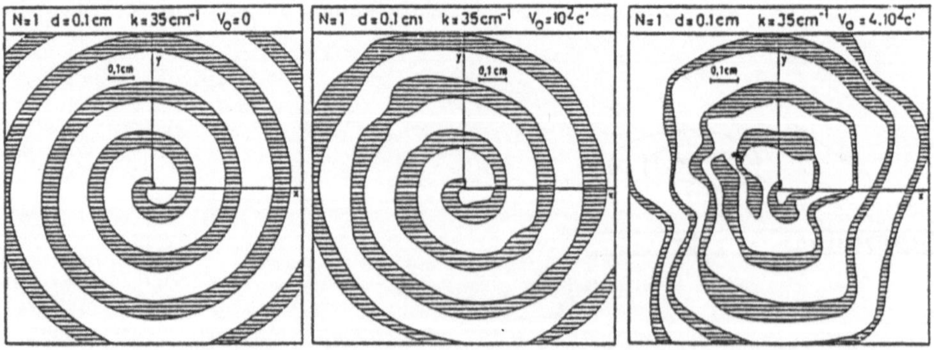

Fig.II: Deformation of a one-armed spiral wave due to the effect of convective rolls.

equation is constructed as an expansion in the diffusion coefficients, we see that at the leading order ($\dot{W}_o = w + (v_z \nabla_z g)\sin W_o$) the temporal oscillations may be suppressed by phase locking above a convective threshold given by $|v_z \nabla_z g| > w$. This implies the coexistence of regions with oscillatory or excitable behaviour within the layer. Although very slow, the chemical diffusion mechanisms are expected to alter this simple picture as they modify the zeroth order phase dynamics according to the following rate equation ($W = W_o + W_1$) :

$$\dot{W}_1 = c'\nabla^2 W_1 + c''(\underline{\nabla}W_1)^2 + \underline{f}(\underline{v},\nabla_z g,\underline{\nabla}W_o).\underline{\nabla}W_1 + h(\underline{\nabla}W_o) + v_z k(W_o,W_1) \quad (8)$$

At the surface, while f vanishes, h contains the print of the convective pattern (through the $(\underline{\nabla}W_o)^2$ term for example) and hence induces spatially dependent frequency variations of the oscillations able to trigger chemical waves in the boundary layer (7).

In this last case the coupling between chemical oscillations and hydrodynamics may lead to a complex spatial structuration of the solution: inhomogeneous phase locking in the center of the layer and wave generation in the boundary layers. The spatial distribution of the excitable regions and the pacemakers should be related to the symmetry of the underlying convective pattern. Although there exists experimental evidence for the existence of such phenomena (8) more quantitative analysis are needed to explore all the possibilities briefly sketched in the present discussion.

1. For a recent review, see for example:
   G.Nicolis and F.Baras, eds.: Chemical Instabilities, Applications to Chemistry, Engineering, Geology and Materials Science, Reidel, Dordrecht, 1983.
2. D.Walgraef, G.Dewel and P.Borckmans: Adv.Chem.Phys.49,311,(1982).
3. R.Ocelli, E.Guazzelli and J.Pantaloni : J.Physique Lett.44,L-567, (1983).
4. C.Vidal and A.Pacault: "Spatial Chemical Structures, Chemical Waves, a Review" in Synergetics (vol.17), H.Haken ed., Springer Verlag, Berlin, 1982, pp.74-99.
5. D.Walgraef, G.Dewel and P.Borckmans: J. Chem.Phys.,78, 3043,(1983).
6. a) G.R.Ivanitsky, V.I.Krinsky, A.N. Zaikin and A.M.Zhabotinsky: Soviet Scientific Review, D, Biology Review, 2, 279-324,(1981).
   b) K.I.Agladze, V.I.Krinsky, and A.M.Pertsov: Nature, 308, 834, (1984).
7. P.Hagan: Adv.Appl.Math 2,400,(1981); SIAM J.Apll.Math. 42, 762, (1982).
8. M.Orban: J.Am.Chem.Soc., 102, 4311, (1980).

Spatial Structures Formed by Chemical Reactions at Liquid Interfaces: Phenomenology, Model Simulations, and Pattern Analysis

D. Avnir[1,3], M.L. Kagan[1,3], R. Kosloff[3], and S. Peleg[2]

Departments of Organic Chemistry[1] and of Computer Sciences[2], and the F. Haber Research Center for Molecular Dynamics[3], The Hebrew University of Jerusalem, Jerusalem 91904, Israel

1. INTRODUCTION. A remarkably wide-scope phenomenon has recently been revealed. Chemical reactions at liquid interfaces proceed in a patterned way; spectacular structures form and grow while matter or energy influx are maintained [1]. Despite its generality and experimental simplicity we could not find descriptions of the phenomenon other than a report by P. Möckel on some photochemical systems [2]. As it turned out, Möckel's observations, which initiated the research project described here, were only the tip of an iceberg; what we subsequently found was beyond any of our expectations.

2. THE PHENOMENON AND ITS GENERALITY. The majority of over 40 reactions we tested formed structures (some special cases in which structures did not form are described below). The initial efforts by us [1,3-6] and by Micheau et al. [7,8] were concentrated on photochemical processes. We found, however, that photochemistry is not a conditional factor: structures are formed also by gases diffusing through a liquid interface and reacting with solutes [5,6,9] and by ground state reactions at liquid/liquid interfaces (e.g., separated by a membrane) [1]. Various solvents were tested and it was found that if strong evaporation, high viscosity, excited state quenching and insolubility of product do not interfere, structure growth is observed.

We found that the phenomenon is quite general regarding the liquid interface: both free and rigid interfaces operate. Selected examples are the interfaces between liquid and air, gas, miscible or immiscible liquids, glass, plastic.

In view of earlier experimental results and models requiring highly complex and non-linear networks of chemical reactions for the formation of (temporal) dissipative structures [10,11], we find it interesting that the phenomenon we investigate shows low sensitivity to the type of reaction and its kinetics. The classes of reactions which produce structure range from complex redox chains [1-4,9], through moderately non-linear reactions of the type $A + B \rightleftharpoons C$ [7,9], to simple first order isomerizations! [7,12]. Examples are the photoreduction of Fe(III) [1,6] and the air oxidation of reduced methylene blue [5] (very complex), protonation of methyl orange [9] (second or quasi-first order), and photochromic isomerization [7]. See Fig. 4.

3. A COMMENT ON THE MECHANISTIC APPROACH. The effort directed towards revealing the underlying mechanism(s) of this phenomenon was split into three directions, which are strongly correlated: a) Testing the response of the structure-growing process to changes in various physical parameters, especially in view of known instability phenomena in liquids [13]. b) Developing models of reaction/diffusion-coupled processes for computer simulation and subsequent experimental testing. c) Developing computerized pattern analysis tools for comparison of structures and for kinetic investigation of structure growth.

4. TESTED PARAMETERS. We describe now, phenomenologically, the effects of parameter changes. Due to space limitations the description is brief; details will be published elsewhere. There were two preliminary questions: first, is a chemical reaction necessary ? Simple diffusion of a dilute solution of dye [1], and irradiation of photoproduct only, suggest a positive answer. Second, are convections a latter stage in the structure building process, or do they pre-

exist ? We showed [1,4-6] that macroscopic convections are driven at a mature stage of the structure growth. No pre-existing convections were detected within the sensitivity of our tests (laser illumination of dusts and deflectometry [4]) when evaporation from the surface was excluded. Micheau et al. have shown [7,8] that if evaporation is allowed, then the pattern formed is a visualization of the evaporation process; this, however, is not the phenomenon we study: we have shown by a variety of experiments that structures form in the absolute absence of evaporation [1,5,6]. The following additional parameters were tested: a) Surface tension. In addition to the various interfaces mentioned above, silicon oil monolayer and addition of surfactants were tested. The structure formation process seems insensitive to this parameter. b) The insensitivity to addition of surfactants seems to exclude also a diffusion-limited aggregation mechanism. c) Depth and viscosity. Profound sensitivity was found at very shallow layers (<1 mm): induction time is increased even to the degree of no-pattern formation. Increase in effective viscosity is probably the reason. Indeed, increase in viscosity at a 1 cm layer produces the same changes [2,12]. A scaling law exists for the average distance between lines as a function of depth [6,7,12]. Structures do not form in a gel [9]. d) Concentration thresholds were searched. We have preliminary indications that these exist, but at very low values. e) Vibrations. It is very difficult to exclude a possibility that the fluctuations which are amplified to a full macroscopic structure are due to minute background vibrations. We went, however, in the other direction: ultrasonic vibrations destroy the structure; light vibrations such as caused by a nearby laboratory magnetic motor, do not seem to interfere. f) Temperature: Rayleigh-number calculations show that in our thin layer experiments, temperature difference should be at least 1 ^{o}C. In practice, the difference is $\ll 1^{o}$C. Furthermore, structure forms also in a stabilized temperature gradient (cooling the bottom, warming to top). g) Gravity. Sensitivity to this parameter is suggested in thin layer experiments. The full picture of the effect of gravity will be obtained from an experiment in space, now in preparation.

5. TWO DIMENSIONAL SIMULATION MODELS WITH A NON-LINEAR DIFFUSION TERM. The accumulation of the experimental data suggests that the investigated phenomenon may indeed be an authentic non-linear reaction/diffusion coupling process. Furthermore, the low sensitivity to the type of chemical reaction suggests that non-linearities in the transport processes are the dominant factors. Therefore, many of our simulation efforts have been directed towards this type of non-linearity. We exemplify the approach with one model; others will appear elsewhere.

We study the reaction $2A \xrightarrow{k} C$ in two dimensions. For low concentration of C, $\mu_c=\mu_c^o + RT \ln C$, so that in Onsager equations $\nabla\mu_c=(RT/C)\nabla C$. To remove the singularity of diffusion rate $\to \infty$ as $C \to 0$, we replace RT/C by RT/(C+α), where LRT/α is the diffusion constant at zero concentration. Using this assumption and Gauss theorem we obtain:

$$\frac{\partial c}{\partial t} = L_{cc} RT \nabla \left(\frac{1}{c+\alpha} \nabla c\right) + kA^2 \tag{1}$$

$$\frac{\partial A}{\partial t} = \frac{L_{AA} RT}{A_o} \nabla^2 A - kA^2 \tag{2}$$

where we assumed $L_{CA}=L_{AC}=0$ and that the reactant A concentration does not change during the period of study. Eq. (2) is the standard linear diffusion case; Eq. (1) is the novel non-linear approach. To the best of our knowledge, no previous attempts were made in coupling non-linear diffusion terms with chemical reactions. Numerical solution of reaction-diffusion problems have been carried out so far mainly by the finite difference method, limiting the studies to one dimension. One of us (R.K.) has recently developed an efficient computational tool which makes multidimensional simulation feasible [14]. The first stage in the numerical solution is surface discretization on an evenly spaced two dimensional grid (64x64 points) with periodic boundary conditions (other boundary conditions could be used as well). Spatial derivatives are calculated using a Fourier

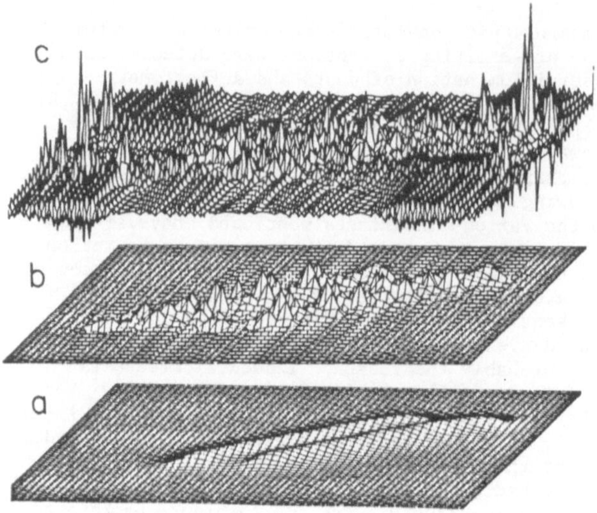

Fig. 1: Simulated pattern growth for 2A → C (see text). Reaction rate: 10^3[mole^{-1} sec^{-1}]; Diffusion rates of A and C: 0.10 [cm^2 sec^{-1}]; α (eq. 1) = [0.01]M; Initial fluctuation: 5%. a) t=0[sec.], b) t=5000[sec.], c) t = 18000[sec.]

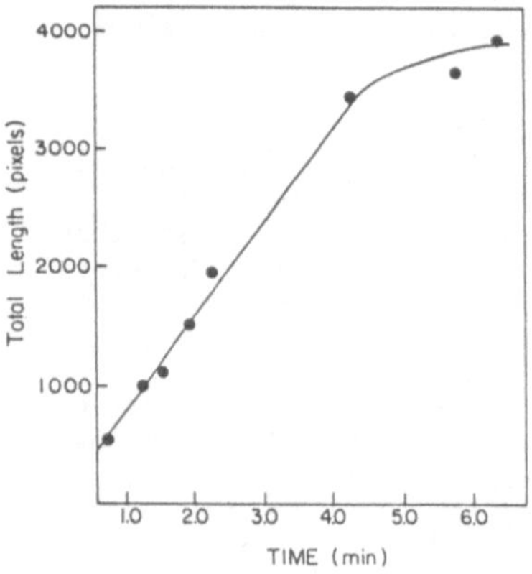

Fig. 2: Total length of pattern skeletonized lines as a function of time

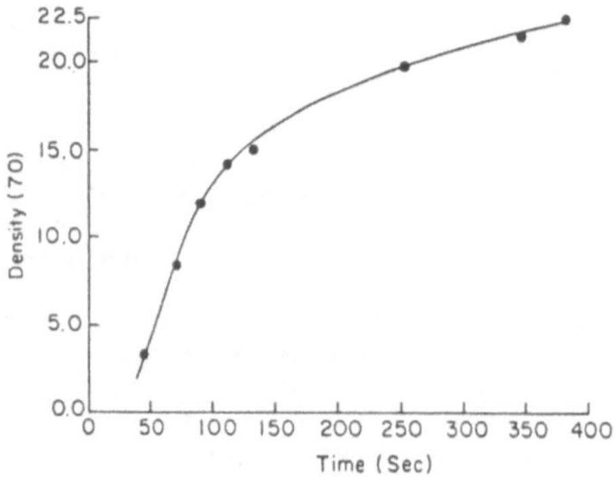

Fig. 3: Growth rate of total product density at shade value of 70

method [14] which gives very accurate values. A first order propagation scheme is used for advancing in time. Initial conditions were chosen such that the concentration of C was zero everywhere and the concentration of A was uniform except for two line-like fluctuations. The pattern-growth is depicted in Fig. 1. Three-dimensional calculations are in progress.

6. COMPUTERIZED IMAGE ANALYSIS OF COMPLEX PATTERNS. The effects of various physical parameters on the process of pattern growth, cannot be fully studied unless tools are developed for quantitative description of spatial structures and their growth kinetics. Computerized image analysis proved to offer solutions [15] some of which are briefly listed in this volume [16]. Two illustrative results are described in Fig. 2 (total length of lines) and Fig. 3 (non-

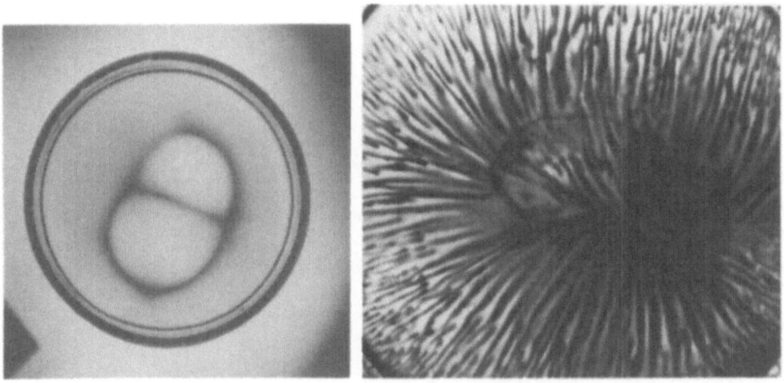

Fig. 4: Simple and complex structures formed by a chemical reaction at an interface. left: photo-oxidation of I^- [2,3]; right: photoreduction of Fe(III) [1].

destructive densitometry), typical of structure formation during photo-oxidations of anilines. Detailed description of the image analysis method will appear elsewhere.

Acknowledgments: Sponsored by the Volkswagen Foundation (under Synergetics). Helpful discussions were carried out with B. Hess, Th. Plesser and S. Müller.

REFERENCES

1. D. Avnir and M. Kagan, Nature, 307, 717 (1984).
2. P. Möckel, Naturwiss. 64, 224 (1977).
3. M. Kagan, A. Levi and D. Avnir, ibid., 69, 548 (1982).
4. D. Avnir, M. Kagan and A. Levi, ibid., 70, 141 (1983).
5. M. Kagan and D. Avnir, Origins of Life, 14, 365 (1984).
6. M. Kagan, E. Meisels, S. Peleg and D. Avnir, Lecture Notes in Biomathematics, 1984, in press.
7. M. Gimenez and J.-C. Micheau, Naturwiss. 70, 90 (1983).
8. J.-C. Micheau, M. Gimenez, P. Borckmans and G. Dewel, Nature 305, 43 (1983).
9. D. Avnir and M. Kagan, Naturwiss. 70, 361 (1983).
10. See, e.g., Springer's Series on Synergetics, M. Haken, editor.
11. See papers in this volume (e.g., I. Epstein, P. Fife).
12. To be published.
13. E.g., "Dynamics and Instability of Fluid Interfaces" ed. M.T.S. Sorensen, Lecture Notes in Physics, 105 (1978).
14. D. Kosloff and R. Kosloff, Comput. Phys. 52, 35 (1983).
15. See also S. Müller and Th. Plesser, Lecture Notes in Biomathematics, 1984, in press.
16. M. Kagan, S. Peleg, A. Tchiprout and D. Avnir, this volume.

Part IV

Chemical Chaos

Chemical Chaos

Harry L. Swinney

Physics Department, The University of Texas, Austin, TX 78712, USA

J.C. Roux

Centre de Recherche Paul Pascal, Université de Bordeaux I, Domaine Universitaire
F-33405 Talence Cêdex, France

1. Introduction

RUELLE [1] suggested more than a decade ago that since nonequilibrium chemical reactions are described by coupled nonlinear differential equations, for some conditions they might exhibit nonperiodic behavior. The nonperiodic behavior that arises from the nonlinear nature of a system rather than from stochastic driving forces is now called chaos, a term that we will define more carefully later.

In 1977 OLSEN and DEGN [2] reported observations of nonperiodic behavior in an enzyme system (peroxidase). Using a celebrated theorem of LI and YORKE [3] ("Period three implies chaos"), Olsen and Degn concluded that the observed nonperiodic behavior was chaos. Unfortunately, however, the Li and Yorke theorem says nothing about the measure of the parameter range of chaos; the chaos predicted by the theorem could be of zero measure and hence unobservable. Thus the observation of period three does not necessarily imply chaos for a physical system.

Soon thereafter observations of nonperiodic behavior in the Belousov-Zhabotinskii (BZ) reaction were reported by SCHMITZ et al. [4], ROSSLER and WEGMANN [5], VIDAL et al. [6], HUDSON et al. [7], SORENSON [8], and NAGASHIMA [9]. By 1981, the time of the last Bordeaux conference [10], observations of nonperiodic behavior in continuously stirred flow tank reactors (CSTR's) had begun to be analyzed in terms of power spectra, phase portraits, and one-dimensional maps, and several papers at the conference concerned theory [11] and experiments [12] on chaos.

However, at that time there still existed considerable healthy skepticism regarding the existence of nonperiodic behavior in well-controlled nonequilibrium chemical reactions. After all, nonperiodic behavior can arise from fluctuations in stirring rate or flow rate, evolution of gas bubbles from the reaction, spatial inhomogeneities due to incomplete mixing, vibrations in the stirring motor, fluctuations in the amount of bromide and dissolved oxygen in the feed, and so on. Any experimental data, no matter how well a system is controlled, will contain some noise arising from fluctuations in the control parameters; therefore, it is reasonable to ask: "Will noise, always present in experiments, inevitably mask deterministic nonperiodic behavior (chaos)?"

The answer is no. We will summarize the large body of evidence, gathered recently by groups in Bordeaux [13-18], Virginia [19-20], and Texas [21-25], that indicates that the nonperiodic behavior observed in chemical reactions is, at least in some cases, chaos, not noise. It is even sometimes possible to separate the experimental noise from the deterministic dynamics. In addition, as we shall describe, low-dimensional deterministic models deduced from the data make it possible to predict the behavior that should occur as a control parameter (e.g., flow rate) is varied, and some of these predictions have been confirmed by experiment.

The quantitative characterization of nonperiodic behavior has required the development of flow reactors with far better control than is necessary for exploratory studies of steady state and oscillating regimes. Positive displacement piston pumps are replacing peristaltic pumps, synchronous stirring motors are replacing voltage-controlled stirrers, and low noise electronics are yielding signals with a precision of 0.1% or better.

Time series strip chart records can be suggestive of chaos, but we have found that it is not possible to distinguish noise from chaos without analyzing long computer records of time series. As an example, we once observed a time series that appeared to be chaotic, but further experiments and data analysis indicated that the nonperiodic behavior arose from stochastic switching between two adjacent periodic states. This kind of nonperiodicity, much discussed in the literature [26-28], is of course not chaos. The length and accuracy of the data record needed to distinguish noise from chaos depend on the particular problem--in particular, on the dimension of the attractor and on the scale of the chaos generating mechanism (the size of the folds in the attractor). In our studies of the BZ reaction we have found that 32000 points spanning 300-600 oscillations usually suffice if the data have a precision of ~0.3% or better. The data requirements can be less stringent if a hallmark of one of the well-established routes to chaos (e.g., period doubling) is observed.

In Sections 2 and 3, respectively, we will describe the characterization of chaos and some of the routes to chaos. Theory and models of chemical chaos will be briefly reviewed in Section 4, and Section 5 contains concluding remarks. The detailed studies of chemical chaos have all been conducted for the BZ reaction, but some evidence for chaos has also been obtained for several other chemical reactions [29-30].

2. Characterization of Chaos

2.1 Power spectra and phase portraits

Around the turn of the century Poincaré and others recognized that much could be learned about dynamical behavior from an analysis of system trajectories in a multi-dimensional phase space in which a single point characterizes the entire system at an instant of time. The set of phase space trajectories for all possible initial conditions (for a given set of control parameter values) forms a phase portrait of the system.

The N-dimensional phase portrait describing the well-stirred (that is, homogeneous) Belousov-Zhabotinskii system could be constructed from measurements of the time dependence of the concentration of all N chemical species in the reaction. Fortunately, such a difficult task is unnecessary - a multi-dimensional phase portrait can be constructed from measurements of a single variable by a procedure proposed by RUELLE [31], PACKARD et al. [32], and TAKENS [33]. The idea is a follows: For almost every observable $B(t)$ and time delay T an m-dimensional portrait constructed from the vectors $\{B(t_k), B(t_k+T), \ldots, B(t_k+(m-1)T)\}$, where $t_k=k\Delta t$, $k=1,2,\ldots,\infty$, will have the same properties (for example, the same spectrum of Lyapunov exponents) as one constructed from measurements of N independent variables, if $m \geq 2N+1$. In principle the choice of time delay T is arbitrary, but in practice time delays of from about one-tenth to two-thirds of a characteristic oscillation time are usually optimum [25].

Time series, power spectra, and phase portraits are shown in Fig. 1 for the BZ reaction for three different flow rates [21]. The power spectra for the periodic states in Figs. 1(a) and (c) contain an instrumentally sharp fundamental frequency and its harmonics, while the spectrum in (b) consists of broadband noise that is well above the instrumental noise level. This spectral noise could arise from either stochastic or deterministic processes. However, at least in principle it should be possible to distinguish stochastic and deterministic processes from the behavior of the power spectrum in the high frequency limit [34]: for stochastic differential equations of order n, $P(\omega) \sim \omega^{-2n}$, while for nonperiodic behavior given by deterministic differential equations, $P(\omega) \sim \exp(-\Gamma\omega)$. To our knowledge this

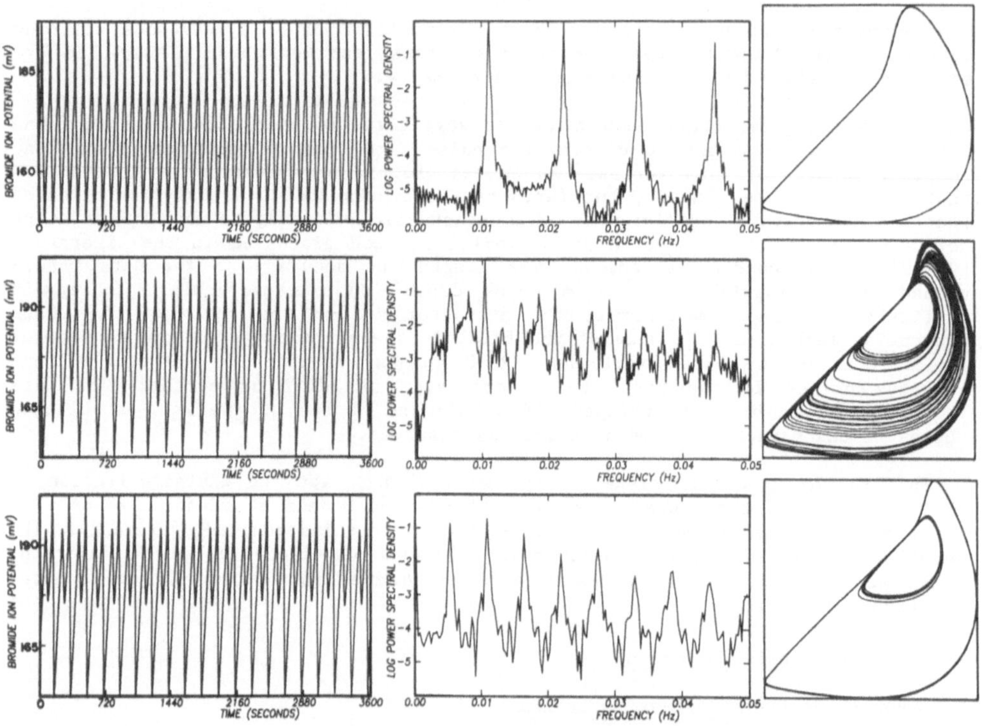

Fig. 1. Data for the BZ reaction at three different residence times τ (reactor volume/total flow rate); the reservoir concentrations, given in [21], were held fixed. (a) τ=0.49 hour; a periodic state with one oscillation per period. (b) τ=0.90 hour; chaos. (c) τ=1.03 hour; a periodic state with 2 oscillations per period. For each τ the graphs show the time dependence of the bromide ion potential, the corresponding power spectrum, and a two dimensional projection of the phase portrait. The phase portraits for the periodic states in (a) and (c) are limit cycles, while the chaotic state in (b) is described by a strange attractor. (From [21].)

test for (deterministic) chaos has not yet been applied to data from nonequilibrium chemical reactions. (However, we note that Fig. 2(b) of [19] looks approximately exponential.)

2.2 Poincaré sections, maps, and attractors

Rather than analyze phase portraits directly it is easier to analyze the lower-dimensional Poincaré section which is formed by the intersection of "positively directed" orbits of an m-dimensional phase portrait with an (m-1)-dimensional hyper-surface. A two-dimensional Poincaré section constructed for a three-dimensional phase portrait of a nonperiodic state is shown on the left-hand side of Fig. 2. (The right-hand side of the figure shows the effect of perturbations, to be discussed shortly.)

The points on the Poincaré section [Fig. 2(b)] lie to a good approximation along a smooth curve. (However, the actual dimension of the Poincaré section must be at least slightly greater than unity because of the fractal nature of attractor [35].) If we parameterize the distance along the curve by a coordinate X, then the coordinate values provide a sequence $\{X_n\}$ which defines a one-dimensional map, $X_{n+1}=f(X_n)$, as shown in Fig. 2(c). The data appear to fall on a single-valued curve, indicating that the system is deterministic, that is, for any X_n, the map determines X_{n+1}.

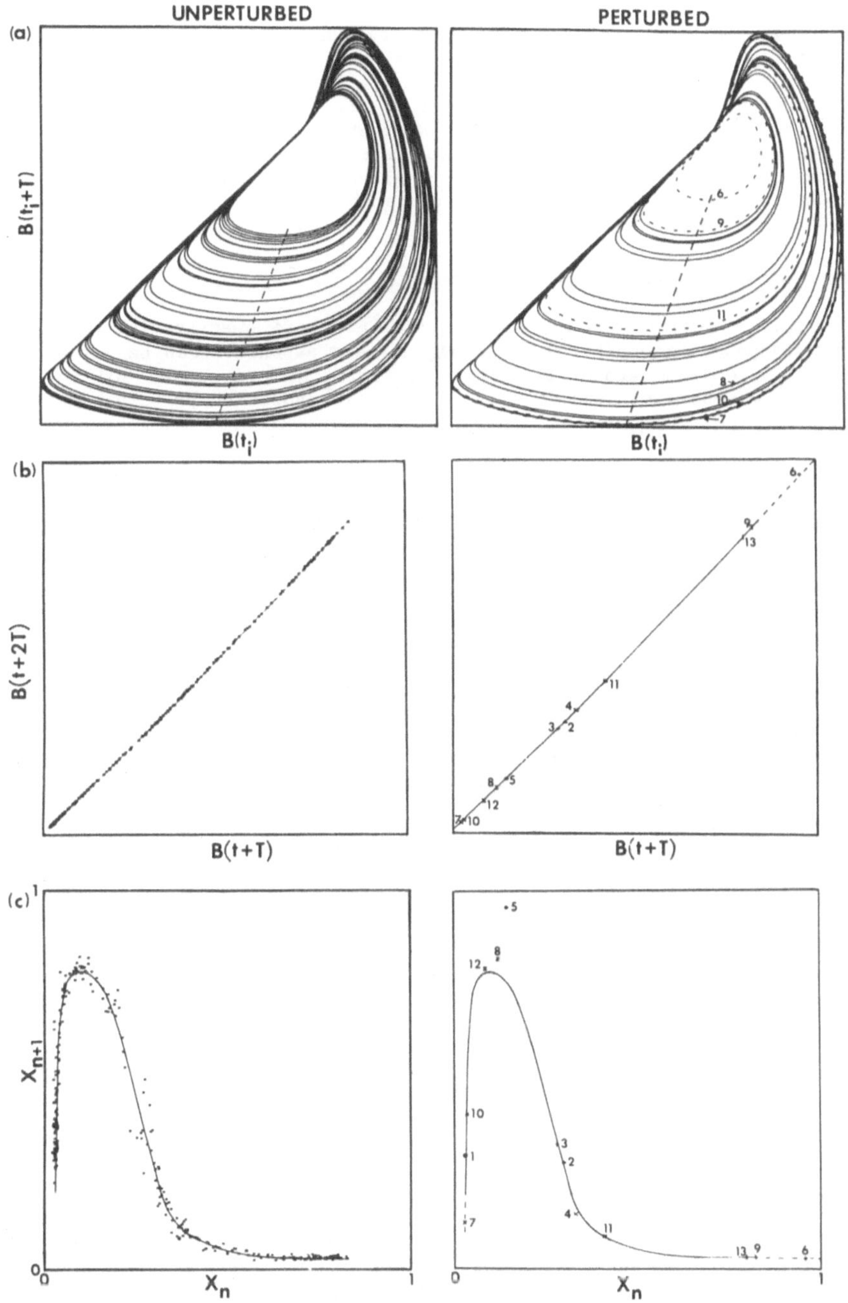

Fig. 2. A chaotic state in the absence of an external perturbation (left-hand side) and with a perturbation (right-hand side): (a) A two-dimensional projection of a three-dimensional phase portrait. (b) A Poincaré section constructed by the intersection of positively directed trajectories with the plane (normal to the paper) passing through the dashed line in (a). (c) a one-dimensional map constructed by plotting as ordered pairs (X_n, X_{n+1}) the successive values of the ordinate of trajectories when they cross the dashed line in (a). (From [22].)

The right-hand side of Fig. 2 illustrates that the post-transient set described by the phase space trajectories is really an attractor: the trajectories rapidly return to this limit set after finite perturbations. The basin of attraction of an attractor is the set of all initial conditions for which the trajectories asymptotically approach the attractor. For the state illustrated in Fig. 2 the trajectories were found to return to the attractor for all perturbations used, so this attractor could be globally attracting [25]. However, in some cases chemical reactions exhibit multistability, where there are two or more disjoint basins of attraction.

2.3 Lyapunov exponents and strange attractors

An attractor in an m-dimensional space is characterized quantitatively by its spectrum of m Lyapunov exponents: $\lambda_1 \geq \lambda_2 \cdots \geq \lambda_m$. These exponents can be defined in terms of the long term evolution of an infinitesimal m-sphere of initial conditions. The m-sphere evolves into an m-ellipsoid whose ith principal axis of length $p_i(t)$ yields the ith Lyapunov exponent [36,37]

$$\lambda_i = \lim_{t\to\infty} \frac{1}{t} \log_2 \frac{p_i(t)}{p_i(o)}, \tag{1}$$

where the λ_i are ordered from largest to smallest. Any continuous time-dependent dynamical system without a fixed point will have at least one zero exponent. Any dissipative dynamical system will have at least one negative exponent, and the post-transient motion of the trajectories will occur on a zero-volume limit set, the attractor.

A strange (or chaotic) attractor is by definition an attractor for which the largest Lyapunov exponent is positive. Then trajectories starting from nearby points will separate exponentially fast as time evolves. Therefore, all information about the initial conditions is rapidly lost, since any uncertainty, no matter how small, will be magnified until it becomes as large as the attractor; thus there is sensitive dependence on initial conditions (RUELLE [38]). Long term predictions about the state of the system are impossible.

The largest Lyapunov exponent for a system described by a one-dimensional map (as in Fig. 2) can be computed from the map:

$$\lambda_1 = \frac{1}{n} \sum_{i=1}^{n} \ln|f'(X_i)|, \tag{2}$$

where $f'(X_i)$ is the derivative of the map at X_i and the sum is over all observed values of X_i, i=1,2,...,n. [In order to have the proper weighting of the sum (by the "invariant distribution"), it is essential that $f'(X_i)$ be computed for the observed values of X_i, not for a uniform mesh in X.] The value for λ_1 computed from the map in Fig. 2 is about 0.6 bits/orbit, definitely positive. Therefore the attractor in Fig. 2 is a strange attractor.

WOLF and SWIFT [36,37] have recently developed a method for computing the non-negative portion of the Lyapunov spectrum directly from a time series. The method does not require the construction of a map from the data, and in fact even when the data are well described by a one-dimensional map, the Wolf and Swift method yields exponent values that are more robust than those obtained from a map. (The difficulty with the map arises in part because $\ln|f'(X_i)|$ is very sensitive to the procedure used to determine the derivative of the map.)

The first step in the Wolf and Swift method is to construct an attractor from the time series, as described in Section 2.1. Then a starting point at time t_o is selected on an arbitrarily chosen fiducial trajectory, as illustrated in Fig. 3. A nearby point is then found; the distance between the two points is called $L(t_o)$. The separation L(t) between the two trajectories is monitored until it becomes large compared to $L(t_o)$ yet still small compared to the size of the folds in the attractor. A new point is then found near the fiducial trajectory; the new point is chosen to be (to the extent possible) in the same direction from the fiducial

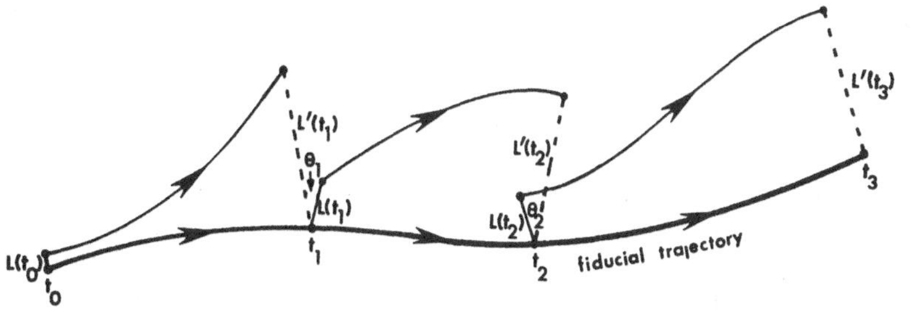

Fig. 3. A schematic diagram illustrating the procedure developed by Wolf and Swift for computing the largest Lyapunov exponent from experimental data (see text and refs. [36,37]).

trajectory as the point that is being replaced; i.e., the replacement step is made with the angle θ_1 and the length $L(t_1)$ simultaneously minimized, as Fig. 3 illustrates. Then after M replacement steps, the estimate for the largest Lyapunov exponent is given by [36, 37]:

$$\lambda = \frac{1}{t_M - t_o} \sum_{k=1}^{M} \log_2 \frac{L'(t_k)}{L(t_{k-1})} . \tag{3}$$

The sum of the two largest exponents, $\lambda_1 + \lambda_2$, is given by a similar expression with lengths L replaced by areas A; replacing lengths with 3-volumes yields $\lambda_1 + \lambda_2 + \lambda_3$, etc.

The Wolf and Swift method has been tested on model systems with known Lyapunov spectra and applied to data for the Belousov-Zhabotinskii chemical reaction and Couette-Taylor flow. See refs. [36-37] for a discussion of the requirements on accuracy and the size of a data file.

2.4 Dimension

Another important property of an attractor is its dimension. It is, loosely speaking, the number of independent degrees of freedom relevant to the dynamical behavior. There are several different definitions that differ mainly in the measure used [35, 39]; these definitions all yield the Euclidean value of dimension for Euclidean objects, but for strange attractors the dimension is in general fractional ("fractal" [40]). As an example of computing the dimension from experimental data, we will describe a procedure for computing the information dimension, d [41]. Let $N(\varepsilon)$ be the number of points in a ball of radius ε about a point x_o on an attractor. For a uniform density of points one would have $N(\varepsilon) \sim \varepsilon^d$. This relation can be used to define dimension,

$$d = \lim_{\varepsilon \to 0} \frac{\log N(\varepsilon)}{\log \varepsilon}, \tag{4}$$

where the result is averaged over several x_o.

The determination of d from slopes of plots of $\log N(\varepsilon)$ vs $\log \varepsilon$ is illustrated in Fig. 4 for data obtained in an experiment [42-43] on fluid flow between concentric rotating cylinders (Couette-Taylor flow). For too large values of ε, $N(\varepsilon)$ begins to saturate at the total number of points in the data file, as can be seen in region C of Fig. 4(b). On the other hand, for ε too small, the algorithm detects the instrumental noise in the data, as can be seen in region A of Fig. 4(b), where the slope approaches the embedding dimension m since random noise fills all dimensions of the subspace. However, the intermediate length scales (region B), where the slope is constant, reflect the fractal structure of the attractor. The dimension of the attractor is then obtained from the asymptotic value of the slopes (in region B) at large embedding dimension, as Fig. 4(c) illustrates.

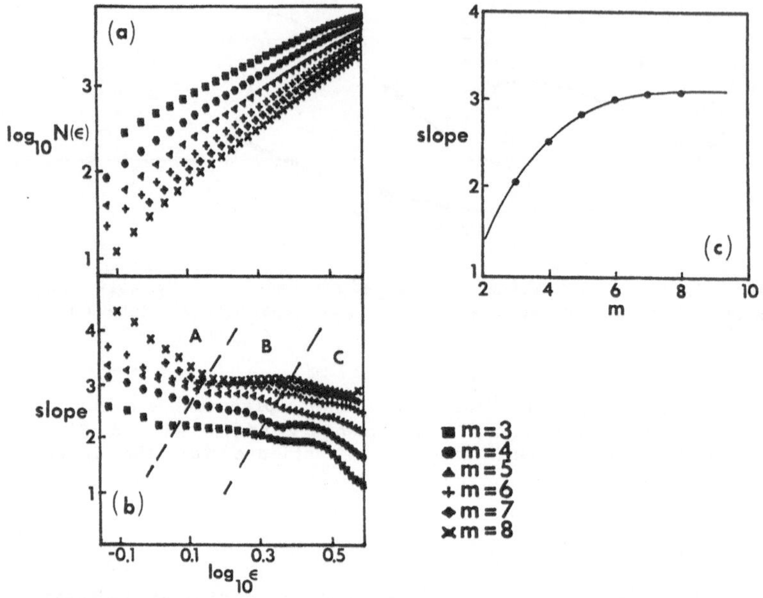

Fig. 4. The computation of the dimension of an attractor is illustrated using velocity data obtained for a weakly turbulent flow in the Couette-Taylor system at $R/R_c=16.0$, where R_c is the Reynolds number for the onset of time-independent Taylor vortex flow. The different curves correspond to different embedding dimensions m. (a) The dependence of N(ε), the average number of points within a ball of radius ε, on ε. (b) The slope of the curves shown in (a). Regions A, B, and C are discussed in the text. (c) The slope in region B as a function of m. The asymptote of the slope at large m is the (information) dimension of the attractor; d=3.1 in this example. (From [43].)

Graphs similar to those in Fig. 4 have also been obtained for a set of nonperiodic data for the BZ reaction (COFFMAN et al. [44]); the value of d was 2.1.

The methods we have described for determining Lyapunov exponents and dimension have been developed very recently. These methods and other methods under development promise to provide a quantitative means for characterizing chaos in future experiments on nonequilibrium chemical reactions.

3. Routes to Chaos

3.1 Types of behavior observed in the BZ reaction

Before describing the different routes to chaos encountered in the BZ reaction, let us describe schematically the general behavior of this reaction deduced from experiments. As a function of two of the control parameters, residence time and acidity, we can distinguish two types of oscillation: (i) quasi-sinusoidal small amplitude oscillations and (ii) large amplitude relaxation-type oscillations. The small amplitude oscillations always appear at the low flow rate boundary between time-independent and time-dependent behavior, as Fig. 5 illustrates, and in some experiments a regime with only small amplitude oscillations is also observed near the high flow rate boundary between time-independent and time-dependent behavior, as the dashed line on the right-hand side of Fig. 5 indicates. In most cases the small and large amplitude oscillations are easily distinguished; for example, see Fig. 1(c).

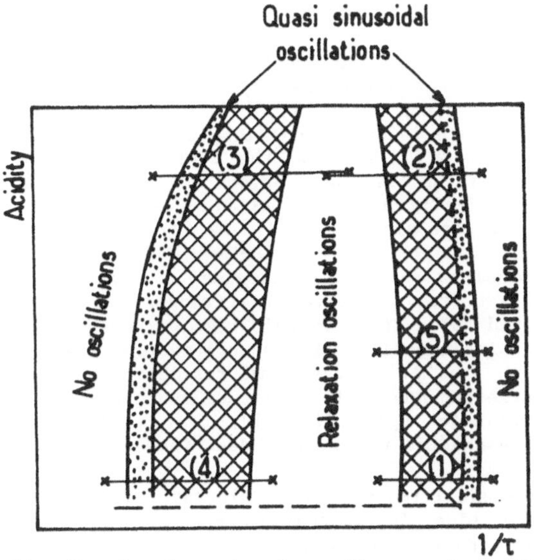

Fig. 5. A schematic phase diagram that illustrates the general dependence of the dynamical behavior of the BZ reaction on acidity and residence time (reactor volume/flow rate). The diagram is a schematic projection of measurements at different bromate, cerium, and malonic acid concentrations onto the acid-residence time plane. Most of the experimental work along path (1) was conducted at Virginia [7,19,20]; along paths (2), (3), and (5) at Bordeaux [13-15 (path 2), 16 (path 3), 52 (path 5)]; and along path (4) at Texas [18,21-25]. See the discussion in the text.

Between the two regions of small amplitude oscillations there exist two regions of what we will call "complex behavior." If the regions of complex behavior are traversed by varying flow rate from either the low or high flow rate side, one ultimately reaches the region of large amplitude relaxation oscillations, the central region of the diagram in Fig. 5. The following three kinds of complex behavior are observed:

(1) Complex periodic behavior. Every state in this regime is a combination of small and large amplitude oscillations, the simplest states being of the form S^pL^q, that is, each period consists of p small amplitude oscillations followed by q large amplitude oscillations. However, far more complex periodic states are also observed [3,6,13,17,18,19,23,24,30]. Perhaps the most complex periodic state of this type observed thus far is one with 26 oscillations per period: $S^3L^3S^2L^2S^1L^2S^2L^2S^1L^2S^3L^3$ [45]. The complex states all appear to be juxtapositions of nearby simpler states. Sequences of complex periodic states have been observed along lines (1), (2), and (4) in Fig. 5.

(2) Chaotic behavior, described by strange attractors, as discussed in Section 2. Chaos has been observed along lines (1), (2), and (5) in Fig. 5.

(3) Some observations of complex behavior do not fall neatly into either category (1) or (2). For very complex states it is not easy to distinguish between chaos and periodic states with a very long period. However, although there will always be some cases where fluctuations in the control parameters and noise from other sources prevent a definitive distinction between chaos and very complex periodic or multiperiodic behavior, the techniques described in Section 2 make it possible to distinguish clearly in many cases between complex periodic behavior, noisy behavior, and (deterministic) chaos.

3.2 Alternating periodic-chaotic sequences

Similar sequences of periodic regimes alternating with chaotic regimes (or noisy complex periodic regimes) have been observed along paths (1), (2), and (4) in Fig. 5 when the residence time is decreased along paths (1) or (2), starting with relaxation oscillations, or when the residence is increased along path (4), again starting with relaxation oscillations. We will describe the experiments along line (4), where the chaotic regime that occurs just beyond the relaxation oscillation region has been well-characterized; chaos has not been observed near the border of the relaxation oscillation region along lines (1) or (2).

When the residence time is increased along path (4), a regime is encountered with irregular time series, broadband spectra, and strange attractors that are essentially two-dimensional; data obtained for a state in this regime are shown in Figs. 1(b) and 2. This regime which contains chaotic states occurs over a range of about 12% in residence time; then with further increase in residence time it yields to a periodic regime with one large and one small amplitude oscillation per period, as shown in Fig. 1(c). This periodic regime is stable over a parameter domain that is about the same size as the chaotic region just described.

Proceeding in this way, we pass through a succession of alternating periodic and chaotic regimes as the residence time is increased: P_1 (the original relaxation oscillations; Fig. 1(a)), C_1 (chaos; Fig. 1(b)); P_2 (two oscillations per period, one with large amplitude and one with small amplitude; Fig. 1(b)); C_2 (chaos); P_3 (three oscillations per period, one with large amplitude and two with small amplitude); C_3, etc. This sequence of regimes is summarized in Fig. 6(a). Successive regimes in the sequence appear to be of roughly comparable width. Beyond C_5 it was not possible to distinguish between periodic and chaotic regimes with the precision available in the experiments [21]. The sequence terminated with the small amplitude quasi-sinusoidal oscillations that occur at the border of the region of oscillations in Fig. 5.

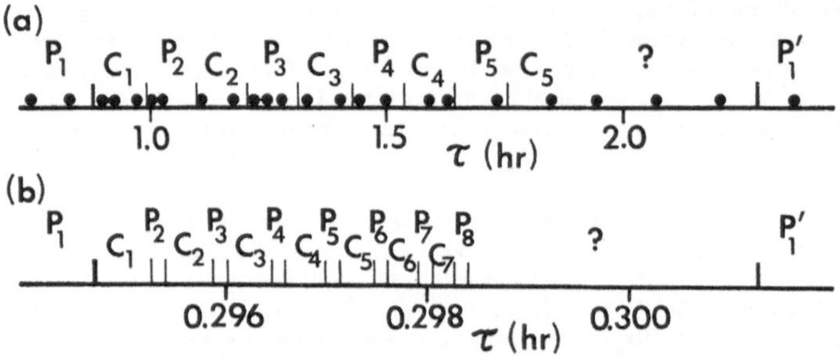

Fig. 6. Alternating periodic and chaotic transition sequences: (a) observed in an experiment on the BZ reaction and (b) found in a four variable reversible Oregonator model. (From [21].)

Within a "chaotic" regime the behavior is <u>not</u> chaotic for every value of the residence time. There are many narrow intervals ("windows") with periodic behavior; see, e.g., the discussion for C_1 in Section 3.4.

The phase space attractors in each successive periodic region contain one more turn in the center than the preceding periodic region [23], and the one-dimensional maps in successive chaotic regions contain one more branch. That is, the map for C_1 is single-valued (as shown in Fig. 2), the map for C_2 is double valued (as shown in COFFMAN et al. [47]), the map for C_3 is tripled-valued, etc.

Each chaotic state C_k consists primarily of what at first glance appears to be a stochastic mixture of the adjacent periodic states P_k and P_{k+1}; thus, for example, C_3 consists of a mixture of P_2 and P_3, as Fig. 6 of ROUX et al. [23] illustrates. However, maps constructed from the time series clearly yield smooth curves, not a scatter of points. These maps indicate that behavior is deterministic, not stochastic. Moreover, it is difficult to imagine stochastic processes that would lead to period doubling, the universal sequence, and tangent bifurcations, yet all of these phenomena associated with chaos are found in one-dimensional maps and in experiments on nonequilibrium chemical reactions, as will now be described.

3.3 Period doubling

One-dimensional maps with a single extremum are predicted to exhibit dynamical behavior that is universal, that is, independent of the details of the map [48-50]. (Proof of universality requires only that the map be basically well-behaved; consult [49] for technical details.) The best known prediction is that the periodic state should, with a change in bifurcation parameter, should lose its stability to a state with twice the period, and the latter state in turn should lose stability to a state with "period 4," and so on. FEIGENBAUM [48-49] showed that the convergence rate for the period doubling sequence is given asymptotically by a universal number, $\delta=4.669\cdots$; that is, the interval in bifurcation parameter over which a state with period 2^{n+1} would occur should be δ times smaller than the interval for the state with period 2^n. The accumulation point for the period doubling sequence at period 2^∞ marks the onset of chaos.

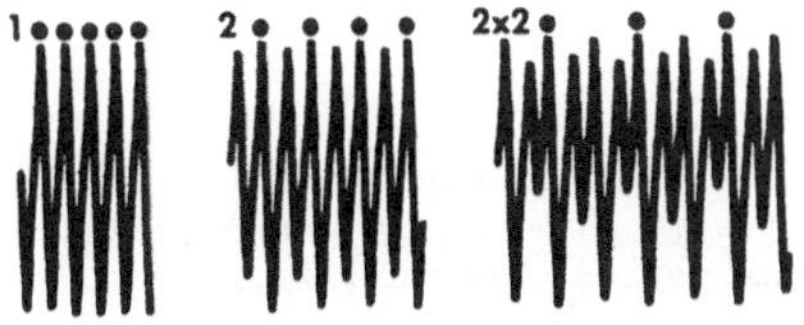

Fig. 7. Period doubling sequence time series with periods T_0(115 s), $2T_0$, 2^2T_0, obtained in experiments on the BZ reaction. The quantity measured was the bromide ion potential. The dots above the time series are separated by one period. (From [24].)

Figure 7 shows the first three states of a period doubling sequence, observed in experiments on the BZ reaction [24]. One further doubling was observed in that reaction, but further doublings beyond period 16 were unobservable because the parameter ranges were too narrow. It is in practice not possible to observe high order doublings in any physical system because the sequence converges extremely rapidly. For example, if a period 2 state were observed over a residence time range of 2%, then the width in residence time of the period 32 state would be about $2\%/\delta^4$ which is only 0.004%!

We have found that period doubling is fairly common in the BZ reaction. Many periodic states lose their stability through period doubling, but often the interval for even the period 2 state is less than 2%. Hence the doubling is easy to miss unless one uses a finer mesh than is usually in experiments on nonequilibrium chemical reactions. For example, if a period 2 state were observed over a 2% range in residence time, then the entire infinite period doubling sequence would occur in a residence time range of only about

$$(2\%) \times \sum_{n=0}^{\infty} \delta^{-n} = 2.455\%. \tag{5}$$

This estimate is approximate, since the convergence rate δ describes in principle only the asymptotic behavior after many doublings, but in practice the ratio of even the first few successive intervals is approximated well by δ [48-50].

3.4 The Universal Sequence

Universality in the period doubling sequence for one-dimensional maps is now well known. Perhaps less well known is the U (universal)-sequence that occurs beyond the accumulation point (2^∞-cycle) of the 2^n-sequence. Years before the universal scaling properties of one-dimensional quadratic maps were discovered by Feigenbaum and others, it was found that one-dimensional maps with a single extremum (not necessarily quadratic) exhibit universal dynamics as a function of the bifurcation parameter (see, e.g. [50]). Beyond the period doubling sequence, which is an infinite sequence of doubling of a periodic state with one oscillation per period, periodic states with K oscillations per period appear for all natural numbers K, and each of these "K-cycles" undergoes its own infinite period doubling sequence, 2^nK. Fig. 8 shows examples of a fundamental 5-cycle, 6-cycle, and 3-cycle (and the first doubling of the 3-cycle) observed in the Belousov-Zhabotinskii reaction.

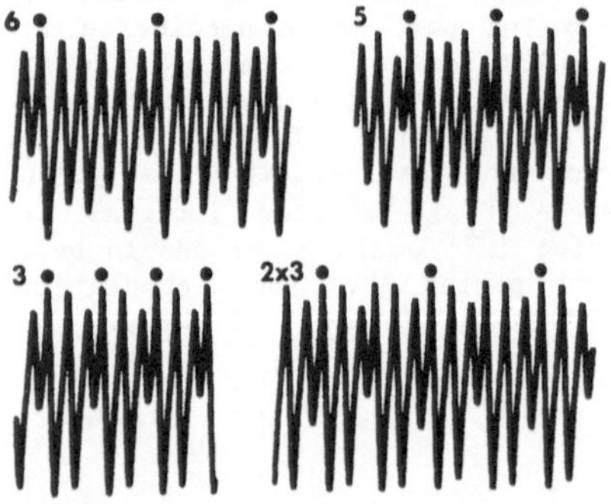

Fig. 8. Some periodic states of the U-sequence observed in experiments on the BZ reaction. The states have periods $6T_o$ (where T_o=115 s), $5T_o$, $3T_o$, and $2x3T_o$. The dots above the time series are separated by one period. (From [24].)

The order in which the periodic states appear as a function of bifurcation parameter is deduced in the theory using only the single-extremum property of the one-dimensional map [50]. The full U-sequence consists of the (infinitely long) list of periodic states allowed by the theory. The larger the fundamental period K, the larger the number of allowed states; there are three distinct allowed 5-cycles, four distinct 6-cycles, and 27 distinct 9-cycles.

The theory also predicts the iteration patterns of the maps--the order of visitation of points on the X-axis. Each iteration pattern is predicted to occur only once, and for a given value of bifurcation parameter not more than one periodic state is stable.

The periodic windows in the U-sequence exist in the "chaotic" regime (e.g., the C_1 regime of Section 3.2) that follows the accumulation point of the period doubling sequence [50]. There are no chaotic intervals, yet the set of bifurcation parameter values for which the behavior is chaotic has positive measure. [For a typical map probably about 85% of the measure (of the bifurcation parameter range in which the U-sequence occurs) corresponds to chaotic states, and 15% to periodic states.]

All the U-sequence states with periods 3, 4, and 5 have been observed in BZ experiments, and some of the U-sequence states with periods 6,7,8,9, and 10 have also been observed [24,47]. Within the experimental resolution the order of occurrence of the periodic states and the observed iteration patterns are in accord with the theory for one-dimensional maps.

3.5 Intermittency

The term intermittency is used to describe a transition from periodic to chaotic behavior characterized by occasional bursts of "noise" [51]. For bifurcation parameter values slightly beyond that corresponding to the onset of the bursts, there are long intervals of nearly periodic behavior between the bursts, but further beyond the transition the time intervals between bursts is shorter. With further change in bifurcation parameter the intervals between bursts decrease until ultimately it is impossible to recognize the regular oscillations of the periodic states.

POMEAU and MANNEVILLE [51] have shown that intermittency can be understood in terms of a tangent bifurcation of a one-dimensional map; at tangency a stable fixed point of the map disappears (or appears, depending on the direction in which the bifurcation parameter is changed.)

Direct evidence for a tangent bifurcation in a chemical experiment was first obtained in Bordeaux and is shown in Fig. 9. The small amplitude oscillations are unstable, increasing steadily in amplitude until at some point they are interrupted by one large amplitude relaxation oscillation. Note that neither the number of small oscillations nor the height of the large one is constant.

Recently it has been shown that some of the U-sequence states lose stability at a tangent bifurcation when the residence time is lowered (see the figure in COFFMAN et al. [47]). Again this is in accord with the theory for one-dimensional maps: U-sequence states should lose stability through period doubling in one direction and a tangent bifurcation in the other direction [50].

3.6 Wrinkles on a torus

Chaos may also occur as a consequence of the destruction of a two-torus that characterizes a quasiperiodic regime with two incommensurate frequencies. Quasiperiodicity in chemistry was recently discovered in experiments on the BZ reaction, and chaos was reached through the development of wrinkles on the torus (ROUX and ROSSI, this volume [52]). ARGOUL et al. [53] (in this volume) have proposed a tentative interpretation of the experiments.

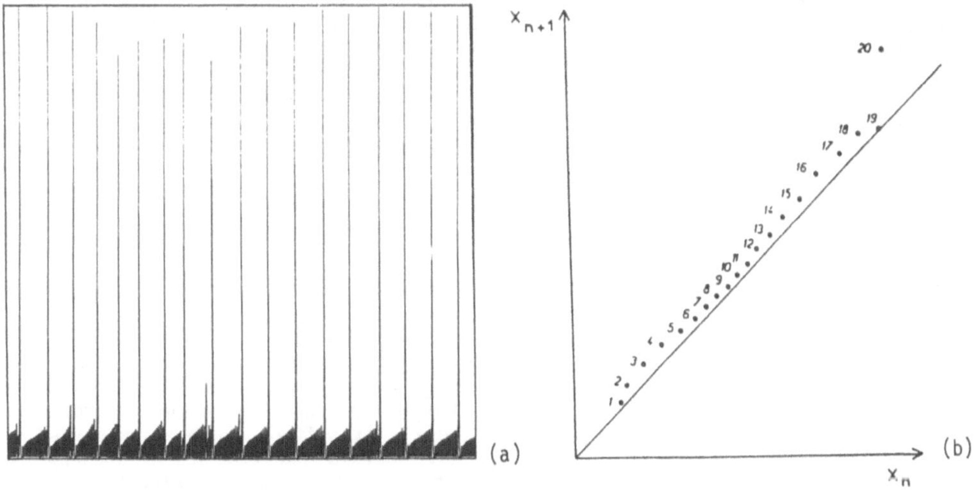

Fig. 9. (a) Time series observed in an intermittent regime in the BZ reaction. (b) Next amplitude map constructed from the successive maxima of the small amplitude oscillations in (a); the map is nearly tangent to the diagonal line given by $X_{n+1}=X_n$. The data were obtained along path (3) in Fig. 5. (From [16,18].)

4. Theory and models

Abstract chemical models exhibiting nonlinear phenomena were proposed more than a decade ago. The Brusselator of PRIGOGINE and LEFEVER [54] has oscillatory (limit cycle) solutions, and the SCHLÖGL [55] model exhibits bistability, but these models have only two variables and hence cannot have chaotic solutions. At least 3 variables are required for chaos in a continuous system, simply because phase space trajectories cannot cross for a deterministic system. As mentioned in the Introduction, the possibility of chemical chaos was suggested by RUELLE [1] in 1973. In 1976 ROSSLER [56], inspired by LORENZ's [57] study of chaos in a 3 variable model of convection, constructed an abstract 3 variable chemical reaction model that exhibited chaos. This model used as an autocatalytic step a Michaelis-Menten type kinetics, which is a nonlinear approximation discovered in enzymatic studies. Recently more realistic biochemical models [58,59] have also been found to exhibit low dimensional chaos.

The detailed chemistry of the BZ reaction was first elucidated (in 1972) by FIELD, KÖRÖS, and NOYES [60]. From the detailed mechanism, which involved more than 20 reactions and as many chemical constituents, FIELD and NOYES [61] then derived a reduced model (the "Oregonator") with only 3 variables. A modified Oregonator (with 7 variables) was then proposed and studied by SHOWALTER et al. [26], who were not successful in their attempt to simulate the observations by SCHMITZ et al. [4] of nonperiodic behavior. SHOWALTER et al. [26] concluded that "the difference between experiment and simulation suggests that the chaotic behavior observed experimentally may result from fluctuations too small to measure in any other way." Similar conclusions have been reached in several other studies [27,28,62,63]. However, abstract models have been developed that display chaos and some of the transition sequences observed in experiments (e.g., see [64-66]).

The only studies of an Oregonator model yielding chaos have been TURNER's [21,67-69] analysis of a four-variable reversible Oregonator and his more recent analysis [70] of a seven-variable Oregonator; the four and seven-variable models yielded essentially the same results. He found an alternating periodic-chaotic transition sequence [21] that is in qualitative accord with that observed in experiments, as Fig. 6(b) illustrates. RINGLAND and TURNER [69] have found that this model has a strange attractor [Fig. 10(a)] and a one-dimensional map [Fig. 10(b)] that appear quite similar to those observed in experiments. The similarity between Figs. 1,2, and 10 is gratifying. However, it should be emphasized that the plots for the model (Fig. 10) were obtained just beyond the accumulation point for the sequence of period doublings of the quasi-sinusoidal oscillations (near the left-hand boundary of Fig. 5), while the plots for the experiments (Figs. 1 and 2) were obtained just beyond the accumulation point of the sequence of period doublings of relaxation oscillations (see Section 3.3). In addition, the chaotic regime beyond the period doubling sequence found in the model [69] spans a much narrower residence time range than the chaotic regime found in the experiments.

It is not yet understood why chaos has proved to be rather elusive in Oregonator models, but some possible reasons are the following:

(1) Even in the simplest nonlinear models the phase diagrams are often extremely complex functions of the multiple control parameters. The numerical studies thus far have examined only a <u>very</u> small fraction of the physically accessible parameter ranges. Furthermore, <u>not</u> every transition sequence should be expected to contain chaotic regimes. Some of the transition sequences observed in the Austin and Bordeaux experiments do <u>not</u> contain chaotic states (e.g., see [45]), while in other cases (with <u>different</u> control parameter values) chaos is observed.

(2) The rate constants in the models are in some cases uncertain by orders of magnitude [63]. Even fairly small changes in the constants in nonlinear models can produce qualitative changes in the dynamics.

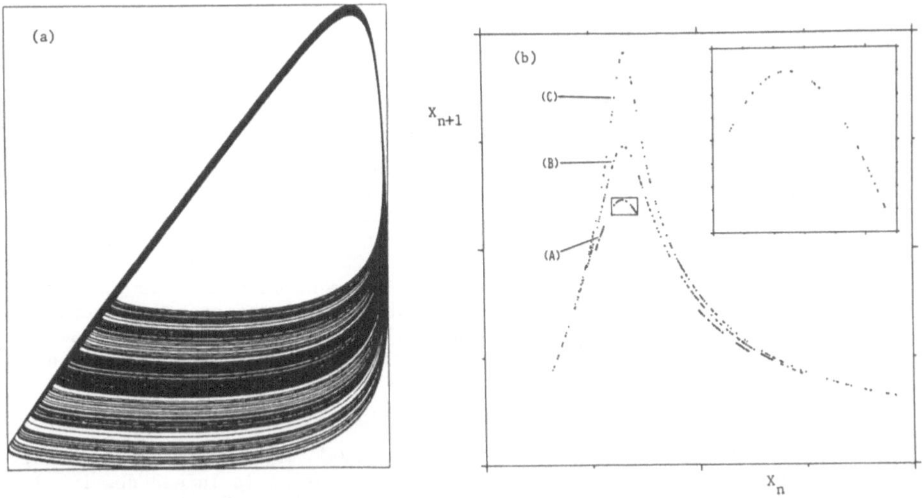

Fig. 10. (a) A two-dimensional projection of a phase portrait obtained from a numerical simulation by RINGLAND and TURNER [69] of a four-variable reversible Oregonator model. This strange attractor was found near the period 3 state of the U-sequence. The inverse residence time was 9.214×10^{-4} s^{-1}; the other parameters are given in TURNER et al. [21]. (b) Next maximum maps at flow rates of (A) 9.2095×10^{-4} s^{-1}, (B) 9.2140×10^{-4} s^{-1}, and (C) 9.2170×10^{-4} s^{-1}.

(3) Experiments have shown that multistability is fairly common in nonequilibrium reactions. Multistability also occurs in models--TURNER [70] discovered (in his numerical exploration of a four-variable reversible Oregonator) a parameter range where there were two stable coexisting states, one periodic and one chaotic. It is hence possible for two investigators studying the same model (with the same constants and control parameters) to use different initial conditions and find quite different behavior. Unfortunately, there is no way known to find all the basins of attraction for a model, even for a given set of constants and control parameters.

(4) The BZ reaction is of course very complex, and the reduced Oregonator-type models might not describe the full range of dynamics occurring in the reaction. Some refinements of the Oregonator are in fact likely warranted [71-73]. Nevertheless, the evidence gathered by many investigators over the past few years in studies of other nonlinear models with 3 or more variables suggests (but by no means proves) that almost any Oregonator will exhibit chaos in some parameter range. If this proves not to be the case, then the Oregonator-type models should be re-examined in view of the experimental evidence for chaos.

5. Concluding Remarks

Evidence that some nonperiodic states observed in nonequilibrium reactions are chaotic includes the following:

*The attractors are not periodic yet appear to have some structure--it is difficult to imagine how stochastic fluctuations could give rise to the smooth nonperiodic structures such as those shown in Figs. 1(b) and 2.

*One-dimensional and higher dimensional maps constructed from the data often yield smooth curves, not a broad distribution of points as might be expected if the nonperiodicity were a consequence of stochastic forcing.

*The largest Lyapunov exponent can be computed from a map or directly from measurements of the divergence rate of nearby trajectories. The two approaches yield the same positive value for the exponent for a given set of data; hence the data are described by a strange attractor.

*While the experimental resolution has to date been insufficient to resolve multiple sheets in the observed attracting (almost two-dimensional) sets, an examination of the attractors reveals clear evidence of the stretching and folding that is a necessary consequence of the exponentially fast separation of nearby trajectories in any strange attractor. .skl

*The strange attractors have small fractional dimension (~2). An unexpected bonus of the determinations of the dimension is that the procedure separates the deterministic dynamical behavior from the stochastic noise (see Section 2.4). Thus the procedure provides an estimate of the experimental noise arising from incomplete mixing, fluctuations in stirring rate, etc., as will be described elsewhere [44].

*Universality in the behavior of one-dimensional maps provides predictive power: If the dynamical behavior of a chemical reaction is indeed described by a one-dimensional map with a single extremum (as in Fig. 2), then the reaction should in general exhibit (with change in control parameter) a period doubling sequence leading to a chaotic regime. This regime beyond the period doubling accumulation point should contain a universal (U) sequence of periodic windows that should have an ordering and corresponding map-iteration pattern given by the theory of one-dimensional maps. Moreover, each periodic window should appear at a tangent bifurcation and disappear through a period doubling sequence. There is evidence from the BZ experiments in support of each of these predictions.

*Another predicted route to turbulence is through the wrinkling of a two-torus. This too has been observed in experiments on the BZ reaction.

Acknowledgments

Our research has been conducted in collaboration with our colleagues in Austin and Bordeaux, including in particular W.D. McCormick, Patrick DeKepper, Annie Rossi, Kerry Coffman, and Jack Swift. The Austin research program is supported by the Department of Energy Office of Basic Energy Sciences and the Robert A. Welch Foundation. The Bordeaux-Austin collaboration is supported by the NSF-CNRS United States-France Cooperative Program. One of us (HLS) acknowledges the support of a Guggenheim Fellowship.

References

1. D. Ruelle, Trans. N.Y. Acad. Sci. II, 35, 66 (1973).
2. L.F. Olsen and H. Degn, Nature 267, 177 (1977).
3. T. Li and J. Yorke, Am. Math. Monthly 82, 985 (1975). This paper was the first to use the term "chaos" with the meaning that it has in the title of the present paper.
4. R.A. Schmitz, K.R. Graziani, J.L. Hudson, J. Chem. Phys. 67, 3040 (1977).
5. O. Rössler, K. Wegmann, Nature 271, 89 (1978); K. Wegmann, O. Rössler, Z. Naturforsch. 33a, 1179 (1978). See also O.E. Rössler, "Chemical Turbulence," in Synergetics, A Workshop, ed. by H. Haken (Springer, 1977), p. 174.
6. C. Vidal, J.C. Roux, A. Rossi, S. Bachelart, C.R. Acad. Sci. Paris 289C, 73 (1979).
7. J.L. Hudson, M. Hart, D. Marinko, J. Chem. Phys. 71, 1601 (1979).
8. P.G. Sorenson, Ann. N.Y Acad. Sci. 316, 667 (1979).
9. H. Nagashima, J. Phys. Soc. Japan 49, 2427 (1980).
10. C. Vidal and A. Pacault, eds., Nonlinear Phenomena in Chemical Dynamics (Springer, Berlin, 1981).

11. See ref. 10: D. Ruelle, p. 30; O.E. Rössler, p. 79.
12. See ref. 10: J.C. Roux and H.L. Swinney, p. 33; J.L. Hudson, J. Mankin, J. McCullough, P. Lamba, p. 44; and C. Vidal. p. 49.
13. C. Vidal, J.C. Roux, S. Bachelart, A. Rossi, Ann. N.Y. Acad. Sci. 357, 377 (1980).
14. J.C. Roux, A. Rossi, S. Bachelart, C. Vidal, Phys. Lett. 77A, 391 (1980).
15. J.C. Roux, A. Rossi, S. Bachelart, C. Vidal, Physica 2D, 395 (1981).
16. Y. Pomeau, J.C. Roux, A. Rossi, S. Bachelart, C. Vidal, J. Phys. Lett. 42, L271 (1981).
17. C. Vidal, S. Bachelart, A. Rossi, J. Phys. (Paris) 43, 7 (1982).
18. J.C. Roux: Physica 7D, 57 (1983).
19. J.L. Hudson and J.C. Mankin, J. Chem. Phys. 74, 6171 (1981).
20. J.L. Hudson, J. Mankin, J. McCullough, P. Lamba, in ref. 10, p. 44.
21. J.S. Turner, J.C. Roux, W.D. McCormick, H.L. Swinney, Phys. Lett. 85A, 9 (1981).
22. J.C. Roux and H.L. Swinney, in ref. 10, p. 33.
23. J.C. Roux, J.S. Turner, W.D. McCormick, H.L. Swinney, in Nonlinear Problems: Present and Future, A.R. Bishop, D.K. Campbell, B. Nicolaenko, eds. (North-Holland, Amsterdam, 1982), p. 409.
24. R.H. Simoyi, A. Wolf, H.L. Swinney, Phys. Rev. Lett. 49, 245 (1982).
25. J.C. Roux, R.H. Simoyi, H.L. Swinney, Physica 8D, 257 (1983).
26. K. Showalter, R.M. Noyes, K. Bar-Eli, J. Chem. Phys. 69, 2514 (1978).
27. R.M. Noyes, in Stochastic Phenomena and Chaotic Behavior in Complex Systems, ed. by P. Schuster (Springer, Berlin, 1984), p. 106; N. Ganapathisubramanian and R.M. Noyes, J. Chem. Phys. 76, 1770 (1982).
28. I.B. Schwartz, Phys. Lett. 102A, 25 (1984).
29. L.F. Olsen, in Stochastic Phenomena and Chaotic Behavior in Complex Systems, ed. by P. Schuster (Springer, Berlin, 1984), p. 116.
30. I.R. Epstein, Physica 7D, 47 (1983); see also the paper by Epstein in ref. 46.
31. D. Ruelle, private communication.
32. N.H. Packard, J.P. Crutchfield, J.D. Farmer, R.S. Shaw, Phys. Rev. Lett. 45, 712 (1980).
33. F. Takens, Lecture Notes in Mathematics, 898, ed. by D.A. Rand and L.S. Young (Springer, Berlin, 1981), p. 366.
34. H.S. Greenside, G. Ahlers, P.C. Hohenberg, R.W. Walden, Physica 5D, 322 (1982).
35. D. Farmer, E. Ott, J. Yorke, Physica 7D, 153 (1983).
36. A. Wolf and J. Swift, in Statistical Physics and Chaos in Fusion Plasmas, ed. by W. Horton and L. Reichl (Wiley, New York, 1984), p. 111.
37. A. Wolf, J. Swift, H.L. Swinney, J. Vastano, "Determining Lyapunov exponents from a time series," submitted to Physica D.
38. D. Ruelle, Ann. N.Y. Acad. Sci. 317, 408 (1978).
39. P. Grassberger and I. Procaccia, Phys. Rev. Lett. 50, 346 (1983).
40. B. B. Mandelbrot, The Fractal Geometry of Nature (Freeman, San Francisco, 1982). See the discussion of chaotic ("fractal") attractors, pp. 193-198.
41. J.D. Farmer, E. Ott, J. Yorke, Physica 7D, 153 (1983).
42. A. Brandstater, J. Swift, H.L. Swinney, A. Wolf, J.D. Farmer, E. Jen, J.P. Crutchfield, Phys. Rev. Lett. 51, 1442 (1983).
43. A. Brandstater, J. Swift, H.L. Swinney, A. Wolf, in Turbulence and Chaotic Phenomena in Fluids, ed. by T. Tatsumi (North-Holland, Amsterdam, 1984); A. Brandstater and H.L. Swinney, in Fluctuations and Sensitivity in Non-equilibrium Systems, ed. by W. Horsthemke and D. Kondepudi (Springer, Berlin, 1984).
44. K. Coffman, W.D. McCormick, J.C. Roux, R. Simoyi, H.L. Swinney, to be published.
45. J. Maselko and H.L. Swinney, Physica Scripta, to appear (1984).
46. Nonequilibrium Dynamics in Chemical Systems, ed. by C. Vidal and A. Pacault (Springer, Berlin, 1984) (this volume).
47. K. Coffman, W.D. McCormick, H.L. Swinney, J.C. Roux, in ref. 46. Also, K. Coffman, W.D. McCormick, J.C. Roux, R. Simoyi, H.L. Swinney, to be published.

48. M.J. Feigenbaum, J. Stat. Phys. 19, 25 (1978).
49. M.J. Feigenbaum, Physica 7D, 16 (1983).
50. P. Collet and J.P. Eckmann, Iterated Maps of the Interval as Dynamical Systems (Birkhauser, Boston, 1980).
51. Y. Pomeau and P. Manneville, Commun. Math. Phys. 74, 189 (1980).
52. J.C. Roux, A. Rossi, in ref. 46.
53. F. Argoul, P. Richetti, A. Arneodo, in ref. 46.
54. I. Prigogine and P. Lefever, J. Chem. Phys. 48, 1695 (1968).
55. F. Schlögl, Z. Phys. 248, 446 (1971).
56. O.E. Rössler, Z. Naturforsch. 31a, 259 (1976); Bull. Math. Biol. 39, 275 (1977).
57. E.N. Lorenz, J. Atmos. Sci. 20, 130 (1963).
58. O. Decroly, A. Goldbeter, Proc. Natl. Acad. Sci. (USA) 79, 6917 (1982).
59. L.F. Olsen, Phys. Lett. 94A, 454 (1982).
60. R.J. Field, E. Körös, and R.M. Noyes, J. Am. Chem. Soc. 94, 8649 (1972).
61. R.J. Field and R.M. Noyes, J. Chem. Phys. 60, 1877 (1974).
62. J. Rinzel, I.B. Schwartz, J. Chem. Phys. 80, 5610 (1984).
63. J.J. Tyson, J. Math. Biol. 5, 351 (1978).
64. K. Tomita, I. Tsuda, Phys. Lett. 71A, 489 (1979).
65. A.S. Pikovsky, Phys. Lett. 85A, 13 (1980); A.S. Pikovsky, M.I. Rabinovich, Physica 2D, 8 (1981).
66. C. Lobry and R. Lozi, in ref. 10, p. 67.
67. J. Turner, Discussion Meeting, Kinetics of Physicochemical Oscillations- (Aachen, September, 1979).
68. J. Turner, in Self-Organization and Dissipative Structures, ed. by W.C. Schieve and P. Allen (University of Texas Press, Austin, 1982), p. 41.
69. J. Ringland and J.S. Turner, "One-dimensional behavior in a model of the Belousov-Zhabotinskii reaction," Phys. Lett., in press (1984).
70. D. Lindberg and J.S. Turner, to be published.
71. Z. Noszticzius, H. Farkas, and Z.A. Schelly, J. Chem. Phys. 80, 6062(1984).
72. R.M. Noyes, J. Chem. Phys. 80, 6071 (1984).
73. J.J. Tyson, J. Chem. Phys. 80, 6079 (1984).

Quasiperiodicity in Chemical Dynamics

J.C. Roux and A. Rossi

Centre de Recherche Paul Pascal, Université de Bordeaux I, Domaine Universitaire F-33405 Talence Cédex, France

Historically, the first description of the transition toward turbulent behavior turbulence predicts chaos to occur after the appearance of complicated dynamics with at least 2 or 3 frequencies (1) (or an infinity according to LANDAU). Although chaotic dynamics were unambiguously identified in chemical systems (2), quasiperiodicity - dynamics with at least two frequencies- was never observed. In this paper we present experimental identifications of such dynamics and of the bifurcations by which they are born and by which they die.

Experimental results

The Belousov-Zhabotinskii reaction was conducted in a CSTR fed at constant temperature (39 C) by the following solutions

$[(CH_2COOH)_2]$ = 0.5 mol/l	$[H_2SO_4]$=1.5 mol/l
$[BrO_3K]$ = 0.036 mol/l	$[H_2SO_4]$=0.75 mol/l
$[Ce_3(SO4)_2]$ = 2.5 10-4 mol/l	$[H_2SO_4]$=0.75 mol/l

The three fluxes were maintained equal, the bifurcation parameter in this study being the overall flow through the reactor, the volume of which is 30cc.(the experiments described in this paper represent a path along line 5 of figure 5 of the paper of SWINNEY et al (2))

The phase portrait and the corresponding Poincare sections presented in the following are constructed using the usual time delay method (3) from a single time series, namely the time variation of the cerium IV monitored spectrophotometrically at 360nm.

At low flow rate the oscillations are nicely periodic, the signal to noise ratio is over 60db as can be measured on the power spectrum. Then the limit cycle becomes unstable for a critical value of the flow rate and gives yield to the regime depicted in figure 1. Two frequencies are evident on the time series as well as in the corresponding power spectra (figure 2), the frequency of the burst and the frequency inside the burst which correspond to the frequency of the limit cycle.

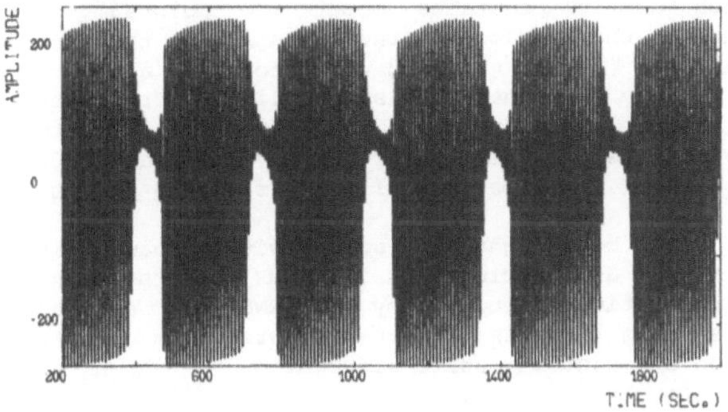

figure 1. Time series of a quasi-periodic regime (about 1/4 of the actual time series is plotted)

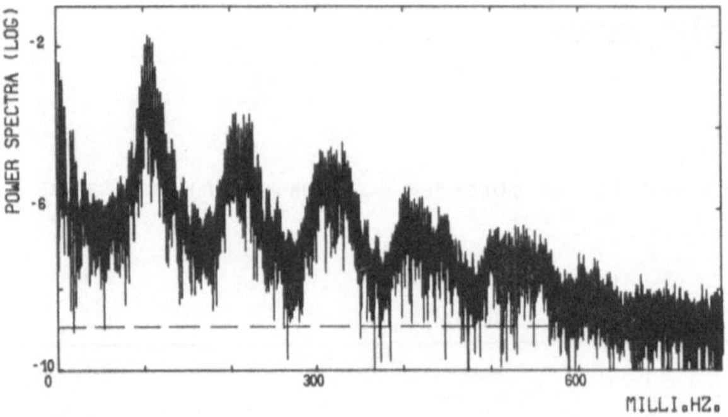

figure 2. Power spectra showing the 107 mHz frequency modulated by the 3.1 mHz frequency. The level of the instrumental noise is indicated by the dotted line.

Description of the trajectories: a wrinkled torus

The reconstructed attractor is shown in figure 3 (for clarity only few trajectories are plotted). Five intersections by planes nearly perpendicular to most of the trajectories (Poincare sections) are depicted. Clearly these sections delimit closed curves: the attractor is a torus as expected for quasiperiodic dynamics.

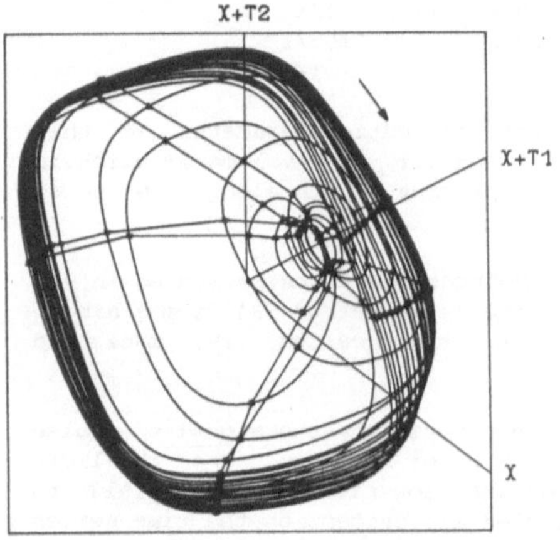

figure 3. Reconstructed attractor T2 = 2T1 = 4 seconds (about 1/20th of the trajectories is plotted)

Successive Poincare sections are plotted figure 4, thus giving an idea of the shape of this torus (each section contains about 700 points). The interesting point to be underlined in the topology of these maps is the tail which appears and develops in sections (2) and (3), and which folds up along the torus (sections (4),(5)) and is finally stretched. This wrinkling occurs every turn inducing some kind of mixing in this part of the attractor; thus the resulting structure of the torus must be fractal. Unfortunately this structure is not evident from figure 4 because of the experimental scatter of the points.

The shape of this torus which can be thought as a sphere with a hole going from north pole to south pole is not as surprising as it seems at first sight. Similar shapes were obtained in theoretical studies by GUCKENHEIMER (4) and LANGFORD (5); they occur as the results of the interactions of two instabilities - a Hopf bifurcation and a steady state bifurcation - which are both generic of these chemical systems (6).

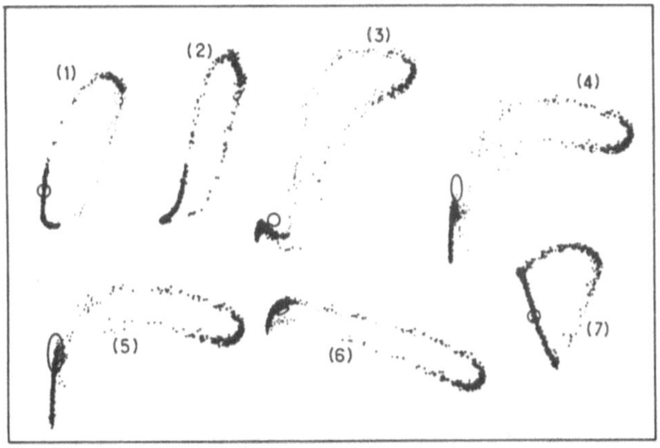

figure 4. 7 successive Poincare sections of the attractor. The center of the torus is always on the right ot the section.

Birth and death of the wrinkled torus

The circled zones in figure 4 represent, at the same scale, the corresponding Poincare sections of the limit cycle that exists prior to the bifurcation toward the torus. The areas of these dashed zones stand for the experimental scatter of the points. It is isotropic almost everywhere except in the part of the limit cycle which corresponds to the stretching of the wrinkle. Furthermore, we must remark that the limit cycle appears to lie on the surface of the torus. This is an indication that the bifurcation leading to the torus is of a saddle-node type rather than of a Hopf type (in this latter case the limit cycle should be inside the torus). This character is also confirmed by the abruptness of the transition and the absence of hysteresis.

Figure 5 is a schematic representation of this type of bifurcation as seen in the Poincare map. The node represents the stable limit cycle that exists before the bifurcation. In one part of this cycle there is a saddle in the vicinity of the cycle. In the direction saddle-node the system will be less contracting, which explains the anisotropic scatter of the experimental intersections of the cycle observed in the sections of figure 4.

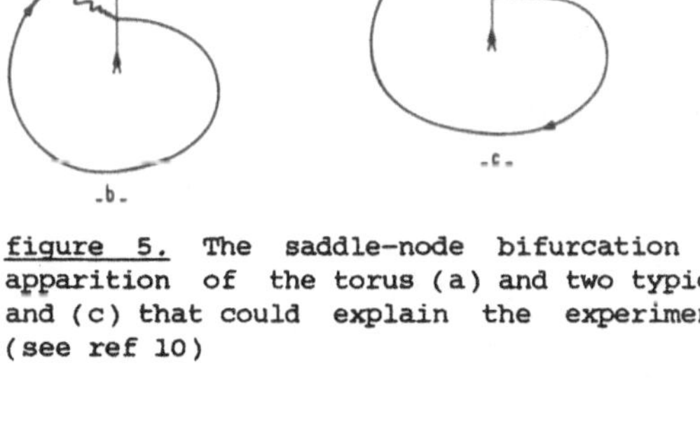

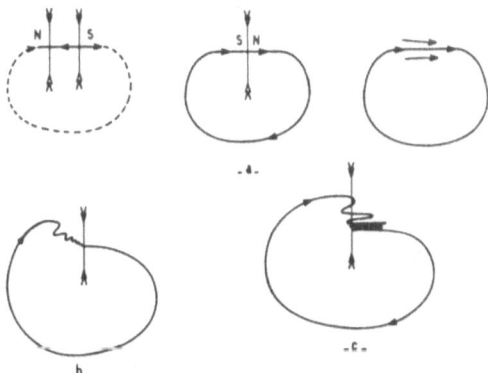

figure 5. The saddle-node bifurcation explaining the apparition of the torus (a) and two typical situations (b) and (c) that could explain the experimental observations (see ref 10)

Bifurcations of this type in diffeomorphisms have been the object of numerical (8) and theoretical (9-10) work. We learn from these studies that in addition to a smooth torus, a saddle-node bifurcation may yield other equally robust situations. In particular two of them appear to be relevant to our work: Either (i) the limit cycle (in the Poincare map) is destroyed when the saddle-node bifurcation occurs, and the disappearance of the saddle-node corresponds to the transition from a regular, periodic, state to a chaotic one, or (ii) the limit cycle is destroyed before the disappearance of the saddle-node and there exists chaos concurrently with the initial periodic orbit (see fig 5c)

The wrinkles observed in our experiments indicate that our data is relevant to one of these two cases. However, it is impossible to distinguish between them on the basis of the present data.

When the flow rate is further increased, the amplitude of the small oscillations decreases and their number increases at the expense of the large one (figure 6). This means that the central "hole" in the torus becomes thinner and thinner and that the system stays longer and longer in it. In the end we reach a stationary state located in the center of the "hole". The disappearance of the torus by this mechanism is consistent with the theoretical interpretation which predicts its existence. (6-7)

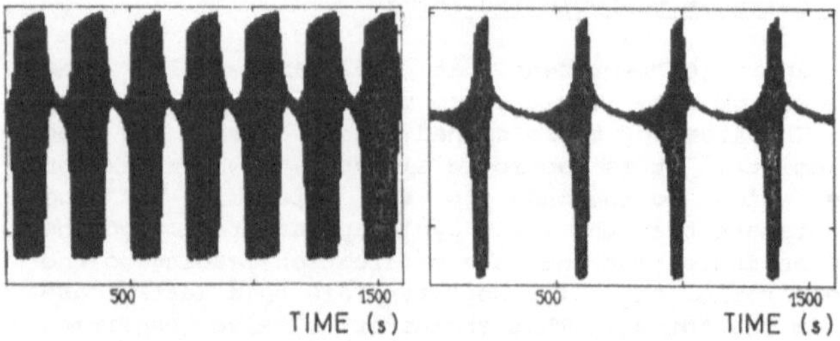

figure 6. Evolution of the dynamics with further increase of the flow rate.

Conclusions

We have presented the first clear evidence of quasiperiodicity in a chemical dynamical system. The shape of the torus as well as its evolution is consistent with the nature of some generic instabilities of this chemical system. Some preliminary experiments in our laboratory indicate that similar tori could be obtained after a Hopf bifurcation of a limit cycle in perfect agreement with the theoretical predictions.

Aknowledgements: The interpretion of the experimental data was enlightened by numerous discussions with A. ARNEODO, C. TRESSER and H.L. SWINNEY. This research is suported by D.R.E.T, contract 83/1332.

References

-1) D. RUELLE and F. TAKENS Commun. Math. Phys. 23 343 (1971)

-2) H. L. SWINNEY and J.-C. ROUX This volume

-3) F. TAKENS Lectures notes in Mathematics ed. by D.A. RAND and L. S. YOUNG (Springer, Berlin - Heidelberg New York 1981) p366

-4) J. GUCKENHEIMER Lectures notes in Mathematics 898 99 (1980)

-5) W. F. LANGFORD in "Nonlinear Dynamics and Turbulence" ed. by I. BARENBLATT, G. IOOSS, D. D. JOSEPH (Pittman 1983) p215

-6) F. ARGOUL, P. RICHETTI, A.ARNEODO This volume

-7) J.C. ROUX, P.RICHETTI, A. ARNEODO, F. ARGOUL J.M.T.A. (to appear)

-8) J. CURRY and J. YORKE Lectures Notes in Mathematics 668 48 (1978)

-9) S. E. NEWHOUSE Ann N. Y. Acad. Sci. 357 292 (1980)

-10) D. G ARONSON, M. A. CHORY, C.R. HALL, R. P. McGEHEE Commun. Math. Phys. 83 303 (1982)

Nonlinear Interactions Between Instabilities Leading to Chaos in the Belousov-Zhabotinsky Reaction

F. Argoul and P. Richetti

Centre de Recherche Paul Pascal, Université de Bordeaux I, Domaine Universitaire
F-33405 Talence Cédex, France

A. Arneodo

Laboratoire de Physique Théorique, Université de Nice, Parc Valrose
F-06034 Nice Cédex, France

The occurrence of well-odered structures in physical systems submitted to some forcing is a manifestation of an instability which affects one of their normal modes. The dissipative structure which appears results from a nonlinear saturation of this instability. Such a transition presents some analogy with phase - transition phenomena[1]. When the value of the control parameter is close to the critical transition value, the temporal evolution of the amplitude of the mode is governed by an "universal" (normal form) first order equation like that given by Landau for the initial development of instability in fluids[2,3]. In more complicated cases, there will exist, in parameter space, polycritical surfaces on which several modes may simultaneously become marginally unstable [4,5]. In neighborhoods of such surfaces, the dynamics is governed by coupled ordinary differential equations for the amplitude of the nearly marginal modes, which generalize the Landau equation to problems of competing instabilities [6-10]. In such cases the temporal evolution of the system may become increasingly complex, e.g. relaxation oscillations, quasiperiodic and even chaotic behavior [5].

Various scenarios to "strange attractor" like behavior have been experimentally observed in the Belousov-Zhabotinsky reaction in an open flow system, i.e. a continuous stirred tank reactor [11-19]. We propose a global interpretation of these transitions to chaos in terms of the competition between three instabilities. In the neighborhood of the polycritical surface we study the normal form which describes this interaction. We limit our investigation to experimental paths which are characteristic of the variety of dynamical behavior one can encounter in this region of parameter space.

We first follow a path along which two among the three instabilities enter the competition, namely an oscillatory instability (Hopf bifurcation) and a stationary (hysteresis) instability whose interaction can give rise to an attracting invariant torus [20,21]. When the control parameter (flow rate) is varied we observe, after several frequency lockings, some stretchings and foldings of the toroidal surface which confirms the a lost of differentiability of the torus into chaos. While the second frequency decreases to zero, the center hole of the torus shrinks to a thin tube, and the trajectory spends more and more of its time nearly stationary in this inner part of the torus before ending on a steady state. Such an evolution of the dynamics is likely to be relevant to explain the history of the fractal torus observed in the Bordeaux experiment [13].

Then we consider a path where the transition to chaos proceeds via the well - known cascade of period-doubling bifurcations [22-24]. As we vary the control parameter, the strange attractor displays a more pronounced spiralling shape. This is characteristic of the chaotic behavior which occurs close to certain homoclinic conditions [5,25,26]. A typical one-dimensional map, associated with these attractors, presents several humps. Maps with double extrema have been extensively studied in the literature. When one varies the parameters in the model, one may encounter other phenomena such as transitions from periodic to chaotic regimes according to the Pomeau-Maneville theory of intermittency [27] and also discontinuous transitions between attractors with hysteresis [28,29]. Experimental evidence of a single humped 1D map have been obtained in the B.Z. reaction in the chaotic regime that immediately follows the few period-doubling bifurcations of the relaxation oscillation detected in a Texas experiment when decreasing the low rate [11]. With further decrease, preliminary results [30] actually suggest that a second hump appears in the reconstructed 1D map which lends strong support to an interpretation of the data in terms of the chaotic behavior which occurs on the way to homoclinicity as predicted by Shil'nikov [26].

We finally emphasize that there is, so far, no experimental evidence of a temporal evolution which does not fit our description in terms of the competition between three instabilities. Therefore it looks very promising to use normal form approach [7-10] to reduce the order of the F.K.N. model [31] by selecting the relevant instabilities which control the dynamics of this chemical reaction.

In consideration of editors requirement to limit the publication to short communication, this paper has to be seen as an extended abstract of a more complete and detailed review to be published in [32].

References

1. S.K.Ma, Modern Theory of critical phenomena (Benjamin, Reading, Mass. (1976)
2. L. Landau, Dokl. Akad. Nauk SSSR 44, 339 (1944)
3. L. Landau and E. M. Lifshitz, Fluid Mechanics, 103 (Pergamon Press, Oxford, 1959)
4. P. H. Coullet and E. A. Spiegel, SIAM J. Appl. Math. 43, 775 (1983)
5. A. Arneodo, P. H. Coullet, E. A. Spiegel and C. Tresser, Asymptotic chaos, preprint (1982) to appear in physicaD
6. H. Haken, Synergetics, 194 (Springer Verlag, NY, 1978)
7. J. Guckenheimer and P. Holmes, Nonlinear Oscillations, Dynamical Systems and Bifurcations of Vector Fields (Springer Verlag, Berlin, 1984)
8. V. I. Arnold, Supplementary chapters to the theory of ordinary differential equations (Nauka, Moscou, 1978)
9. H. Poincare, Thesis-1879, Oeuvres (Gauthier-Villars Paris, 1928), p.1-69.
10. G. O. Birkhoff, in Dynamical Systems, A. M. S. Publications, Providence (1927)
11. J. C. Roux and H. L. Swinney, in Nonlinear Phenomena in Chemical dynamics, A. Pacault and C.Vidal eds (Springer, Berlin, 1981), p.38. R. H. Simoyi, H. L. Swinney and A. Wolf, Phys. Rev. Lett. 49, 245 (1982)
12. Y. Pomeau, J. C. Roux, A. Rossi, S. Bachelart and C. Vidal, J. Phys. Lett. 42, L271 (1981)

13. J. C. Roux and A. Rossi in this volume
14. R. A. Schmitz, K. R. Graziani and J. L. Hudson, J. Chem. Phys, 71, 3040 (1977)
15. J. L. Hudson, M. Hart and D. Marinko, J. Chem. Phys, 71, 1601 (1979)
16. J. L. Hudson and J. C. Mankin, J. Chem. Phys. 74, 6171 (1981)
17. C. Vidal, J. C. Roux, A. Rossi and S. Bachelart, Ann. N. Y. Acad. Sci. 357, 377 (1980)
18. J. C. Roux, J. S. Turner, W. D. McCormick and H. L. Swinney, in Nonlinear Problems: Present and Future, A. R. Bischop ed (North-Holland, Amsterdam, 1982), p.409.
19. J. C. Roux, R. H. Simoyi and H. L. Swinney, Physica 8D, 251 (1983)
20. W. F. Langford, in Nonlinear Dynamics and Turbulence, ed G.Iooss and D. D. Joseph, Pitman Press, (1982)
21. W. F. Langford. Proceeding of the conference on Numerical Methods for Bifurcation Problems.Univ.Dortmund(1983)
22. M. J. Feigenbaum, J. Stat. Phys. 19, 25 (1978); 21, 669 (1979)
23. P. Coullet and C. Tresser, J. de Physique, colloque 39, 5 (1978)
24. C. Tresser and P. Coullet, C. R. Acad. Sc. Paris, 287, 577 (1978)
25. A. Arneodo, P. H. Coullet and C. Tresser, Comm. Math. Phys. 79, 573 (1981); J. Stat. Phys. 27, 171 (1982)
26. L. P. Shil'nikov, Sov. Math. Dokl. 6, 163 (1965); Math. USSR. Sbornick, 10, 91 (1970)
27. Y. Pomeau and P. Manneville, Comm. Math. Phys. 74, 189 (1980)
28. C. Tresser, P. Coullet and A. Arneodo, J. Phys. Lett. 41, 243 (1980)
29. S. Fraser and R. Kapral, Phys. Rev. A25, 3223 (1982)
30. K. Coffman, H. L. Swinney, J. C. Roux and W. D. McCormick, Private Communication.
31. R. J. Field, E. Koros and R. M. Noyes, J. Am. Chem. Soc, 94, 8649 (1972)
32. J. C. Roux, P. Richetti, A. Arneodo and F. Argoul preprint 1984, to be published in Journal de Mécanique et de Techniques Appliquées.

Part V

Noise Effects

Noise Induced Transitions

Werner Horsthemke

Center for Studies in Statistical Mechanics, Department of Physics
University of Texas, Austin, TX 78712, USA

1. Introduction

In this paper I will deal with the effect of external random perturbations, "noise", on chemical systems and other open nonlinear systems. As a concrete example let us consider a CSTR. This is an open system and as such subject to external constraints, namely the concentrations of the chemical species in the feed streams, the flow rate, the stirring rate, the temperature, and the incident light intensity in the case of a photochemical reaction. These external constraints characterize the state of the environment of the open system and will, in general, fluctuate more or less strongly. Such environmental fluctuations are particularly important for natural systems; here random fluctuations are always present and their amplitude is not necessarily small as in laboratory systems. In the latter systems the experimenter will of course try to minimize the effect of random perturbations, though it is impossible to eliminate noise completely. Clearly, random external noise is ubiquitous in open systems, but this fact by itself would hardly warrant a systematic study of the effects of external fluctuations. The question is whether noise is more than a mere nuisance we have to live with. Is there any hope of finding interesting physics? The intuitive, and wrong, answer would be negative: The system averages out rapid fluctuations and the only trace of external noise would be a certain fuzziness in the state of the system. Of course, if the state of the system becomes unstable, the fluctuations initiate the departure from the unstable state. Then the dynamics of the system take over and the system evolves to a new stable state. Besides these trivial effects of random fluctuations, external noise can give rise to unexpected and interesting phenomena. Noise can change the stability properties of a system, namely stabilize or destabilize a steady state. This effect was first theoretically predicted in studies on oscillations in radio circuits [1], on the survival of populations [2], and on oscillating enzyme systems [3]. Even more surprisingly, external noise can create *new* states which *never* exist under deterministic environmental conditions [4]. Clearly, there is interesting physics to be found in a study of noise-induced phenomena.

The organization of this paper is as follows: First I will discuss the modeling of nonlinear systems coupled to a noisy environment. Then I will highlight the main theoretical results and I will conclude by mentioning problems currently under study and experimental results on noise-induced transitions. More details on most of the aspects of noise-induced phenomena discussed here can be found in the recent monograph [5].

2. Modeling Systems with External Noise

For the sake of clarity I will discuss the effect of external noise in as simple a case as possible. I will therefore assume that the system has the following three properties: i) It is spatially homogeneous. This corresponds to the limit of fast transport. ii) The system is macroscopic and can be described by intensive variables. This corresponds to the thermodynamic limit and implies that any finite size effects, such as internal fluctuations, can be neglected. iii) The state of the system can be described by one variable. This is only a point of mathematical convenience; explicit analytical results can be obtained for one variable systems.

These three properties imply that the evolution of the system can be modeled by the following kinetic equation:

$$\dot{x}(t) = f(x,\lambda) = h(x) + \lambda g(x) \quad . \tag{1}$$

Here x denotes the state of the system at time t, for instance the concentration of a chemical species in the reactor. λ is the external parameter which we will let fluctuate later on. The other external constraints are not explicitly written in (1). In most applications f is nonlinear in x but linear in λ. This is the only case I will consider in the following. Nonlinear external parameters can be treated as is shown in [5]. Associated with (1) is a typical time scale, τ_{macro}, which is characteristic of the macroscopic evolution of the system.

In order to take into account the influence of external noise, we will take (1) as our starting point and replace λ by a random process. In other words, (1) describes the system for a deterministic environment and is obtained as the amplitude of the external noise is made to vanish. The fact that the external parameter becomes a stochastic process implies that the state of the system at a given instant of time is a random variable. The state is no longer characterized by a simple number but by a probability distribution. To model the external noise, we assume that the environment has the following three properties: i) It is stationary, i.e. in particular

$$\langle\lambda_t\rangle = \bar{\lambda}, \quad \langle(\lambda_t - \bar{\lambda})(\lambda_{t+\tau} - \bar{\lambda})\rangle = C(\tau) \quad . \tag{2}$$

Here $\langle\ \rangle$ denotes the mathematical expectation or average. Stationarity is assumed in order to assess the effects of noise separately from any effects due to a systematic evolution of the environment. Furthermore, this assumption is fulfilled in most applications to a good degree of approximation. We write the external noise as

$$\lambda_t = \bar{\lambda} + Z_t \quad , \tag{3}$$

where Z_t is now a process with mean value zero. ii)a) If the noise is not deliberately put on the system by an experimenter, we will assume that Z_t is Gaussian distributed. This is justified by the Central Limit Theorem: The external parameter represents the cumulative effect of a large number of small, additive contributions, which are at most weakly coupled. b) If the experimenter wishes to study the effects of external noise by deliberately randomizing an external constraint, then the so-called dichotomous noise is often used. In this case, Z_t takes on only two values,

$$Z_t \in \{\Delta_-,\Delta_+\} \quad . \tag{4}$$

Such a noise is easy to generate electronically. iii) Since we are interested in macroscopic systems, we will observe the system usually only on macroscopic time scales. It is then reasonable to assume that Z_t is a Markov process. Furthermore, it has been argued that a non-Markovian noise will not introduce any essentially new physics into the problem [5]. Properties i) - iii) uniquely specify the noise process. In the case of ii)a) we find, in light of DOOB's theorem [6], that Z_t is given by a stationary Ornstein-Uhlenbeck process, i.e. it obeys the following Langevin equation:

$$\dot{Z}_t = -\gamma Z_t + \sigma\xi_t \quad . \tag{5}$$

This equation was originally introduced to model the velocity of a Brownian particle. ξ_t is Gaussian white noise,

$$\langle\xi_t\rangle = 0, \quad \langle\xi_t\xi_{t+\tau}\rangle = \delta(\tau) \quad . \tag{6}$$

It is a very irregular process with no memory at all. As indicated by the fact that its correlation function is a Dirac delta function, i.e. a generalized function, Gaussian white noise is a generalized stochastic process. The use of generalized random processes can be avoided and we can work entirely in the framework of ordinary stochastic processes by exploiting the fact

$$\int_0^t \xi_s ds = W_t \quad . \tag{7}$$

Here W_t is the Wiener process, which describes the position of a Brownian particle, and which is characterized by

$$\langle W_t \rangle = 0, \quad \langle W_t W_s \rangle = \min(t,s) \quad , \tag{8}$$

and

$$p(w,t|u,s) = [2\pi(t-s)]^{-\frac{1}{2}} \exp\left\{-\frac{1}{2}\frac{(w-u)^2}{t-s}\right\} \quad . \tag{9}$$

$p(w,t|u,s)$ is the probability density to find a value w for W_t knowing that $W_s = u$.

Multiplying (5) formally by dt and using (7), we obtain the stochastic differential equation (SDE)

$$dZ_t = -\gamma Z_t dt + \sigma dW_t \quad . \tag{10}$$

The probability density of Z_t obeys the following evolution equation:

$$\partial_t p(z,t|z_0,0) = -\partial_z(-\gamma z)p(z,t|z_0,0) + \frac{1}{2}\partial_{zz}\sigma^2 p(z,t|z_0,0) \quad . \tag{11}$$

This type of equation is known as a Fokker-Planck equation. The stationary solution of (11) is given by

$$p_s(z) = [2\pi(\sigma^2/2\gamma)]^{-\frac{1}{2}} \exp\left\{-\frac{1}{2}\frac{z^2}{(\sigma^2/2\gamma)}\right\} \quad . \tag{12}$$

If the Ornstein-Uhlenbeck process is started with this probability density, then it is a stationary process and

$$\langle Z_t \rangle = 0, \ \langle Z_t Z_{t+\tau} \rangle = \frac{\sigma^2}{2\gamma} e^{-\gamma|\tau|} = C(\tau) \quad . \tag{13}$$

The correlation time, which characterizes the memory of the environment, is given by

$$\tau_{cor} = \gamma^{-1} \quad . \tag{14}$$

The power spectrum of Z_t is

$$S(\nu) = \frac{1}{2\pi}\int e^{-i\nu\tau} C(\tau) d\tau = \frac{\sigma^2}{2\pi}\frac{1}{\nu^2+\gamma^2} \quad . \tag{15}$$

Thus in case ii)a) we have

$$\dot{X}_t = h(X_t) + (\bar{\lambda} + Z_t) g(X_t) \quad , \tag{16}$$

or multiplying by dt:

$$dX_t = [h(X_t) + \bar{\lambda} g(X_t)]dt + Z_t g(X_t) dt \quad , \tag{17}$$

$$dZ_t = -Z_t dt + \sigma dW_t \quad . \tag{18}$$

Here time has been rescaled, so that $\gamma = 1$. (The case ii)b) will be discussed later.) The joint probability density $p(x,z,t)$ for the system and its environment obeys the following Fokker-Planck equation:

$$\partial_t p(x,z,t) = -\partial_x[h(x) + \bar{\lambda} g(x) + zg(x)]p(x,z,t) - \partial_z(-z)p(x,z,t) + \frac{\sigma^2}{2}\partial_{zz}p(x,z,t) \quad . \tag{19}$$

After the system has been coupled to the environment for a sufficiently long time, it will, in general, settle down to a steady state. To describe this state, we have to find the stationary solution, i.e. $\partial_t p = 0$, of (19). It turns out that this is, in most cases, an intractable problem. However, headway can be made by realizing that often the external noise is very rapid, i.e.

$$\tau_{cor} << \tau_{macro} \quad . \tag{20}$$

It is then tempting to pass to the idealization of $\tau_{cor} = 0$. In other words, we would like to approximate the real system and environment by an approximate one in which the external noise has no memory. Obviously, such an approximation would only be any good if the essential features of the real system are preserved. This requires some circumspection. Similarly to taking the thermodynamic limit of a macroscopic system, we have to replace the real system and environment by a series of equivalent systems and environments in which τ_{cor} goes to zero. As shown by BLANKENSHIP and PAPANICOLAOU [7], this can be done by speeding up the noise, namely considering it on the faster time scale t/ε^2, $Z_t \to Z_{t/\varepsilon^2}$, with $\varepsilon \to 0$ and by scaling up the amplitude of the noise by a factor $1/\varepsilon$. This amounts to replacing (17) and (18) by

$$dX_t = [h(X_t) + \bar{\lambda} g(X_t)]dt + \frac{1}{\varepsilon} Z_t g(X_t)dt \quad , \tag{21}$$

$$dZ_t = -\varepsilon^{-2} Z_t dt + \varepsilon^{-1} \sigma dW_t \quad . \tag{22}$$

The Fokker-Planck equation (19) reads then:

$$\partial_t p^\varepsilon(x,z,t) = -\partial_x [h(x) + \bar{\lambda} g(x)] p^\varepsilon(x,z,t) - \frac{1}{\varepsilon} \partial_x z g(x) p^\varepsilon(x,z,t) + \frac{1}{\varepsilon^2} [\partial_z z + \frac{\sigma^2}{2} \partial_{zz}] p^\varepsilon(x,z,t) \quad . \tag{23}$$

The limit $\varepsilon \to 0$ is known as the Gaussian white noise limit, since we have for the power spectrum $S^\varepsilon(\nu)$ of $\varepsilon^{-1} Z_{t/\varepsilon^2}$:

$$S^\varepsilon(\nu) = \frac{\sigma^2}{2\pi} \frac{1}{\varepsilon^4 \nu^2 + 1} \to \frac{\sigma^2}{2\pi} \quad . \tag{24}$$

The form of (23) suggests that we try the following perturbation expansion:

$$p^\varepsilon(x,z,t) = p_0(x,z,t) + \varepsilon p_1(x,z,t) + \varepsilon^2 p_2(x,z,t) + \cdots \quad . \tag{25}$$

This perturbation expansion is known as the wide band perturbation expansion [8,5]. I will skip the technical details here and just present the main results. It turns out that to lowest order the joint probability density factorizes

$$p_0(x,z,t) = p(x,t) p_s(z) \quad , \tag{26}$$

as it should, since the system and environment are independent at the same instant of time in the white noise case. $p_s(z)$ is given by (12) and $p(x,t)$ obeys the Fokker-Planck equation

$$\partial_t p(x,t) = -\partial_x [h(x) + \bar{\lambda} g(x) + \frac{\sigma^2}{2} g'(x) g(x)] p(x,t) + \frac{\sigma^2}{2} \partial_{xx} g^2(x) p(x,t) \quad . \tag{27}$$

(Prime denotes derivative with respect to x.) The stationary solution of (27) is given by

$$p_s(x) = N g^{-1}(x) \exp\left\{\frac{2}{\sigma^2} \int^x \frac{h(y) + \bar{\lambda} g(y)}{g^2(y)} dy\right\} \quad . \tag{28}$$

Here N is a normalization constant and $p_s(x)$ exists if N is finite.

3. Noise-Induced Transitions

The phenomenon of nonequilibrium transitions for nonfluctuating environmental conditions is by now a familiar one and well understood [9,10]. A transition corresponds to a qualitative change in the state of the system. For instance, a steady state may lose stability for a certain value of the external parameter and new branches of stable states bifurcate. In direct analogy we will say that a nonequilibrium transition occurs in a system with noise, if the state of the system changes qualitatively. As discussed above, the state of the system is a random variable and can thus be characterized by a probability distribution. In order to determine when a transition occurs, we need to monitor $p_s(x)$ for qualitative changes. To do so, we need to find a suitable indicator. As we deal with a probability distribution, moments come to mind. Consider however the simple Landau equation

$$\dot{x} = \lambda x - x^3 = f(x,\lambda) \quad . \tag{29}$$

Here $x \in [0,\infty)$ and $\lambda \in (-\infty,+\infty)$. This equation is frequently encountered in modeling equilibrium and nonequilibrium critical phenomena. In the deterministic case we have for the steady states $\bar{x}$

$$f(\bar{x},\lambda) = 0 \quad : \quad \bar{x} = 0, \quad \text{and } \bar{x} = \pm\sqrt{\lambda} \quad \text{if } \lambda > 0.$$

A stability analysis shows that $\bar{x} = 0$ is stable for $\lambda < 0$ and loses its stability at $\lambda_c = 0$, where two new stable branches $\bar{x} = \pm\sqrt{\lambda}$ bifurcate. Let us now take into account other rapid degrees of freedom in the system by adding a white noise term

$$\dot{X}_t = \lambda X_t - X_t^{\,3} + \sigma\xi_t \quad . \tag{30}$$

We have

$$p_s(x) = N \exp\left\{\frac{2}{\sigma^2}\left(\lambda \frac{x^2}{2} - \frac{x^4}{4}\right)\right\} \quad . \tag{31}$$

It is easily verified that for $\lambda < \lambda_c = 0$, $p_s(x)$ consists of a single peak centered on $x = 0$, whereas for $\lambda > 0$, $p_s(x)$ consists of two peaks centered on $+\sqrt{\lambda}$ and $-\sqrt{\lambda}$. This transition is however <u>not</u> reflected in the moments of $p_s(x)$. We have

$$\langle x^{2n+1} \rangle = 0 \quad \text{for all } \lambda \quad , \tag{32}$$

and $\langle x^{2n} \rangle$ are infinitely often differentiable with respect to the bifurcation parameter λ. Clearly, the transition of the system cannot be detected by monitoring the moments. Moments are *not* a reliable indicator of nonequilibrium transitions. Furthermore, moments are a bad choice also on general theoretical grounds, since moments often do not characterize a probability distribution uniquely [11]. It is obvious from the above example that the appropriate indicator of a transition are the extrema of the probability density: i) They reflect the qualitative features of $p_s(x)$, for instance if $p_s(x)$ is single-humped or multi-humped. ii) It can be shown that x_m converges towards $\bar{x}$ as the noise is turned off; see below. iii) The maxima are the most probable values and are preferentially observed in an experiment; they correspond, so to speak, to the "phases" of the system.

In the white noise limit the extrema of the system are given by

$$[h(x_m) + \bar{\lambda} g(x_m)] - \frac{\sigma^2}{2} g'(x_m) g(x_m) = 0 \quad . \tag{33}$$

We have to distinguish two types of noise: i) additive noise, $g(x) \equiv \text{const}$. In this case the effect of the external noise does not depend on the state of the system. Since $g' = 0$, (33) reduces to the equation for the deterministic steady states $\bar{x}$. Thus

$$x_m \equiv \bar{x} \quad \text{for all } \sigma^2 \quad .$$

This gives further support to our identification of x_m with the "phases" of the system. ii) multiplicative noise, $g(x) \not\equiv \text{const}$. In this case the effect of the external noise is modulated by the state of the system. For small noise intensities σ^2 the term $g'g$ in (33) may be neglected and we find that $x_m \approx \bar{x}$. However, if σ^2 becomes larger and if $g'g$ is sufficiently nonlinear compared to h and g, then the extrema x_m can be very different, in number and location, from the deterministic steady states. Since this change in the behavior of the system arises without any changes in the systemic parameters but simply by varying the noise intensity, we have called this phenomenon a noise-induced transition.

Noise-induced transitions have been studied theoretically in quite a few physical and chemical systems, namely the optical bistability [12,13,5], the Freedricksz transition in nematics [14,15,16,5], the superfluid turbulence in helium II [17], the dye laser [18,19], in photochemical reactions [20], the van der Pol-Duffing oscillator [21] and other nonlinear oscillators [22]. Here I will present a very simple model which exhibits a noise-induced critical point. The so-called genetic model was first discussed in [4]. I will not describe its application to population genetics in this paper, see [5] for this aspect, but use a chemical model reaction scheme:

$$A + X + Y \rightleftarrows 2Y + A^* \quad ,$$
$$B + X + Y \rightleftarrows 2X + B^* \quad , \tag{34}$$

where A, B, A^*, and B^* are assumed to be in large excess. The total number N of X and Y particles remains constant in the scheme (34) and we find for the fraction of X particles in the system the following kinetic equation

$$\dot{x} = \frac{1}{2} - x + \lambda x(1-x) \quad , \tag{35}$$

with $x \in [0,1]$ and $\lambda \in (-\infty,+\infty)$. The steady states of (35) are

$$\bar{x} = \frac{1}{2\lambda}\,[\lambda - 1 + \sqrt{1+\lambda^2}\,] \quad . \tag{36}$$

In other words, for each value of the external parameter λ, which is a combination of the concentrations of the major reactants and the kinetic constants, there is exactly one steady state; this state is globally stable. Thus the genetic model (35) does not display any transition for deterministic external constraints. Assume now that the concentrations of A and B fluctuate rapidly, which in turn implies that λ fluctuates. In the Gaussian white noise idealization, we obtain from (33) the following equation for the extrema of the stationary probability density:

$$\frac{1}{2} - x_m + \lambda x_m(1-x_m) - \frac{\sigma^2}{2}\,x_m(1-x_m)(1-2x_m) = 0 \quad . \tag{37}$$

Consider the case that $\lambda = 0$. Then in the deterministic situation $\bar{x} = \frac{1}{2}$. From (37) we find for the system with external noise that

$$x_{m_0} = \frac{1}{2}\,, \quad \text{and} \quad x_{m\pm} = \frac{1}{2}\,[1 \pm \sqrt{1-(4/\sigma^2)}\,] \quad \text{for} \quad \sigma^2 \geq \sigma_c^{\,2} = 4 \quad . \tag{38}$$

The system with external white noise has a noise-induced critical point at $\lambda_c = 0$, $\sigma_c^2 = 4$, $x_c = 1/2$. For $\sigma^2 < 4$, the stationary probability density has a single peak centered on $x_{m_0} = 1/2$. At $\sigma^2 = 4$, this maximum becomes a double maximum and for $\sigma^2 > 4$ it splits in two; the probability density becomes double-humped. The external noise has induced bistable behavior, which does not exist for deterministic external constraints. Like equilibrium critical behavior, the noise-induced critical point is characterized by critical exponents. According to our discussion above, the order parameter is given by

$$m = x_{m_+} - \bar{x} \quad . \tag{39}$$

We have

$$m \sim \sqrt{\sigma^2 - \sigma_c^{\,2}} \qquad \text{for } \lambda = \lambda_c = 0 \quad , \text{ i.e. } \beta = \frac{1}{2} \quad , \tag{40}$$

$$m \sim \lambda^{1/3} \qquad \text{for } \sigma^2 = \sigma_c^{\,2} = 4 \quad , \text{ i.e. } \delta = 3 \quad , \tag{41}$$

and

$$\left.\frac{\partial m}{\partial \lambda}\right|_{\lambda=0}(\sigma^2) \sim |\sigma^2 - \sigma_c^{\,2}|^{-1} \quad , \quad \text{i.e. } \gamma = \gamma' = 1 \quad . \tag{42}$$

We find the classical or Landau values for the critical exponents, which is expected, since our model excludes any spatial inhomogeneities. To characterize the dynamics near the noise-induced critical point, define a time t_c at which an initially single-humped probability density develops a double maximum and becomes double-humped. It has been shown [5] that

$$t_c \sim -\ell n(\sigma^2 - \sigma_c^2) \quad \text{for } \sigma^2 > \sigma_c^2 ,$$

i.e. t_c diverges logarithmically as the critical point is approached.

4. Dichotomous Noise and Poisson White Noise

Let us now consider the case ii)b), i.e. the two-state noise case. This noise is also known as the random telegraph signal. It is characterized by the following, so-called, Master equation

$$\frac{d}{dt}\begin{pmatrix} P(\Delta_-,t) \\ P(\Delta_+,t) \end{pmatrix} = \begin{pmatrix} -\beta & \alpha \\ \beta & -\alpha \end{pmatrix}\begin{pmatrix} P(\Delta_-,t) \\ P(\Delta_+,t) \end{pmatrix} . \tag{43}$$

α is the average frequency of transitions from Δ_+ to Δ_- and β is the average frequency of transitions from Δ_- to Δ_+. Let T_{Δ_+} and T_{Δ_-} denote the random waiting times in state Δ_+ and Δ_-, respectively. These random times are exponentially distributed, i.e.

$$p_{T_{\Delta_+}}(t) = \alpha e^{-\alpha t} , \quad p_{T_{\Delta_-}}(t) = \beta e^{-\beta t} . \tag{44}$$

The stationary solution of (43) is

$$p_s(\Delta_+) = \frac{\beta}{\gamma} , \quad p_s(\Delta_-) = \frac{\alpha}{\gamma} \tag{45}$$

with

$$\gamma = \alpha + \beta . \tag{46}$$

If the dichotomous noise is started with the probability (45), then it is stationary and

$$\langle Z_t \rangle = \frac{1}{\gamma} (\alpha\Delta_- + \beta\Delta_+) , \tag{47}$$

$$C(\tau) = \frac{\alpha\beta}{\gamma^2} (\Delta_+ - \Delta_-)^2 \exp(-\gamma|\tau|) . \tag{48}$$

Since we have to impose $\langle Z_t \rangle = 0$, according to (3), we find that

$$\alpha\Delta_- = -\beta\Delta_+ \tag{49}$$

has to hold. In [5] the equation for the probability density of a system subjected to dichotomous Markov noise has been derived:

$$\partial_t p(x,t) = -\partial_x f(x,\bar{\lambda}) p(x,t) + \alpha\beta\gamma^{-2} (\Delta_- - \Delta_+)^2 \, \partial_x g(x) \int_{-\infty}^{t} dt' \exp\left\{-[\gamma + \partial_x (f(x,\bar{\lambda}) + Ig(x))](t - t')\right\} \partial_x g(x) p(x,t') , \tag{50}$$

where $I = \gamma^{-1}(\beta\Delta_- + \alpha\Delta_+)$ and recall that $f(x,\bar{\lambda}) = h(x) + \bar{\lambda} g(x)$. Further, the operator ∂_x acts on everything to the right of it. The memory kernel in (50) indicates that the system is described by a non-Markovian process X_t. In fact, it holds that X_t is Markov if and only if Z_t is white. Contrary to the case of Ornstein-Uhlenbeck noise, (19), the stationary solution of (50) can be found in an exact explicit way:

$$p_s(x) = N \frac{g(x)}{(f(x)+\Delta_+ g(x))(f(x)+\Delta_- g(x))} \exp\left\{-\gamma \int^{x} \frac{f(y)}{(f(g)+\Delta_+ g(y))(f(y)+\Delta_- g(y))} dy\right\} \tag{51}$$

for $x \in U$ with $U = [\bar{x}(\bar{\lambda} + \Delta_-), \bar{x}(\bar{\lambda} + \Delta_+)]$ and $p_s(x) \equiv 0$ for $x \notin U$.

The white noise limit can be discussed directly in this case. The speeding up of the noise corresponds to letting γ go to infinity and the scaling up of the amplitude to letting Δ_+ go to infinity. If the noise is symmetric, i.e.

$$-\Delta_- = \Delta_+ \equiv \Delta \quad , \qquad \alpha = \beta = \frac{\gamma}{2} \quad , \tag{52}$$

then the Gaussian white noise limit, already discussed in section 2, is obtained if Δ^2/γ = const = $\sigma^2/2$. In the asymmetric case, Poisson white noise is obtained if $\Delta_+ \to \infty$, $\alpha \to \infty$ such that Δ_+/α = const = w [23]. The requirement (49) implies that $\Delta_- = -\beta w$. Note that the area of a Δ_+ pulse, $\Delta_+ \cdot T_{\Delta_+}$, is exponentially distributed, according to (44). Thus in the white noise limit, we have the following picture: The noise spends essentially all the time in the state $\Delta_- = -\beta w$. With an average frequency of β, this state is interrupted by a Dirac delta spike with a weight that is a random variable. The weights are exponentially distributed with mean value w. The Poisson white noise limit of (50) is

$$\partial_t p(x,t) = -\partial_x f(x,\bar{\lambda})p(x,t) + \beta w^2 \partial_x g(x)[1 + w\partial_x g(x)]^{-1}\partial_x g(x)p(x,t) \quad , \tag{53}$$

whose stationary solution is given by

$$p_s(x) = N[f(x,\bar{\lambda}) - \beta w g(x)]^{-1} \exp\left\{-\frac{1}{w}\int^x \frac{f(y,\bar{\lambda})}{g(y)[f(y) - \beta w g(y)]}\,dy\right\} \tag{54}$$

for $x \in U$. The equation for the extrema of (54) reads

$$[h(x_m) + \bar{\lambda}g(x_m)] - \beta w^2 g'(x_m)g(x_m) + w[h'(x_m) + \bar{\lambda}g'(x_m)]g(x_m) = 0 \quad . \tag{55}$$

Since Poisson white noise is bounded from below by $-\beta w$, contrary to Gaussian white noise, it is particularly useful in modeling situations where the parameter should remain positive for physical reasons. This can be achieved by imposing that $\bar{\lambda} > \beta w$. To illustrate the application of Poisson white noise, consider the simple photochemical model

$$\begin{aligned} &A + X \rightleftarrows 2X \\ &X + h\nu \rightarrow C \quad . \end{aligned} \tag{56}$$

Here again A is in large excess and C is immediately removed from the reactor. With suitable scaling, the kinetic equation for the concentration of X reads

$$\dot{x} = \lambda x - x^2 - kI(1 - e^{-\alpha x}) \quad . \tag{57}$$

I is the incident light intensity and α is the absorption coefficient times the sample thickness. Let us briefly discuss the steady states of (57) for nonfluctuating light intensity. $\bar{x} = 0$ is a solution of $\dot{x} = 0$ for all values of I. For large enough I, $\bar{x} = 0$ is stable. This is the expected result, since X is photochemically degraded. As I is decreased, $\bar{x} = 0$ becomes unstable and the intermediate can accumulate in the reactor. $\bar{x} = 0$ loses stability at

$$I_0^d = \frac{\lambda}{k\alpha} \quad . \tag{58}$$

At this point a new branch of nontrivial steady states bifurcates. A bifurcation analysis shows that this new branch can emerge subcritically or supercritically, depending on the value of λ. For $\lambda > \lambda_c^d$ with

$$\lambda_c^d = \frac{2}{\alpha} \quad , \tag{59}$$

the bifurcation is subcritical, i.e. the new branch exists for values of I greater than I_0^d. These states are unstable till the branch turns around at some I_u and becomes stable. We have thus for $I_0^d < I < I_u$ bistability; a nonzero stable steady state $\bar{x}$ coexists with the state $\bar{x} = 0$. For $\lambda < \lambda_c^d$ the bifurcation is supercritical; i.e. the new branch exists for values of I smaller than I_0^d. This branch is stable. Let us now study the effect of light intensity fluctuations on the loss of stability of the trivial branch $\bar{x}=0$ and on the point where the bifurcation changes from subcritical to supercritical behavior. Obviously, I should always be a positive quantity. Thus, Poisson white noise is more adequate than Gaussian white noise to model rapid fluctuations in the incident light intensity. Let $\bar{I}$ denote the mean value. Then to guarantee positivity, we require

$$\beta w < \bar{I} \quad .$$

Applying (55) to our model (57), we find that $x_m = 0$ ceases to be a maximum at

$$I_0 = \frac{\lambda}{k\alpha} - \frac{\beta w^2 k\alpha}{1-kw\alpha} \tag{60}$$

$$\approx I_0^d - \beta w^2 k\alpha + O(w^3) \quad . \tag{61}$$

The transition changes from subcritical to supercritical at

$$\lambda_c = \frac{2}{\alpha}\,[1 - 3kw\alpha - \beta w^2 k^2\alpha^2 + 2k^2w^2\alpha^3]\,[1 - kw\alpha]^{-1}\,[1 - 2kw\alpha]^{-1} \tag{62}$$

$$\approx \lambda_c^d - 2\beta w^2 k^2 \alpha^2 - 36k^2w^2\alpha + O(w^3) \quad . \tag{63}$$

It is interesting to compare these results to those we obtain if we use the, in this case "unphysical", Gaussian white noise:

$$I_0 = I_0^d - \frac{\sigma^2}{2} k\alpha \quad , \qquad \lambda_c = \lambda_c^d - \sigma^2 k^2 \alpha^2 \quad . \tag{64}$$

We see, as was to be expected from the form of (55), that qualitatively the same result is obtained, as long as the noise intensities are small. (We expect that at large noise intensities the particularities of the nature of the noise will be more pronounced.) In fact, Gaussian white noise underestimates the shift in I_0 and λ_c. It follows easily from the above expressions that in general Poisson white noise converges to Gaussian white noise, as $\beta \to \infty$, $w \to 0$ such that $\beta w^2 = \text{const} = \sigma^2/2$.

5. Conclusions

So far we have discussed noise-induced transitions only in the white noise limit. Naturally, the question arises if these noise-induced phenomena are robust. In other words, are essentially the same phenomena observed for τ_{cor} small, but nonvanishing? The answer is positive. Noise-induced transitions are *not* an artifact of white noise; they occur also for colored noise, i.e. for environments with nonzero correlation times. This has been established by various techniques, namely the wide band perturbation expansion [8,5], the approximate Fokker-Planck operator techniques [24], and approximate renormalized equations of evolution for Gaussian noise [25]. These methods are perturbation expansions and are limited to small τ_{cor}. In order to explore the dependence of noise-induced phenomena on the correlation time for a wider range the dichotomous Markov noise has been used [26,27,5], since the stationary probability density can be calculated exactly for any value of the correlation time and any value of the noise intensity. Not surprisingly, the transition behavior is even richer under colored noise than under white noise. Besides studies on the influence of colored noise, the following problems are currently investigated: i) the dynamics of the system [28-33], ii) systems with two or more variables [34-38], iii) the interaction between internal and external fluctuations [39,40], iv) the influence of fluctuations and stirring in a CSTR [41,42], v) the influence of external noise in spatially distributed systems [43,44], vi) the effect of periodic versus random variations [45,46], and vii) the effects of noise on deterministic chaos [47].

Let me conclude by listing the experimental studies on noise-induced transitions and by expressing the hope that their number will increase in the future. At present the theoretical part of the field of noise-induced transitions is more developed than its experimental side. There is a clear need for more experimental work. The first experiment in which a noise-induced transition was observed, was carried out by KABASHIMA, KAWAKUBO and coworkers [48]. They studied a nonlinear electrical circuit. The influence of fluctuating illumination on photochemical reactions was investigated experimentally by P. DEKEPPER and W. HORSTHEMKE [49] and by J. C. MICHEAU et al. [50]. Experiments on the effect of fluctuating electric fields on electrohydrodynamic instabilities in nematics have been conducted by various groups [51-53]. The experimental findings on the dye laser [54] were interpreted as noise-induced effects [18,19]. The first measurement of the critical exponents of a noise-induced critical point were carried out by MOSS and coworkers on a nonlinear electronic circuit [55].

Acknowledgement: This work was supported in part by grant NSF INT 8115672.

1. P.I. Kuznetsov, R.L. Stratonovic, V.I. Thikhonov: "The Effect of Electrical Fluctuations on a Valve Oscillator", in Nonlinear Transformations of Stochastic Processes, ed. by P.I. Kuznetsov, R.L. Stratonovic, V.I. Thikhonov (Pergamon, Oxford 1965) p. 223
2. R.M. May: Stability and Complexity in Model Ecosystems (University Press, Princeton, NJ 1973)
3. H.S. Hahn, A. Nitzan, P. Ortoleva, J. Ross: Proc. Natl. Acad. Sci. USA 71, 4067 (1974)
4. L. Arnold, W. Horsthemke, R. Lefever: Z. Phys. B29, 367 (1978)
5. W. Horsthemke, R. Lefever: Noise-Induced Transitions. Theory and Applications in Physics, Chemistry, and Biology. Springer Series in Synergetics 15 (Springer, Berlin 1984)
6. J.L. Doob : Ann. Math. 43, 351 (1942)
7. G. Blankenship, G.C. Papanicolaou: SIAM J. Appl. Math. 34, 437 (1978)
8. W. Horsthemke, R. Lefever: Z. Phys. B40, 241 (1980)
9. H. Haken: Synergetics. An Introduction. 2nd ed. (Springer, Berlin 1977)
10. G. Nicolis, I. Prigogine: Self-Organization in Nonequilibrium Systems (Wiley, New York 1977)
11. Yu.V. Prohorov, Yu.A. Rozanov: Probability Theory (Springer, Berlin 1969)
12. A.R. Bulsara, W.C. Schieve, R.F. Gragg: Phys. Lett. A68, 294 (1978)
13. J. de la Rubia, M.G. Velarde: Phys. Lett. A69, 304 (1978)
14. H. Brand, A. Schenzle: J. Phys. Soc. Jpn. 48, 1382 (1980)
15. M. San Miguel, J.M. Sancho: Z. Phys. B43, 361 (1981)
16. W. Horsthemke, C.R. Doering, R. Lefever, A.S. Chi: Phys. Rev. A (to appear)
17. F. Moss, G.V. Welland: Phys. Rev. A25, 3389 (1982)
18. R. Graham, M. Höhnerbach, A. Schenzle: Phys. Rev. Lett. 48, 1396 (1982)
19. S.N. Dixit, P.S. Sahni: Phys. Rev. Lett. 50, 1273 (1983)
20. R. Lefever, W. Horsthemke: Proc. Natl. Acad. Sci. USA 76, 2490 (1979)
21. K.A. Wiesenfeld, E. Knobloch: Phys. Rev. A26, 2946 (1982)
22. W. Ebeling, H. Engel-Herbert: Physica 104A, 378 (1980)
23. C. van den Broeck: J. Stat. Phys. 31, 467 (1983)
24. J.M. Sancho, M. San Miguel: Z. Phys. B36, 357 (1980)
25. K. Lindenberg, B.J. West: Physica 119A, 485 (1983)
26. K. Kitahara, W. Horsthemke, R. Lefever: Phys. Lett. A70, 377 (1979)
27. K. Kitahara, W. Horsthemke, R. Lefever, Y. Inaba: Prog. Theor. Phys. 64, 1233 (1980)
28. A. Schenzle, H. Brand: Phys. Rev. A20, 1628 (1979)
29. M. Suzuki, K. Kaneko, F. Sasagawa: Prog. Theor. Phys. 65, 828 (1981)
30. J.M. Sancho, M. San Miguel, S.L. Katz, J.D. Gunton: Phys. Rev. A26, 1589 (1982); J.M. Sancho, M. San Miguel, H. Yamazaki, T. Kawakubo: Physica 116A, 560 (1982)
31. P. Hänggi, P. Riseborough: Phys. Rev. A27, 3379 (1983); P. Talkner, P. Hänggi: Phys. Rev. A29, 768 (1984)
32. E. Ben-Jacob, D.J. Bergman, B.J. Matkowsky, Z. Schuss: Phys. Rev. A26, 2805 (1982)
33. E. Cortes, K. Lindenberg: Physica 123A, 99 (1984)
34. R. Graham: Phys. Rev. A25, 3234 (1982)
35. R. Lefever, J.W. Turner: in Fluctuations and Sensitivity in Nonequilibrium Systems, ed. by W. Horsthemke and D.K. Kondepudi, Springer Proceedings in Physics 1 (Springer, Berlin 1984)

36. E. Knobloch, K.A. Wiesenfeld: J. Stat. Phys. 33, 611 (1983)
37. F.J. de la Rubia, J. Garcia Sanz, M.G. Verlarde: in Nonlinear Stochastic Problems, ed. by R. Bucy, J.M.F. Moura (Reidel, Dordrecht 1983)
38. K.H. Hoffmann: Z. Phys. B49, 245 (1982)
39. M. San Miguel, J.M. Sancho: Phys. Lett. 90A, 455 (1982); M.A. Rodriguez, L. Pesquera, M. San Miguel, J.M. Sancho: preprint 1984
40. W. Horsthemke, R. Lefever: Phys. Lett. A (submitted)
41. L. Hannon: in Fluctuations and Sensitivity in Nonequilibrium Systems, ed. by W. Horsthemke and D.K. Kondepudi, Springer Proceedings in Physics 1 (Springer, Berlin 1984)
42. W. Horsthemke, L. Hannon: J. Chem. Phys. (in press); see also this volume
43. A.S. Mikhailov, L. Schimansky-Geier, W. Ebeling: Phys. Lett. 96A, 453 (1983)
44. A.S. Mikhailov: Phys. Lett. 73A, 143 (1979); Z. Phys. B41, 277 (1981); A.S. Mikhailov, I.V. Uporov: Sov. Phys. JETP 52, 989 (1980)
45. K. Lindenberg, V. Seshadri, B.J. West: Phys. Rev. A22, 2171 (1980)
46. C.R. Doering, W. Horsthemke: J. Stat. Phys. (Feb. 1985)
47. For a review, see: J.P. Eckmann: Rev. Mod. Phys. 53, 643 (1981)
48. S. Kabashima, T. Kawakubo: Phys. Lett. 70A, 375 (1979); S. Kabashima, S. Kogure, T. Kawakubo, T. Okada: J. Appl. Phys. 50, 6296 (1979)
49. P. DeKepper, W. Horsthemke: C. R. Acad. Sci. Paris C287, 251 (1978)
50. J.C. Micheau, W. Horsthemke, R. Lefever: J. Chem. Phys. (Sept. 1984)
51. S. Kai, T. Kai, M. Takata, K. Hirakawa: J. Phys. Soc. Jpn. 47, 1379 (1979)
52. T. Kawakubo, A. Yanagita, S. Kabashima: J. Phys. Soc. Jpn. 50, 1451 (1981)
53. H.R. Brand, S. Kai, S. Wakabayashi: preprint 1984
54. K. Kaminishi, R. Roy, R. Short, L. Mandel: Phys. Rev. A24, 370 (1981)
55. J. Smythe, F. Moss, P.V.E. McClintock: Phys. Rev. Lett. 51, 1062 (1983)

External Noise in Finite Systems

M. San Miguel [1,2*], M.A. Rodriguez [3], L. Pesquera [3], and J.M. Sancho [2]

[1] Physics Department, Temple University, Philadelphia, PA 19122, USA
[2] Departamento de Fisica Teorica, Universidad de Barcelona, Diagonal 647 E-Barcelona 28, Spain
[3] Departamento de Fisica Teorica, Universidad de Santander, Av. los Castros E-Santander, Spain

1. Introduction

Stochastically driven systems exhibit a variety of interesting nonequilibrium effects. These have been recently reviewed by HORSTHEMKE and LEFEVER [1] and also addressed by other authors in this workshop. In this contribution we focus our attention on the role played by the internal fluctuations of a system driven by an external noise [2,3,4]. External noise effects are usually studied in the thermodynamic limit in which internal fluctuations become negligible. This procedure assumes that the external driving noise completely dominates the fluctuations in the system. Nevertheless, a framework in which internal and external fluctuations are simultaneously considered is necessary to calculate finite size effects. Within such a framework a better understanding of the physical contents of "noise induced transition" phenomena [1] is obtained by investigating how changes in a stationary distribution induced by external noise are smoothed out by internal fluctuations. A major novel outcome of the unified theory of internal and external fluctuations presented here is the existence of "crossed-fluctuation" contributions which couple the two independent sources of randomness in the system.

2. External Noise in Master Equations

We consider a homogeneous system with internal fluctuations modeled by a one-step Markovian master equation.

$$\frac{\partial P(N,t)}{\partial t} = [(E^{-}-1)\, W_{+}(N) + (E^{+}-1)W_{-}(N)]P(N,t) \equiv \Gamma\left(\frac{\partial}{\partial N},N\right)P(N,t) \tag{2.1}$$

where $E^{\pm} = \exp \pm \frac{\partial}{\partial N}$. We assume extensive transition probabilities $W_{+}(N) = Vr_{+}(x)$, $W_{-}(N) = Vr_{-}(x)$ with $x = N/V$ and V the volume of the system. In the thermodynamic limit a deterministic description emerges

$$\dot{x}(t) = r_{+}(x) - r_{-}(x). \tag{2.2}$$

External noise is usually modeled by considering the existence of random parameters in (2.2). In a more detailed description we model external noise by random parameters in the transition probabilities of (2.1):

$$W_{+}(N) = W_{+,0}^{o}(N) + \alpha(t)W_{+,1}(N) = W_{+,o}(N) + \xi_{+}(t)W_{+,1}(N) \tag{2.3}$$

$$W_{-}(N) = W_{-,0}^{o}(N) + \beta(t)W_{-,1}(N) = W_{-,o}(N) + \xi_{-}(t)W_{-,1}(N) \tag{2.4}$$

where $\alpha(t) = \bar{\alpha} + \xi_{+}(t)$, $\beta(t) = \bar{\beta} + \xi_{-}(t)$ and $\xi_{+}(t)$, $\xi_{-}(t)$ are specified random processes. Eq. (2.1) with (2.3)-(2.4) is well defined as a master equation with stochastic transition probabilities as long as $W_{+}(N)$ and $W_{-}(N)$ are positive for all realizations of $\xi_{+}(t)$ and $\xi_{-}(t)$. This positivity requirement on $\alpha(t)$, $\beta(t)$ restricts the possible choices of $\xi(t)$. In particular, Gaussian white noise cannot be consistently considered. A consistent choice for $\xi_{+}(t)(\xi_{-}(t))$ is a dichotomic Markov process [3] with amplitude $\Delta_{+}(\Delta_{-})$ and inverse relaxation time $\lambda_{+}(\lambda_{-})$. The

* Permanant address.

positivity of $\alpha(t)$ ($\beta(t)$) is guaranteed whenever $\bar{\alpha} \geq \Delta_+$ ($\bar{\beta} \geq \Delta_-$). Another possible choice [4] is given by a Poisson white noise of zero mean: $\xi(t) = z^w(t) - \lambda \bar{w}$. The Poisson white noise $z^w(t)$ consists of a sequence of delta peaks at random points in time. The average time difference between two peaks is controlled by λ and the amplitude w of the peaks is distributed according to a probability density $\rho(w)$ with mean value $\bar{w}$. The process $\xi(t)$ is bounded from below, $\xi(t) \geq \lambda \bar{w}$. When $\xi_+(t)$ ($\xi_-(t)$) is given by a zero mean Poisson white noise characterized by $\lambda_+, \bar{w}_+$ ($\lambda_-, \bar{w}_-$) the stochastic master equation is meaningful for $\bar{\alpha} \geq \lambda_+ \bar{w}_+$, $\bar{\beta} \geq \lambda_- \bar{w}_-$

With the substitutions (2.3) (2.4) in (2.1), P(N,t) becomes a functional of $\xi(t)$. A unified description of fluctuations is given by a probability $\bar{P}(N,t)$ defined as the average of P(N,t) over the realizations of $\xi(t)$. The equation satisfied by $\bar{P}(N,t)$ is obtained averaging (2.1). In the case of a dichotomic Markov process $\xi_+(t)$ (with $\xi_-(t)=0$) we obtain an integrodifferential equation for $\bar{P}(N,t)$ [3]

$$\partial_t \bar{P}(N,t) = \Gamma_0 \bar{P}(N,t) + \Delta_+^2 \Gamma_{+,1} \int_0^t \exp[-(\lambda_+ + \Gamma_0)(t-t')] \Gamma_{+,1} \bar{P}(N,t')\, dt' \tag{2.5}$$

where Γ_0 is defined as Γ in (2.1) but with $W_+(N), W_-(N)$ replaced by $W_{+,0}(N), W_{-,0}(N)$ respectiveley and $\Gamma_{+,1} = (E^- - 1) W_{+,1}(N)$. In the thermodynamic limit $\Gamma_0 \to -\partial_x (r_{+,0}(x) - r_-(x))$, $\Gamma_{+,1} \to -\partial_x r_{+,1}(x)$ and (2.5) reduces to the standard description of external dichotomic noise [1]. When considering white Poisson noise processes for $\xi_+(t)$ and $\xi_-(t)$, $\bar{P}(N,t)$ still satisfies a Markovian master equation [4]

$$\begin{aligned} \partial_t \bar{P}(N,t) = \Gamma_0 \bar{P}(N,t) &+ \lambda_+ [\{e^{w_+ \Gamma_{+,1}}\}_{av} - \bar{w}_+ \Gamma_{+,1} - 1]\, \bar{P}(N,t) \\ &+ \lambda_- [\{e^{w_- \Gamma_{-,1}}\}_{av} - \bar{w}_- \Gamma_{-,1} - 1]\, \bar{P}(N,t) \end{aligned} \tag{2.6}$$

where Γ_0, $\Gamma_{+,1}$ are defined as above and $\Gamma_{-,1} = (E^+ - 1) W_{-,1}(N)$. The average $\{\ldots\}_{av}$ is over the realizations of the amplitudes $w_{+,-}$. An important difference between (2.6) and (2.1) is that as a result of the consideration of external noise, (2.6) includes nonvanishing effective transition probabilities $\bar{W}(N \to N \pm n)$ for any step size n. For example, for a multiplicative linear noise of the form $W_{\pm,1}(N) = a_\pm N$ we find[4]

$$\bar{W}(N \pm 1 \to N) = W_\mp (N \pm 1) + \lambda_\mp [\{\pm e^{-w_\mp a_\mp N} (1 - e^{\mp w_\mp a_\mp})\}_{av} (N \pm 1) - \bar{w}_\mp a_\mp (N \pm 1)] \tag{2.7}$$

$$\bar{W}(N \mp n \to N) = \lambda_\mp [\{(\pm)^n e^{-\bar{w}_\mp a_\mp N} (1 - e^{\mp w_\mp})^n\}_{av} (N \pm 1) \ldots (N \pm n)/ n!\,],\; n > 1 \tag{2.8}$$

The effect of white external noise in the master equation is then twofold: First, a modification of the one-step transition probabilities (2.7), and second the introduction of new transition probabilities for $n > 1$. In the thermodynamic limit, (2.6) becomes a partial differential equation for $\bar{P}(x,t)$ which involves derivatives of all orders with respect to x. It has the same form than (2.6) with the replacements $\Gamma_0 \to -\partial_x (r_+(x) - r_-(x))$, $\Gamma_{+,1} \to -\partial_x r_{+,1}(x)$, $\Gamma_{-,1} \to \partial_x r_{-,1}(x)$ [4]. An expansion around this limit permits a systematic calculation of finite size effects. Alternative formulations of (2.5) and (2.6) can be given in terms of the generating function and Poisson representation associated with (2.1).

A most interesting consequence of the unified description of fluctuations given by (2.6) shows up explicitly when considering the equations satisfied by the moments of $\bar{P}(N,t)$. For the sake of clarity, and although such equations can be written quite generally, here we restrict ourselves to the two first moments for the example considered in (2.7)-(2.8) and with $\lambda_+ = 0$. Inclusion of a fluctuating parameter in W_+ gives similar results. In terms of the variable x=N/V we find[4]

$$d_t \langle x \rangle = \langle r_{+,0}(x) - r_{-,0}(x) \rangle + \lambda_- (\{e^{-w_- a_-}\}_{av} - \bar{w}_- a_- - 1) \langle x \rangle \tag{2.9}$$

$$\begin{aligned} d_t \langle x^2 \rangle = 2 \langle x (r_{+,0}(x) - r_{-,0}(x)) \rangle &+ \lambda_- (\{e^{-2w_- a_-}\}_{av} - 2\bar{w}_- a_- - 1) \langle x^2 \rangle \\ + V^{-1} \langle r_{+,0}(x) + r_{-,0}(x) \rangle &- \lambda_- V^{-1} (\{e^{-2w_- a_-} - e^{-\bar{w}_- a_-}\}_{av} + \bar{w}_- a_-) \langle x \rangle \end{aligned} \tag{2.10}$$

Eq. (2.9) is the same that one obtains in the thermodynamic limit and it contains no contributions from internal fluctuations. The first term corresponds to the deterministic limit and the second one is the external noise contribution. The first two terms in the r.h.s. of (2.10) have the same origin than in (2.9) but there are additional contributions from internal fluctuations which scale with V^{-1}. The third term exists in the absence of external noise. The fourth term is a "crossed-fluctuation"contribution which couples external noise to the internal fluctuations of the system. This coupling effect is of special relevance for small systems and can only be obtained through a unified description of fluctuations. It can be seen that in general "crossed-fluctuations" contributions exist in the equation for $\langle x\rangle$ whenever $r_{+,1}(x)$ or $r_{-,1}(x)$ are nonlinear functions of x. In the equation for $\langle x^2\rangle$ in general "crossed-fluctuations" contributions only vanish for additive external noise. In this sense such contributions are a consequence of multiplicative (state dependent) external noise in a finite system. We note that this coupling effect also exists for external dichotomic Markov noise [3] and in the white noise limit of the theory [2]. It seems to be a consequence of any reasonable unified theory of fluctuations. We finally remark that this effect is not obtained by simply adding a new noise term to the stochastic differential equation which describes external noise in the thermodynamic limit.

3. Creation and Annihilation Process with Source

As an illustrative example of the general theory we consider a model of a nonequilibrium process with creation and annihilation of particles and a source term. Mathematical details are given elsewhere [4]. The model is defined by the chemical reactions [5]

$$A + X \rightleftharpoons C \qquad\qquad B + X \longrightarrow 2X \tag{3.1}$$

The chemical process is described by (2.1) with $W_+(N) = \alpha V + \gamma N$, $W_-(N) = \beta N$. The same mathematical model has also been used in the context of maser amplification [6] and nuclear reactor modeling [7]. We consider random fluctuations of the annihilation parameter β as given by (2.4) with $\xi(t)$ being a Poisson white noise.

The stationary distribution of the model is shown in Fig. 1 in several circumstances. In the absence of external noise one has a smooth function defined for all values of $N \geqslant 0$. This is rather different of the situation with external noise in the thermodynamic limit: We first consider the case $\bar\beta - \gamma - \lambda\bar w > 0$ for which the stochastic damping coefficient $\beta + \xi(t) - \gamma$ is always positive. The stationary distribution is then defined in the interval $(0,x_0)$, $x_0 = \alpha/(\bar\beta - \gamma - \lambda\bar w)$, and it exhibits a transition at $\lambda = \lambda_c = \bar\beta - \gamma - \lambda\bar w$. At this point in parameter space, the stationary distribution $\bar P_{st}(x_0)$ changes its value from 0 to ∞ (Fig. 1). This transition is modified by internal fluctuations in a system of volume V. The distribution is then defined for all valus of $x \geqslant 0$ and no sharp changes of $\bar P_{st}(x)$ are found. For any finite value of V the distribution goes smoothly to zero as $x \to \infty$ irrespective of the value of λ . However, for a fixed and large value of V a smooth but important change of $\bar P_{st}(x)$ still exists for $\lambda \simeq \lambda_c$. The height of the distribution at its maximum becomes significantly smaller when λ becomes larger than λ_c . If the value of V is decreased, internal fluctuations become dominant and destroy any remnant of the transition. For small values of V the stationary distribution is practically independent of the value of λ .

When the parameters of the system are such that $\bar\beta - \gamma - \lambda\bar w < 0$, a stationary distribution only exists for $\bar\beta - \gamma > 0$. This condition is the same in the thermodynamic limit than for a finite system. Also in both cases the stationary distribution is defined in the interval $(0,\infty)$ and no transition is found. Although the system reaches a steady state the fact that the damping coefficient has negative realizations manifests itself in the divergence of the stationary moments $\langle x^m\rangle_{st}$ for $m > m_0 = (\bar\beta - \gamma)/\bar w(\gamma - \bar\beta + \lambda\bar w)$. Again, the value of m_0 is independent of the system size. The formal expressions for $\langle x^m\rangle$ (any m) are the same for $\bar\beta - \gamma - \lambda\bar w \gtrless 0$. The mean value and relative fluctuations are given by

$$\langle x\rangle_{st} = \alpha/[\bar\beta - \gamma - \lambda\bar w + \lambda\bar w/(1+\bar w)] \tag{3.1}$$

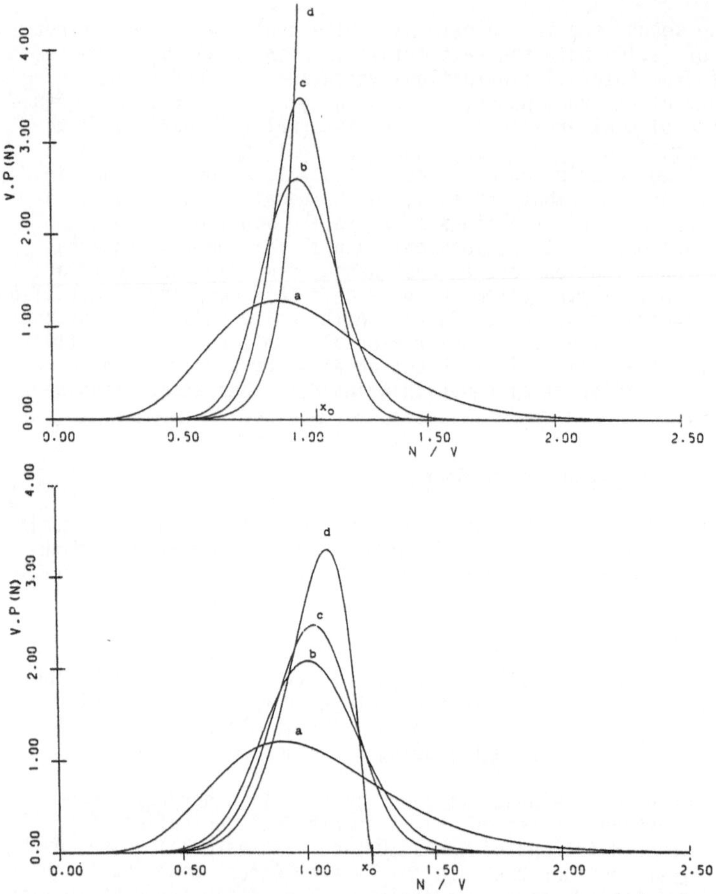

Fig. 1. Stationary distribution for $\alpha=0.1, \bar{\beta}=1, \gamma=0.9, \bar{w}=0.1$. a) V=100, b) V=500, c) V=1000, d) $V=\infty$. Above $\lambda=0.05$ ($\bar{\beta}-\gamma-\lambda\bar{w}>\lambda$). Below $\lambda=0.2$ ($\bar{\beta}-\gamma-\lambda\bar{w}<\lambda$).

$$\frac{\langle x^2\rangle_{st}-\langle x\rangle_{st}^2}{\langle x\rangle_{st}^2} = \frac{\bar{\beta}}{\alpha V} + \frac{\lambda\bar{w}^2}{(1+\bar{w})[(1+2\bar{w})(\bar{\beta}-\gamma) - 2\lambda\bar{w}^2]} + \frac{2\lambda^2\bar{w}^4 - \lambda\bar{w}^2[(1+2\bar{w})\bar{\beta} - 2(1+\bar{w})\gamma]}{\bar{\alpha}V(1+\bar{w})[(1+2\bar{w})(\bar{\beta}-\gamma) - 2\lambda\bar{w}^2]} \tag{3.2}$$

The mean value is independent of V. The first and second terms in the r.h.s. of (3.2) are the contributions from internal and external fluctuations respectively. The third term is the "crossed-fluctuation" contribution discussed earlier in general.

We finally note that similar conclusions about the interplay of internal and external fluctuations are obtained when considering fluctuations in the source (α) or creation (γ) parameters. Also in these cases a transition is smeared out by internal fluctuations. In these two cases the probability distribution in the thermodynamic limit is defined for values of x larger than a boundary value. A change of behavior at this boundary is found for a critical value of λ. Fluctuations of α and γ turn out to have qualitatively the same effects. They are qualitatively different from the previous case of β-fluctuations in which $P_{st}(x)$ was defined for $x < x_0$. It is interesting to note that if one considers external gaussian white noise the situation is very different. Because the realizations of gaussian noise are unbounded it is not possible to distinguish fluctuations of β from those of γ in the thermodynamic limit. The only difference then is between additive (α) or multiplicative (β,γ) noise. We also remark that none of the transitions described above exists in the gaussian white noise case. Nevertheless, we expect that the white noise limit of the theory accounts for the main finite size effects in the calculation of averaged quantities like mean values and relative fluctuations.

Acknowledgement: One of us (MSM) acknowledges financial support from NSF grant #DMR-8312958.

References

1. W. Horsthemke and R. Lefever: Noise Induced Transitions (Springer Verlag, Heidelberg, 1983).
2. M. San Miguel and J.M. Sancho: Phys. Lett. 90 A, 455 (1982).
J.M. Sancho and M. San Miguel in "Recent Developments in Nonequilibrium Thermodynamics". Eds. J. Casas-Vazquez and D. Jou. (Lecture Notes in Physics, Springer Verlag, Heidelberg, 1984).
3. J. M. Sancho and M. San Miguel: J. Stat. Phys. (1984).
4. M. A. Rodriguez, L. Pesquera, M. San Miguel and J.M. Sancho: "Master Equation Description of External Poisson White Noise in Finite Systems" (Preprint 1984).
5. C.W. Gardiner and S. Chaturvedi: J. Stat. Phys. 17, 429 (1977).
6. N.G. Van Kampen: Stochastic Processes in Physics and Chemistry (North Holland, Amsterdam, 1983).
7. M.M.R. Williams: Random Processes in Nuclear Reactors (Pergamon Press, Oxford, 1974).
M.A. Rodriguez, M. San Miguel and J.M. Sancho: Ann. Nucl. Energy 10, 263 (1983).

Hopf Bifurcation and Ripple Induced by External Multiplicative Noise

René Lefever and John Wm. Turner

Chimie Physique II, Université Libre de Bruxelles C.P. 231
B-1050 Bruxelles, Belgium

1. Introduction

The influence of external noise on nonequilibrium systems can advance or delay the onset of oscillatory behaviour. Without drawing much attention, this phenomenon was apparently first described in the field of radio engineering [1-3]. KUTZNETSOV et al. [1] in a paper on the valve oscillator remark that the amplitude of the oscillations tends to zero if the intensity of the noise exceeds a certain threshold. Conversely, studying numerically the effect of substrate input noise on an oscillatory enzymatic reaction, HAHN et al. [4] found that it may induce quasi-periodic behaviour under conditions where oscillations do not occur according to the deterministic equations.

More recently, while the interest in noise induced transitions is increasing (see e.g. [5-7] and the references cited therein), similar observations have been made by KABASHIMA et al. [8,9] and by DE KEPPER. In the first case, the system studied is an electrical parametric oscillator, the oscillations of which can be suppressed by turning on a random pumping current; in the second case, it is a light sensitive chemical oscillator (Brigg-Rauscher reaction) in which the light intensity fluctuations advance the occurrence of the oscillatory regime.

On the theoretical side, the influence of external multiplicative noise on a Hopf bifurcation has been investigated by EBELING and ENGEL-HERBERT [11] on various abstract model systems. These authors have shown that the direction in which the bifurcation is shifted by the noise depends on the model considered or on the parameter which is perturbed. The dynamics of fluctuations near a Hopf bifurcation with fluctuating control parameter has been analyzed by GRAHAM [12]. In this work, the external noise is added to the dimensionless parameters which appear in the universal normal form equation describing near a Hopf bifurcation, the deterministic dynamics on a long time scale. In that case, regarding the amplitude of the oscillations, it decreases with the intensity of the noise.

The results we summarize in this short note demonstrate, for a given chemical model (Brusselator) and a given fluctuating control parameter, that the direction of shift of a Hopf bifurcation induced by external noise can be changed by acting on the speed of the noise relative to the speed of rotation of the system on its deterministic limit cycle. We use for this purpose the perturbation scheme, which as indicated recently [13] provides a method for solving in a systematic way the Fokker-Planck equation describing the kind of problem at hand. We conclude the note by commenting, in the light of the results mentioned above, on the conditions under which pure noise induced transitions (ripples) occur in this system.

2. Influence of external noise speed on the direction of shift of a Hopf bifurcation

We consider the Brusselator chemical reaction scheme

$$A \longrightarrow X,\ 2X+Y \longrightarrow 3X,\ B+X \longrightarrow Y+D,\ X \longrightarrow E. \tag{1}$$

Its evolution equations read

$$\dot{X} = A-X+X^2Y-BX \qquad (2)$$

$$\dot{Y} = BX-X^2Y \qquad (3)$$

and admit the unique steady state solution

$$X_s = A, \; Y_s = B/A$$

which undergoes a Hopf bifurcation at

$$B = B_c = 1+A^2.$$

At the bifurcation point, the linear period of oscillation is equal to

$$T = A^{-1}.$$

We now examine the behaviour of this system when the parameter B fluctuates. Let us assume that

$$B_t = B_c+\varepsilon^2A^2(I+\frac{z_t}{A^{1/2}K}) \qquad (4)$$

where ε is a smallness parameter expressing the distance of the average $\langle B\rangle$ with respect to the deterministic bifurcation point, i.e.

$$\langle B\rangle-B_c = \varepsilon^2A^2I \qquad (5)$$

with I = 1, 0, -1 according to whether B_t is on the average above, at or below the deterministic Hopf bifurcation point. z_t is a colored noise with zero mean value. We take it to be the Ornstein-Uhlenbeck process given by the stochastic differential equation

$$dz_t = -\frac{\gamma z_t}{K^2} + \frac{\sigma}{K}dW_t. \qquad (6)$$

The correlation time and variance of the noise z_t are respectively

$$\tau_{cor} = K^2/\gamma \text{ and } Var = \sigma^2/(2\gamma). \qquad (7)$$

The scaling by the smallness parameter K in (4,6) is the appropriate procedure for taking the white noise limit of the noise or for studying its neighbourhood [5,14]. In the following we shall work in this limit, i.e. $K \rightarrow 0$, and without loss of generality let $\gamma = 1$. Introducing the time scale

$$\tau = At \qquad (8)$$

in which the unit time corresponds for $B = B_c$ to one period of the rotation in the deterministic system and introducing polar coordinates through the transformation

$$\begin{aligned} X &= A+\varepsilon Au^{1/2}\cos\theta, & Y &= \langle B\rangle/A+\varepsilon u^{1/2}(\sin\theta-A\cos\theta) \\ &= A+\varepsilon Au^{1/2}c & &= \langle B\rangle/A+\varepsilon u^{1/2}(s-Ac), \end{aligned} \qquad (9)$$

(2,3,6) can be rewritten as

$$\begin{aligned} d_\tau u &= 2\{\varepsilon[c^3(\tfrac{1}{A}-A)u^{3/2}+2c^2su^{3/2}]+\varepsilon^2(c^2IAu-c^4Au^2+c^3su^2)+\varepsilon^3(c^3IAu^{3/2})- \\ &\quad -\frac{z}{A^{1/2}K}(\varepsilon cAu^{1/2}+\varepsilon^2c^2Au)\} \\ &= \varepsilon f_1+\varepsilon^2f_2+\varepsilon^3f_3-z(A^{1/2}K)^{-1}(\varepsilon g_1+\varepsilon^2g_2) \\ &= f-z(A^{1/2}K)^{-1}g \end{aligned} \qquad (10)$$

$$\begin{aligned} d_\tau\theta &= -1+\varepsilon[c^2s(A-\tfrac{1}{A})u^{1/2}-2cs^2u^{1/2}]+\varepsilon^2(-csIA+c^3sAu-c^2s^2u)+\varepsilon^3(-c^2sIAu^{1/2})+ \\ &\quad +\frac{z}{A^{1/2}K}(\varepsilon\frac{sA}{u^{1/2}}+\varepsilon^2csA) \end{aligned}$$

$$= -1+\varepsilon h_1+\varepsilon^2 h_2+\varepsilon^3 h_3+ \frac{z}{A^{1/2}K}(\varepsilon\tilde{g}_1+\varepsilon^2\tilde{g}_2)$$

$$= h+z(A^{1/2}K)^{-1}\tilde{g} \tag{11}$$

$$dz_\tau = -z_\tau/(AK^2)d\tau+\sigma/(A^{1/2}K)dW_\tau . \tag{12}$$

The Fokker-Planck equation giving the probability density of the triple (u,θ,z) is

$$\partial_\tau p(u,\theta,z) = [-\partial_u(f- \tfrac{z}{\eta}g)-\partial_\theta(h+ \tfrac{z}{\eta}\tilde{g})+ \frac{1}{\eta^2}(\partial_z z+ \tfrac{\sigma^2}{2}\partial_{zz})] p(u,\theta,z), \quad \eta = A^{1/2}K \tag{13}$$

In the following we shall analyze the stationary properties of (13) under conditions where the time scale corresponding to the relaxation towards the limit cycle is longer than the other two characteristic time scales of the system, i.e.

$$\tau_{macro} = (\varepsilon^2A^2)^{-1}>>\tau_{cor} \text{ and } T.$$

We shall more particularly be interested in the effect of the speed of the external noise on the noise-induced Hopf bifurcation. We consider successively the two cases where the correlation time of the noise is shorter than the period of oscillation T or longer.

2.1 The correlation time of the noise is shorter than the period of oscillation

In this case

$$\tau_{macro}>> T>>\tau_{cor} \text{ or } (\varepsilon^2A^2)^{-1}>>\frac{1}{A}>>K^2 \tag{14}$$

so that

$$\eta = A^{1/2}K = (\tau_{cor}/T)^{1/2}<<1 \tag{15}$$

is a smallness parameter. We write the stationary equation of (13) in term of the powers of η

$$\frac{1}{\eta^2}(\partial_z z+\frac{\sigma^2}{2}\partial_{zz})p_s(u,\theta,z) = [\frac{z}{\eta}(-\partial_u g+\partial_\theta\tilde{g})+\partial_u f+\partial_\theta h]\, p_s(u,\theta,z) \tag{16}$$

and expand the solution of (16) as

$$p_s(u,\theta,z) = p_o'(u,\theta,z)+\eta p_1'(u,\theta,z)+\dots . \tag{17}$$

The perturbation scheme set up in this manner is exactly the wide band perturbation method described elsewhere [5,14]. To the lowest order in η, the solution of (16) can straightforwardly be written as

$$p_o'(u,\theta,z) = p_s(z)p_s''(u,\theta) \tag{18}$$

where $p_s(z)$ is the stationary probability density of the Ornstein-Uhlenbeck process and $p_s''(u,\theta)$ is the solution of

$$\{\frac{\sigma^2}{2}(\partial_{uu}g^2-2\partial_{\theta u}g\tilde{g}+\partial_{\theta\theta}\tilde{g}^2)-\partial_u[f+ \frac{\sigma^2}{2}((\frac{\partial g}{\partial u})g-(\frac{\partial g}{\partial\theta})\tilde{g})] -\partial_\theta[h+ \frac{\sigma^2}{2}((\frac{\partial\tilde{g}}{\partial\theta})\tilde{g}-(\frac{\partial\tilde{g}}{\partial u})g] \}p_s'' = 0. \tag{19}$$

(18,19) are intuitively evident: the limit $\eta \longrightarrow 0$ amounts to replacing z_t by white noise in (10,11).

To solve (19) we now make use of the smallness parameter which remains at our disposal and expresses the slowness of the radial motion towards the limit cycle. We thus write (19) and its solution in the form of an expansion in ε:

$$(F_o+\varepsilon F_1+\varepsilon^2F_2+\varepsilon^3F_3+\varepsilon^4F_4)p_s''(u,\theta) = 0, \tag{20}$$

$$p_s''(u,\theta) = p_0''(u,\theta)+ \varepsilon p_1''(u,\theta)+\dots \quad (21)$$

After solving (20,21) up to the order ε^2 (calculations therefore must be pushed up to the order ε^4), one conclydes that the reduced stationary probability

$$p_0(u) = \int_0^{2\pi} p_s''(u,\theta)d\theta \quad (22)$$

is given by (for A = 1)

$$p_0(u) = \{1+\varepsilon^2[2C+Q(u)] \}\phi(u) \quad (23)$$

where

$$\phi(u) = N\exp\frac{1}{\sigma^2}(Iu- \frac{3u^2}{8}), \quad (24)$$

$$Q(u) = \frac{77}{24}u+ \frac{4I}{\sigma^2}u^2+(\frac{25I^2}{36\sigma^4} - \frac{239}{72\sigma^2})u^3- \frac{63I}{64\sigma^4} + \frac{57}{160\sigma^4}u^5 \quad (25)$$

and

$$C = - \frac{1}{2}\int_0^{\infty} Q(u)\phi(u)du. \quad (26)$$

In the limit $\sigma^2 \ll 1$ and with I = 1, (23) admits a maximum u_m, located at

$$u_m \simeq \frac{4}{3} - \frac{760}{81}\varepsilon^2 + \frac{\sigma^2}{2}(\frac{77}{36}\varepsilon^2). \quad (27)$$

Accordingly, under the conditions (14) the external noise on B advances the onset of the oscillatory regime as compared to the deterministic situation.

2.2 The correlation time of the noise is longer than the period of oscillation

Inequalities (14) are now replaced by

$$\tau_{macro} \gg \tau_{cor} \gg T \text{ or } (\varepsilon^2 A^2)^{-1} \gg K^2 \gg \frac{1}{A}, \text{ with } A \gg 1. \quad (28)$$

The appropriate parameter for expanding the Fokker-Planck equation and its solution is now

$$\eta' = \varepsilon A. \quad (29)$$

At the lowest order in η', the solution is circularly uniform, i.e. one has

$$p_s(u, ,z) = \frac{p_0'(u,z)}{2\pi}$$

where $p_0'(u,z)$ is the solution of

$$\partial_u [2\varepsilon(Iu- \frac{3u^2}{4} - \frac{z}{\eta} u)] p_0'(u,z) - \frac{1}{\eta^2}(\partial_z z+ \frac{\sigma^2}{2}\partial_{zz})p_0'(u,z) = 0. \quad (30)$$

Since

$$\varepsilon \ll \frac{1}{\eta} \text{ and } \frac{1}{\eta} = \frac{T}{\tau_{cor}} \ll 1,$$

we look for a solution of (30) in the form of an expansion in powers of η. This yields at the lowest order

$$p_0'(u,z) = p_s(z)\phi(u) \quad (31)$$

where $\phi(u)$ is simply the solution of the Fokker-Planck equation of the one variable Verhulst model, namely

$$\cdot[2\varepsilon\frac{\sigma^2}{2}\partial_u u\partial_u u - \partial_u(Iu - \frac{3u^2}{4})]\phi(u) = 0.$$

It is well known that in that case the bifurcation is delayed by a shift proportional to the intensity of the noise.

3. Comment on noise induced ripples

We have shown using a Galerkin type of resolution scheme, that when the kinetic constant of the trimolecular step in (1) fluctuates, new pure noise induced transitions become possible. For the sake of clarity let us recall for example that when A = 2, the extrema of the probability density behave as represented in figure 1. u_m is plotted as a function of σ^2 and for the values of ε^2 indicated. The curves labelled 1 correspond to the situation typical above the Hopf bifurcation. For small intensities (curve in the lower left corner) one sees that the noise suppresses the extremum corresponding to the usual limit cycle; this is the same behavior as in 2.2. Its amplitude which is equal to one (in normalizing with respect to the deterministic limit cycle, i.e. 4/3) when $\sigma^2 = 0$, rapidly drops to zero for $\sigma^2 = 0.1$. Between 0.1 and ≈0.6, no other extremum is found. For $\sigma^2 > 0.6$, a new branch of physically acceptable extrema appears. The probability density exhibits a ripple suggesting that a new kind of sustained periodic regime sets in. This branch of solutions does not disappear when ε^2 is equal to zero or even negative (but in absolute value smaller than B_c). As ε^2 diminishes this pure noise induced branch of extrema shifts to lower values of σ^2, reaches the abscissa and finally the direction of the bifurcation reverses.

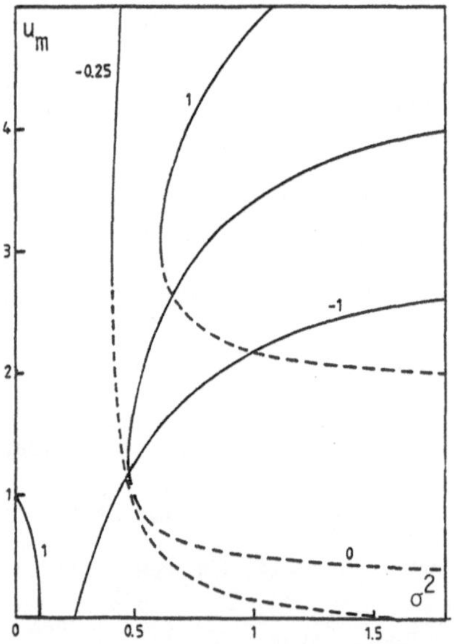

Figure 1: extrema of the probability density when the trimolecular constant fluctuates. The values of ε^2 are indicated on the curves.

It would be possible to verify these predictions using the perturbation expansions of the preceding section. Owing to their length, such calculations however are hardly feasible when the trimolecular constant fluctuates. Nevertheless the results above indicate conditions for which the Galerkin procedure employed in [13] can be expected to converge rapidly towards the exact solution. Indeed, this procedure is based on the assumption that to lowest order the stationary probability density is nearly circularly uniform. We saw in 2.2 that this is the case when A >> 1 and when the correlation time of the noise is longer than the period of the deterministic limit cycle.

References

1. P. I. Kutznetsov, R. L. Stratonovich, V. I. Thikhonov: "The effect of electrical fluctuations on a valve oscillator," in Nonlinear transformations of stochastic processes, ed. by P. I. Kutznetsov, R. L. Stratonovich, V. I. Thikhonov (Pergamon, Oxford 1965) p. 223
2. I. N. Amiantov, V. I. Thikhonov: "The response of typical nonlinear elements to normally fluctuating inputs," ibid p. 175
3. R. L. Stratonovich, P. S. landa: "The effect of noise on an oscillator with fixed excitation," ibid p. 259

4. H. S. Hahn, A. Nitzan, P. Ortoleva, J. Ross: Proc. Natl. Acad. Sci. USA 71, 4067 (1974)
5. W. Horsthemke, R. Lefever: Noise induced transitions. Theory and applications in physics, chemistry and biology. Springer Ser. Synergetics, vol. 15 (Springer, 1979)
6. A. Schenzle, H. Brand: Phys. Rev. A20, 1628 (1979)
7. J. M. Sancho, M. San Miguel, S. L. Katz, J. D. Gunton: Phys. Rev. A26, 1589 (1982)
8. S. Kabashima, T. Kawakubo: Phys. Lett. A70, 375 (1979)
9. S. Kabashima, S. Kogure, T. Kawakubo, T. Okada: J. Appl. Phys. 50, 6296 (1979)
10. P. De Kepper, private communication
11. W. Ebeling, H. Engel-Herbert: Physica 104A, 378 (1980)
12. R. Graham: Phys. Rev. A25, 3234 (1982)
13. R. Lefever, J. Wm. Turner: "Sensitivity of a Hopf bifurcation to external multiplicative noise", to appear in Lect. Notes Phys. (1984)
14. W. Horsthemke, R. Lefever: Z. Phys. B40, 241 (1980)

Inhomogeneities Induced by Imperfect Stirring in a C.S.T.R. Effect on Bistability

J. Boissonade, J.C. Roux, H. Saadaoui, and P. De Kepper

Centre de Recherche Paul Pascal, Domaine Universitaire
F-33405 Talence Cêdex, France

1. Introduction

During the past ten years the C.S.T.R. has been considered as the fundamental tool to study temporal behaviour of isothermal *homogeneous* chemical reactions maintened far from equilibrium. The control parameters generally are the concentrations in the input feed and the pumping rate $k_E = \frac{1}{\tau}$, where τ is the residence time of the reactor. The steady states are characterized by the concentrations of all species in the reactor. Bistability [1] and various types of oscillations [2] have been the basic phenomena explored by this technique. A system is *bistable* when there are two different steady states for the same set of control parameter values. The bistability range extends necessarily on a finite range of k_E values since at very large flows τ is short with respect to the reaction time and the only steady state has the concentrations of the feed. The set of steady states which extends to $k_E = \infty$ will be refered as the "flow branch", and the other one as the "thermodynamic branch" (we include exotic cases [3]). Figure 1 displays a typical bistability in the range $k_i < k < k_s$ (S or Z shape as well as positive or negative slopes of both branch are possible according to the specific system and the species chosen to characterize the steady state). Recent experiments [4,5,6] have shown that, even at stirring rate usually considered as sufficient to guarantee homogeneity the bistability of the chlorite iodide reaction is very sensitive to this stirring rate. The two main outcome of these experiments are : i) When the stirring rate is decreased the critical switching value k_s from the thermodynamic branch to the flow branch is dramatically shifted to lower values of the flow and the thermodynamic branch moves towards the flow branch as shown on Fig.1 of ref.[6] ii) When performing the supercritical perturbation the transition does not occur immediately, but the system remains on the initial branch during a time statistically distributed and long compared to the residence time. This could suggest that the initial state is metastable. Various interpretations have been proposed to this effect of imperfect mixing. At first, it was suggested [4] that, if one considers turbulent mixing as an enhanced diffusion, these results are an experimental evidence for a nucleation process, in agreement with stochastic theories initially developed by PRIGOGINE and NICOLIS [7] and NITZAN et al. [8] (for more references see ref. [4] and [5]). In this theoretical frame, transition occurs when a fluctuation has a sufficient size to drive the whole system out of the initial state. Large diffusion rates favour damping of spontaneous fluctuations, increasing the required critical size. These theories

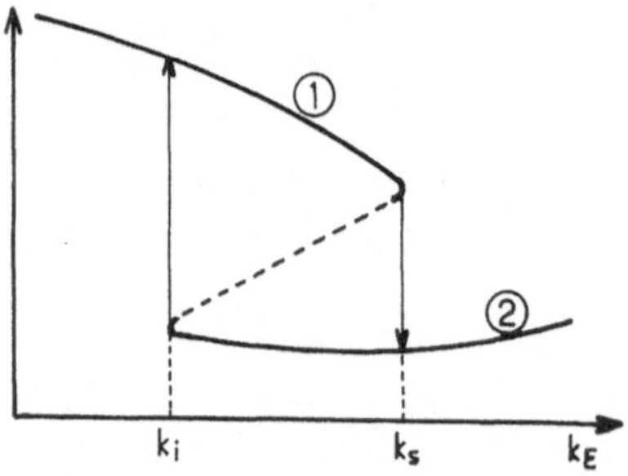

Fig. 1 Typical bistability in a C.S.T.R.
① : thermodynamic branch
② : flow branch

predict a decrease of the effective bistability range when diffusion decreases and a metastability related to the time for a critical fluctuation to form. Such interpretations could be supported by recent experiments by MENZINGER [9] but do not account for the shift of the thermodynamic branch and other experimental features reported in ref. [5]. HORSTHEMKE and HANNON [10,11] interpret the results with a different stochastic theory, based on the dynamics of turbulent eddies in the reactor.

Besides these stochastic interpretations, deterministic interpretations are presently developed: GRAY [12], KUMPINSKY and EPSTEIN [13] , propose systemic approaches, commonly used in chemical engineering : several ideal reactors are coupled by conservative flows with expandable coefficients, so that by-passes or dead zones may be taken into account. NICOLIS and FRISCH [14] use a quasi-Semenov equation in the limit of large diffusion coefficients and obtain a renormalization of k_E. DEWEL et al. [15] use a phenomenological theory of turbulent mixing to study surface effects produced by the feed of the reactor.

With the exception of nucleation theory, all the interpretations are related to inhomogeneous perturbations by the feed. We propose a very simple interpretation of the feed effect supported by numerical simulations.

2. General Scheme

We shall consider that the velocity field can be described as the superposition of a convective velocity field and a diffusion expressing the rate of turbulent mixing. For perfect mixing the diffusion coefficient D goes to infinity,so that the system is homogeneous. The injection has generally a small diameter d_i. Except for small flow rates,the velocity field is extremely large in the vicinity of the inlet,so that the convective transport is large in regard to the diffusive part. If we consider a small volume, typically of size d_i, localized around the port, the concentration inside are those of the feed or at least this small volume can be considered as a small reactor with very large flow rates. In both cases, the stationary state of this volume belongs to the flow branch. In some respect it plays the role of a fluctuation in the nucleation theory. If the turbulent diffusion is not two large, this small "nucleus" tends to expand (or equivalently d_i is longer than the critical size) and drive the bulk into the stationary state on the flow branch. When the stirring rate is decreased, the flow branch is favoured. One could speak of a "forced nucleation" but this term is somehow misleading since the process is purely deterministic and the local perturbation due to the feed is permanent. The validity of the conclusions implies that the input flow is large enough to define the small volume above. This is certainly the case close to k_s, where the flow becomes dominant compared to the reaction processes on a macroscopic scale. The transition is shifted to lower values of the flow rate. For the reverse transition at k_i, where the flow is often very small and where the reaction process becomes dominant no definite conclusion can be drawn. We shall now develop these ideas on simple models and show that localized injection of reactants induces effects in qualitative agreement with the experiments.

3. The Model

We use a three variables model, proposed previously [16] to account for a competition between an autocatalytic step an a non autocatalytic one :

$$\begin{aligned} B + X &\xrightarrow{k_1} 2X \\ D + X &\xrightarrow{k_2} \text{Products} \end{aligned} \qquad \text{(M.1)}$$

If B,D,X are the concentrations inside the reactor, B_0, D_0, X_0 the concentrations in the feed, and using mass action kinetics,the homogeneous evolution equations written:

$$\frac{dX}{dt} = (k_1 B - k_2 D) X + k_E (X_0 - X)$$

$$\frac{dB}{dt} = - k_1 B X + k_E (B_0 - B) \qquad (1)$$

$$\frac{dD}{dt} = - k_2 D X + k_E (D_0 - D)$$

It was shown [16] that there are two stable stationary solutions for a proper choice of the control parameter values. With $X_0 = 20$, $B_0 = 62.5$, $D_0 = 750$, the system is bistable for $0 \leqslant k_E < 1.68$[1]. These values are used in the rest of the work. One and two dimensional model-reactors were used. In a one dimensional reactor of length ℓ, the convection velocity is all along equal to $\frac{\ell}{\tau}$. If the space is discretized into N identical cells, the equation for species C in cell i is :

$$\frac{dC_i}{dt} = R_i + D(C_{i-1} + C_{i+1} - 2C_i) - k_E^{'} (C_i - C_{i+1}) + \text{source terms} \qquad (2)$$

where R_i is the reaction term for an homogeneous system, D a renormalized diffusion coefficient, and $k_E^{'} = N\, k_E$. To take into account the *local* character of the feed the source terms are also localized and equal to i) $k_E^{'} C_0$ in cell 1 (input cell), ii) $- k_E^{'} C_N$ in cell N (output cell), iii) 0 elsewhere (Fig. 2a). In our numerical experiments N = 16. In a multidimensional reactor, the convection velocity depends on the geometry of the reactor and on the stirring process. In order to produce a qualitative description of the turbulent mixing, we represent convection and turbulent diffusion by a generalized non-uniform diffusion process. The two dimensional model-reactor is represented by a square array of N cells, with periodic boundary conditions (N = 16 in the computations) and source terms defined as previously within the input and output cells according to Fig. 2b. Equation (1) becomes here :

$$\frac{dC_i}{dt} = R_i + D_j \sum_j C_j - 4D_i C_i + \text{Source terms} \qquad (3)$$

where the j are the indexes of the first neighbouring cells and the D_i are the generalized diffusion coefficients deduced from the linear conservation equations :

$$4D_i - \sum_j D_j = \text{Flow source terms} \qquad (4)$$

They take the form $D_i = d + \alpha_i k_E$ where α_i are constants depending only on the cell and d is an expandable parameter giving the rate of turbulent diffusion.

Fig. 2a Model of 1-D reactor

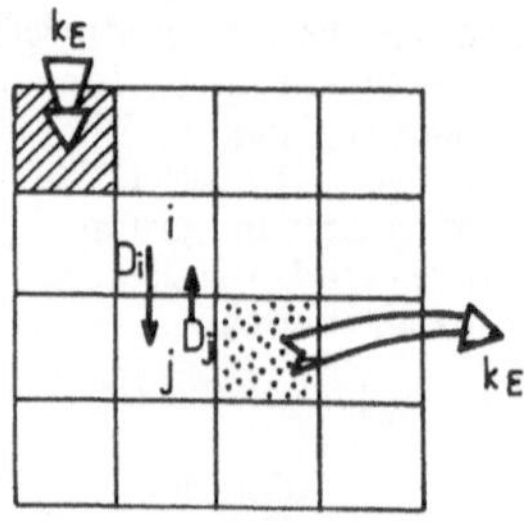

Fig. 2b Model of 2-D reactor

The dynamical equations of both reactors were solved on a computer for various mixing rates and the stationary states carefully determined. The states of the system were characterized by the concentrations averaged over all the cells.

[1] *Due to absence of reverse reactions and absorbing boundary limit at* $X = 0$, *bistability extends to* $k_E = 0$.

4. Results

For the one-dimensional reactor, the computed bistability range as a function of k_E is displayed on figure 3 for several values of the diffusion coefficient D. The results are in perfect qualitative agreement with the experiments : the transition point k_S is shifted to lower flow values when D decreases. Moreover, close to the transition point, stationary states on the thermodynamic branch deviates from the homogeneous case values, a behaviour not predicted by nucleation theory.

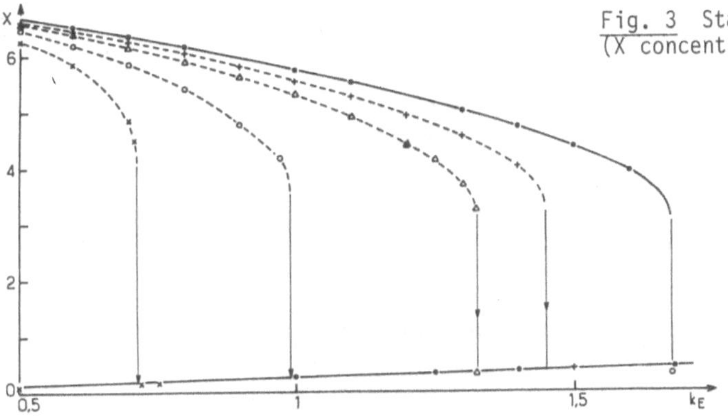

Fig. 3 Stationary state. Model M.1 (X concentration - 1-D reactor)

—●— D = ∞
+ D = 500
△ D = 250
o D = 100
X D = 50

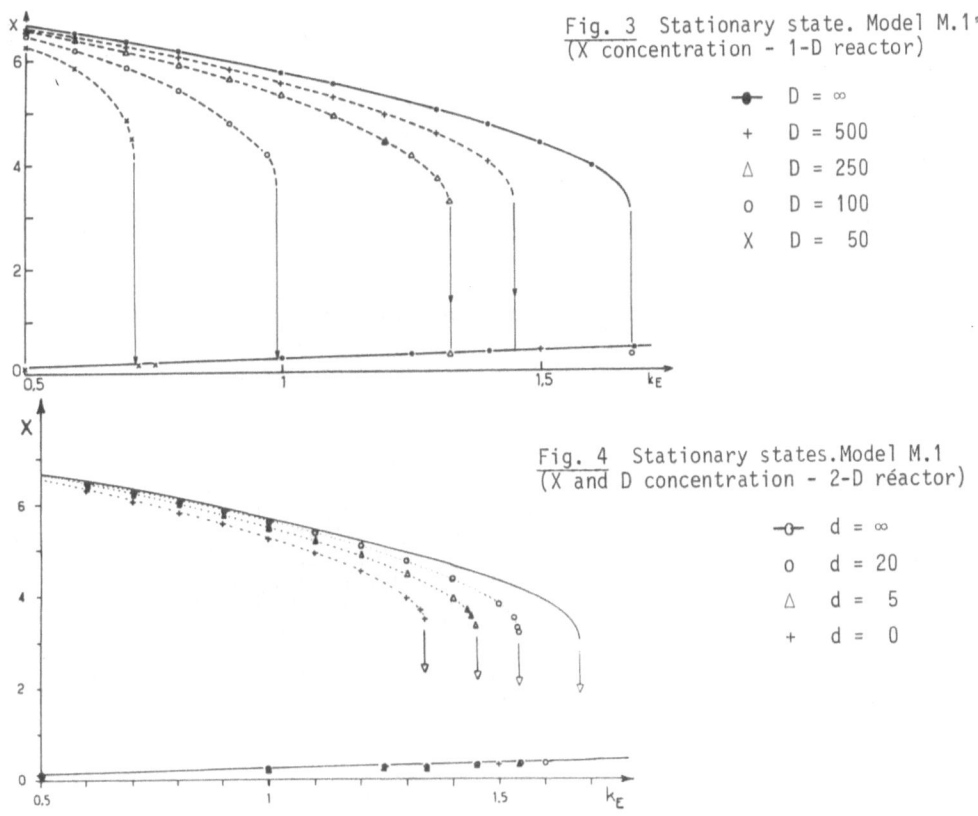

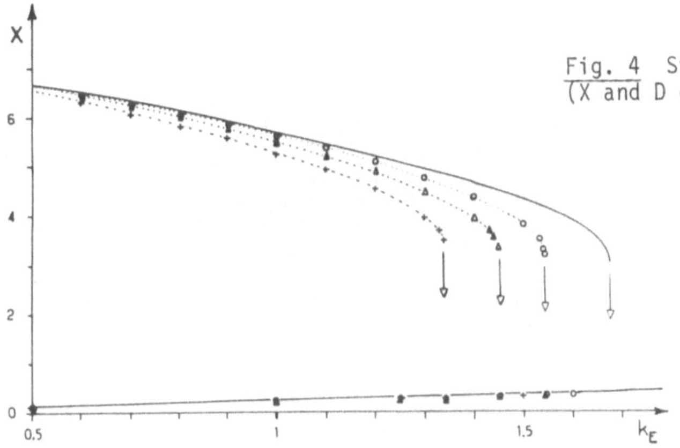

Fig. 4 Stationary states.Model M.1 (X and D concentration - 2-D réactor)

—o— d = ∞
o d = 20
△ d = 5
+ d = 0

On figure 4 analog results are obtained for the two dimensional reactor where d is varied from 0 to ∞. As could be expected the feed effect is found less dramatic than in the one dimensional reactor, since the introduced perturbation is more easily damped with increasing dimensionality. Computations have also been performed in the one-dimensional reactor for the reaction

$$A + 2X \rightleftarrows 3X \qquad (M.2)$$

where A and was supposed to be in large excess. The homogeneous equation is a cubic rate law :

$$\frac{dX}{dt} = k_+ X^2 - k_- X^3 + k_E (X_0 - X) \qquad (5)$$

With $k_+ = k_- = 9$ and $X_0 = 0.03$ this system is bistable for $0.98 < k_E < 2.42$ .The stationary states for D = ∞ and D = 100 are represented on figure 5. Similar results to those of model M.1 were obtained. Moreover the reverse transition at k_i was also found shifted to lower flow values.

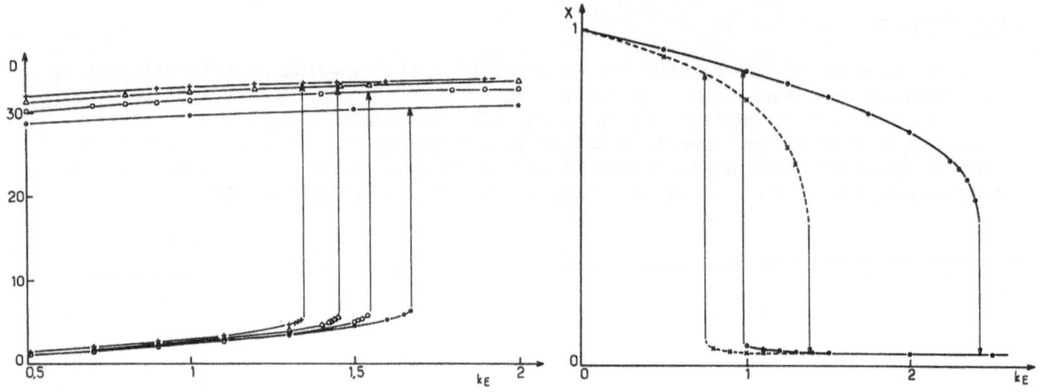

Fig. 5 Stationary states. Model M.2 (1D reactor)

— D = ∞

--- D = 100

A major argument in favour of a stochastic nucleation theory was the experimental observation of metastable states. DEWEL et al. [17] have shown that, due to critical slowing down, after a slightly supercritical perturbation the concentrations values first decay very slowly and present a long plateau close to the values of the initial state before switching rapidly to the final stable state. This type of relaxation has also been observed in our simulations: we present on Fig. 6 examples of the response to two supercritical perturbation obtained for model M.1 with $d = 20$. For these conditions the transition point is $k_s = 1.542$. The initial state was $k_E = 1.53$, starting on the thermodynamic branch. A jump to $k_E = 1.543$ at $t = 0$ induces a plateau (curve II) longer than 30 τ before switching. Such long relaxation times have been also observed experimentally by PIFER et al. [18]. If the flow jump is slightly larger i.e. to $k_E = 1.55$, the plateau disappears and the system switches rapidly (curve I). Thus, less than 1 % change in the values of the flow may induce dramatic variations in the relaxation times. Since, typically, in stirring experiments [4-6], the pumping rate is not more reliable, a rather large distribution is produced in these relaxation times in repeated experiments but this apparent "metastability" does not imply any nucleation or stochastic process.

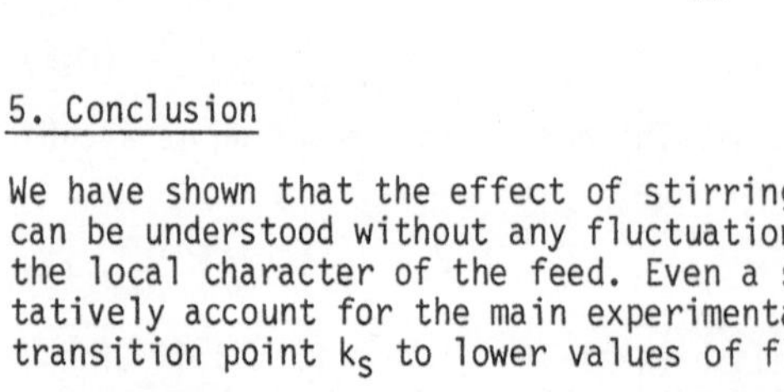

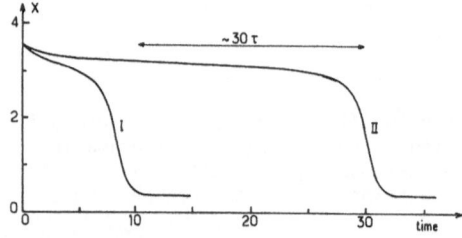

Fig. 6 Relaxation to a supercritical perturbation (X concentration ; 2-D reactor)

II $k_E \to 1.543$

I $k_E \to 1.55$

5. Conclusion

We have shown that the effect of stirring on bistable chemical systems in a C.S.T.R. can be understood without any fluctuation process. Inhomogeneities originate in the local character of the feed. Even a simple reaction-diffusion scheme can qualitatively account for the main experimental facts, in particular the shift of the transition point k_s to lower values of flow when the stirring rate is decreased.

We have mentioned that the argument for a "forced nucleation" process falls at low flows, in particular for the reverse transition at k_i and that no general conclusion can be drawn without the complete kinetical equations. Recent experiments [13] support this statement : a shift of the reverse transition point k_i to larger flows was found in the minimal bromate oscillator.

All these results show that the non ideality of real flow reactors has to be considered carefully in experimental work on "homogeneous" reactions. The "mixing rate" cannot be simply dismissed as a small side-effect but has sometimes to be retained as a control parameter.

This work has been supported by grant 831332 of the Direction des Recherches, Etudes et Techniques.

References

1. P. De Kepper and J. Boissonade, chapter 7 in ref. [2]
2. Oscillations and traveling waves in chemical systems, Ed. R.J. Field and M. Burger, Wiley, in press
3. P. Gray and S.K. Scott, Chem. Eng. Sci. 38, 29 (1983)
4. J.C. Roux, P. De Kepper and J. Boissonade, Phys. Lett. 97A, 168 (1983)
5. J.C. Roux, H. Saadaoui, P. De Kepper and J. Boissonade, in "Fluctuations and Sensitivity in non equilibrium systems", Ed. W. Horsthemke and D.K. Kondepudi, Springer Proc. Phys. 1, 70 (1984)
6. H. Saadaoui, J.C. Roux, P. De Kepper and J. Boissonade, this volume (1984)
7. I. Prigogine and G. Nicolis, Proc. 3rd Int. Conf. "From Theoretical Physics to Biology", p. 89, Karger (1973)
8. A. Nitzan, P. Ortoleva and J. Ross, Faraday Symp. Chem. Soc. 9, 241 (1975)
9. M. Menzinger, this volume (1984)
10. W. Horsthemke and L. Hannon, J. Chem. Phys., in press (1984)
11. W. Horsthemke and L. Hannon, this volume (1984)
12. P. Gray, this conference, short communication, unpublished (1984)
13. E. Kumpinsky and I. Epstein, preprint (1984)
14. G. Nicolis and H. Frisch, submitted to Phys. Rev. A (1984)
15. G. Dewel, P. Borckmans and D. Walgraef, preprint
P. Borckmans, G. Dewel and D. Walgraef, this volume (1984)
16. J. Boissonade, J. Chim. Phys. 73, 740 (1976)
17. G. Dewel, P. Borckmans and D. Walgraef, preprint and this volume (1984)
18. T. Pifer , N. Ganapathisubramanian and K. Showalter, this volume (1984)

Turbulent Mixing and Nonequilibrium Chemical Instabilities: The Effect of Reactant Streams in a CSTR

W. Horsthemke and L. Hannon

Center for Studies in Statistical Mechanics, Department of Physics
University of Texas, Austin, TX 78712, USA

1. Introduction

We use a phenomenological model to study the effects of imperfect mixing in the continuous flow stirred tank reactor (CSTR). In the limit of an infinitely fast stirring rate (perfect mixing), macroscopic spatial homogeneity is achieved throughout the reactor. Of course, realistic stirring rates are necessarily finite, in which case, various mechanisms can give rise to spatial inhomogeneities which may not be completely eliminated by stirring (imperfect mixing). In this paper, we examine reactant streams in the role of breaking stirring-induced homogeneity, and, the consequent effects on the state of the chemical reaction. We find results similar to those presented in our earlier paper [1] where we considered a different mechanism in this role, namely fluctuations.

We consider behaviour in the Schlögl model [2] as a function of flow rate (inverse residence time). This result is in better agreement with experimental findings [3] than that presented in our previous paper [1]. We conclude that the interaction between reactant streams and finite stirring is the dominant mechanism underlying the phenomena observed in experiment [3].

2. The N-Variable CSTR Model

The N-variable model for chemical reactions in a CSTR is [1,4]

$$\frac{\partial}{\partial t} p(\underline{x},t) = -\sum_{n=1}^{N} \frac{\partial}{\partial x_n} f_n(\underline{x}) p(\underline{x},t) + \frac{\sigma^2}{2} \sum_{n=1}^{N} \sum_{m=1}^{N} \frac{\partial^2}{\partial x_n \partial x_m} h_{nm}(\underline{x}) p(\underline{x},t)$$

$$+ 2\beta^2 \{ d\underline{x}' \int d\underline{x}'' \delta[\tfrac{1}{2}(\underline{x}' + \underline{x}'') - \underline{x}] \, p(\underline{x}',t) p(x'',t) - p(\underline{x},t) \}$$

$$- \alpha\{p(\underline{x},t) - p_o(\underline{x},t)\} \quad , \tag{1}$$

where $\underline{x}$ is an N-dimensional vector, the elements of which are concentrations of the individual chemical species, x_n, and $p(\underline{x},t)$ is the N-variable probability density for species concentrations. The first term on the right hand side describes chemical kinetics according to the deterministic rate equations

$$\frac{\partial}{\partial t} x_n = f_n(\underline{x}) \quad , \quad n = 1,\ldots,N \quad . \tag{2}$$

The next term reflects the existence of fluctuations in the system (σ^2 is the fluctuation strength, $h_{nm}(\underline{x})$ are elements of the $N \times N$ diffusion matrix). The third term uses a coalescence-dispersion mechanism (for details, see [1,4]) to model turbulent mixing (2β is the inverse characteristic mixing time). The last term describes the flow of reactants into and out of the CSTR (α is the inverse residence time; $p_o(\underline{x},t)$ is the probability density for input reactant stream concentrations).

Again we find it advantageous to work with the characteristic function, $q(\underline{k}) = \langle\exp(i\underline{k}\cdot x)\rangle$, which, in steady state, is governed by the equation

$$-\sum_{n=1}^{N}(-ik_n)f_n\left(\frac{1}{i}\frac{\partial}{\partial \underline{k}}\right)q(\underline{k}) + \frac{\sigma^2}{2}\sum_{n=1}^{N}\sum_{m=1}^{N}(-ik_n)(-ik_m)h_{nm}\left(\frac{1}{i}\frac{\partial}{\partial \underline{k}}\right)q(\underline{k})$$
$$- 2\beta[q(\underline{k}) - q^2(\tfrac{1}{2}\underline{k})] - \alpha[q(\underline{k}) - q_o(\underline{k})] = 0 \quad . \tag{3}$$

The solution to (3) in the infinitely fast stirring limit, $\beta\to\infty$, is

$$q(\underline{k}) = \exp(i\underline{k}\cdot\underline{x}_s) \quad \text{or} \quad p(\underline{x}) = \delta(\underline{x}-\underline{x}_s) \quad , \tag{4}$$

where $\underline{x}_s$ is determined by the condition

$$f_n(\underline{x}_s) - \alpha[x_{sn} - x_{on}] = 0 \ , \quad x_{on} = \int d\underline{x}\ x_n p_o(\underline{x}) \ , \quad n = 1,\ldots,N \quad . \tag{5}$$

We expand about this solution in the smallness parameter, β^{-1}, to obtain solutions for rapid stirring (β large but finite)

$$q(\underline{k}) = \exp(i\underline{k}\cdot\underline{x}_s)[1 + \sum_{\ell=1}^{\infty}\frac{1}{\beta^\ell}\phi_\ell(\underline{k})] \quad . \tag{6}$$

We substitute (6) into (3) and equate terms of like order in the smallness parameter to obtain

$$2[\phi_1(\underline{k}) - 2\phi_1(\tfrac{1}{2}\underline{k})] = -\alpha[1 - \exp(-i\underline{k}\cdot\underline{x}_s)q_o(\underline{k})]$$
$$- \sum_{n=1}^{N}(-ik_n)f_n(\underline{x}_s) + \frac{\sigma^2}{2}\sum_{n=1}^{N}\sum_{m=1}^{N}(-ik_n)(-ik_m)h_{nm}(\underline{x}_s) \quad , \tag{7a}$$

$$2[\phi_{\ell+1}(\underline{k}) - 2\phi_{\ell+1}(\tfrac{1}{2}\underline{k})] = -\alpha\phi_\ell(\underline{k}) + 2\sum_{\ell'=1}^{\ell}\phi_{\ell'}(\tfrac{1}{2}\underline{k})\phi_{\ell-\ell'+1}(\tfrac{1}{2}\underline{k})$$
$$- \sum_{n=1}^{N}(-ik_n)\sum_{r_1=0}^{R_1}\cdots\sum_{r_N=0}^{R_N}\frac{(-i)^{r_1+\ldots+r_N}}{r_1!\ldots r_N!}f_n^{(r_1,\ldots,r_N)}(\underline{x}_s)\phi_\ell^{(r_1,\ldots,r_N)}(\underline{k})$$
$$+ \frac{\sigma^2}{2}\sum_{n=1}^{N}\sum_{m=1}^{N}(-ik_n)(-ik_m)\sum_{r_1=0}^{R_1}\cdots\sum_{r_N=0}^{R_N}\frac{(-i)^{r_1+\ldots+r_N}}{r_1!\ldots r_N!}h_{nm}^{(r_1,\ldots,r_N)}(\underline{x}_s)$$
$$\times\ \phi_\ell^{(r_1,\ldots,r_N)}(\underline{k}) \quad . \tag{7b}$$

We next expand the functions, $\phi_\ell(\underline{k})$, as power series in the variables $(-ik_n)$.

$$\phi_\ell(\underline{k}) = \sum_{j_1=0}^{\infty}\cdots\sum_{j_N=0}^{\infty}\gamma_{\ell,j_1,\ldots,j_N}(-ik_1)^{j_1}\ldots(-ik_N)^{j_N} \quad . \tag{8}$$

We substitute (8) into (7a) and (7b), equate terms of like order in $(-ik_n)$, and solve for the unknowns, $\gamma_{\ell,j_1,\ldots,j_N}$. These coefficients enter into the moments of the finite stirring probability density via the relationship

$$\langle x_1^{j_1}\ldots x_N^{j_N}\rangle = \left(\frac{1}{i}\frac{\partial}{\partial k_1}\right)^{j_1}\cdots\left(\frac{1}{i}\frac{\partial}{\partial k_N}\right)^{j_n}q(\underline{k})\Big|_{\underline{k}=0} \quad . \tag{9}$$

For example, finite stirring will lead to a shift in the mean concentration of the jth chemical species away from the steady state, x_{sj}, of the global rate equation (5)

$$\langle x_j\rangle = x_{sj} - \sum_{\ell=1}^{\infty}\frac{1}{\beta^\ell}\gamma_{\ell,0,\ldots,1,\ldots,0} \quad ,$$

where $\gamma_{\ell,0,\ldots,1,\ldots,0}$ is the coefficient of the term in the power series for $\phi_\ell(\underline{k})$ which is first order in k_j and zeroth order in all other components of $\underline{k}$.

In the remainder of this paper, we make the following assumptions for simplicity. First, we assume that the chemical reactions under consideration can be adequately described using one intermediate species, the remaining chemical species being in excess. Second, we assume that fluctuations in the reactant feed stream are negligible. This allows us to represent the probability density for feed concentrations by a delta function. Finally, since we wish to specifically study the effects of reactant streams in breaking homogeneity, we omit the term reflecting fluctuations.

3. The Schlögl Model

The Schlögl Model [2] is a generic model for bistability in a chemical reaction described by the evolution of one intermediate species. The reaction scheme

$$A = 2X \underset{k_1'}{\overset{k_1}{\rightleftarrows}} 3X \quad , \quad B \underset{k_2'}{\overset{k_1}{\rightleftarrows}} X \quad , \tag{10}$$

leads to the kinetic rate equation

$$\dot{x} = -k_1'x^3 + k_1ax^2 - k_2'x + k_2b = f(x) \quad . \tag{11}$$

For this model, we assume the feed concentration probability density to be

$$p_o(x) = \delta(x - x_o) \quad . \tag{12}$$

Substituting (11) and (12) into (7a,b), we obtain to first order

$$\gamma_{10} = 0 \quad ,$$

$$\gamma_{11} = \frac{f^{(2)}(x_s)\gamma_{12} - f^{(3)}(x_s)\gamma_{13}}{f^{(1)}(x_s) - \alpha} \quad ,$$

$$\gamma_{1n} = \left(\frac{2^{n-1}}{2^n - 2}\right)\frac{\alpha}{n!}(x_s - x_o)^n \quad ,$$

$$-k_1'x_s^3 + k_1ax_s^2 - (\alpha + k_2')x_s + \alpha x_o - k_2b = 0 \quad . \tag{13}$$

We compare these results to the first order results presented in our previous paper [1] where we analyzed the effects of a different mechanism, fluctuations, for breaking homogeneity

$$\gamma_{10} = 0 \quad ,$$

$$\gamma_{11} = \frac{f^{(2)}(x_s)\gamma_{12}}{f^{(1)}(x_s)} \quad ,$$

$$\gamma_{12} = \frac{\sigma^2}{2}h(x_s) \quad ,$$

$$\gamma_{1n} = 0 \quad , \quad n \geq 2 \quad ,$$

$$-k_1'x_s^3 + k_1\,ax_s^2 - k_2'x_s + k_2b = 0 \quad . \tag{14}$$

The two sets of results (13) and (14) are similar in that they both predict a decrease in the area of the bistable region as well as a downward shift and finite variance for states on the upper branch as a function of the stirring rate.

In Fig. 1, we plot the mean and variance (calculated to fifth order)

$$\langle x \rangle = x_s - \sum_{\ell=1}^{5}\frac{\gamma_{\ell 1}}{\beta^\ell} \quad , \quad \langle (x - \langle x \rangle)^2 \rangle \simeq \sum_{\ell=1}^{5}\frac{\gamma_{\ell 2}}{\beta^\ell} \tag{15}$$

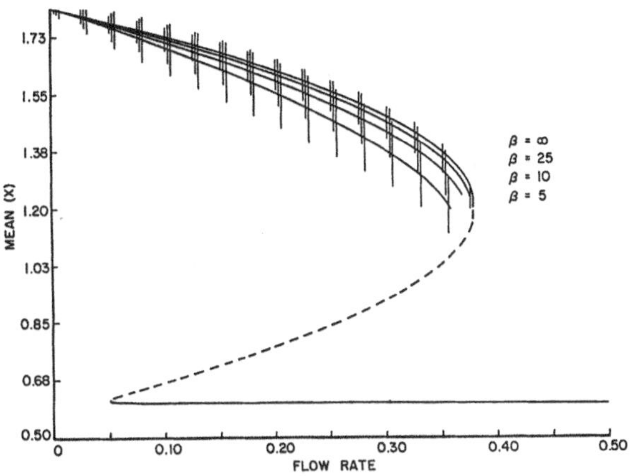

Fig. 1. Schlögl Model: Stationary state mean and variance vs. flow rate ($k_1^1 = 1$, $k_1a = 3$, $k_2^1 = 2.5$, $k_2b = 0.636$, $x_o = 0.602$) $\beta = \infty$, 50, 25, 10, 5

of the stationary states as a function of the flow rate, α, for several values of the mixing rate, 2β. The reactant stream model (11) exhibits features which are closer to experimental findings [3] than the fluctuation model (1). Model II shows no stirring-related changes on the lower (flow) branch, whereas Model I shows small upward shifts and variance. Model II predicts that the variance for states on the upper (thermodynamic) branch will increase as the transition point is approached whereas Model I predicts that the variance will show a small decrease as the transition point is approached.

4. Conclusions

The Schlögl model [2] yields results that are closer to experimental findings [3] when the homogeneity breaking mechanism is reactant flows than when it is fluctuations. We conclude that the observed phenomena are essentially the effects of reactant streams and finite stirring rates in the CSTR.

Acknowledgements

W.H. would like to thank J.C. Roux, P. De Kepper, J. Boissonade, G. Dewel, D. Walgraef, P. Borkmans, and G. Nicolis for stimulating discussions on different aspects of and viewpoints on the effects of stirring in chemical reactions. This work was partially supported by NSF grant INT 81-15672 (W.H.) and the International Paper Co., J. Stanford Smith Fellowship (L.H.).

References

1. W. Horsthemke, L. Hannon: J. Chem. Phys. (to appear)
2. F. Schlögl: Z. Physik 248, 446 (1971); 253, 253 (1972)
3. J.C. Roux, P. Dekepper, J. Boissonade, Phys. Lett. 97A, 168 (1983)
4. R.L. Curl; A.I.Ch.E.J. 9, 175 (1973); J.S. Evangelista, S. Katz, R. Shinnar, A.I.Ch.E.J. 15, 843 (1969)

Part VI

Stochastic Analysis

Stochastic Aspects of Nonequilibrium Transitions in Chemical Systems

G. Nicolis, F. Baras, and M. Malek Mansour

Faculté des Sciences de l'Université Libre de Bruxelles, Campus Plaine, CP 226
B-1050 Bruxelles, Belgium

1. Introduction

One of the principal preoccupations of the present meeting is the study of the transitions occurring in nonlinear chemical systems far from equilibrium. Experimental evidence provides us with an impressive number of examples, and brings out clearly the diversity of these phenomena, from bistability and sustained oscillations to chaotic dynamics and the formation of spatial patterns in an initially uniform system.

The theory of dynamical systems, which was marked by spectacular developments since the 1960's, establishes that there exist large classes of nonlinear equations whose solutions show precisely this kind of behavior. Whence a growing feeling about the inevitability of complex transition phenomena in physical chemistry, and about the suitability of macroscopic physics - of which the theory of dynamical systems is the principal tool - to handle them.

Now, bifurcation or any other transition implies that, at some stage of its development, a macroscopic system becomes capable of generating, propagating and sustaining reproducible relationships between its constitutive parts, extending over its entire dimensions. How can this extraordinary coherence arise ? Certainly not by the random collisions of the molecules of reacting species, which is the usual picture afforded by classical chemical kinetics in the presence of short-range forces. Nor can phase-transition related phenomena be invoked since, as a rule, the systems of interest are in a well-defined, thermodynamically stable phase of matter. The theory of dynamical systems is unable to handle this major question. An enlarged description of nonequilibrium transitions is thus needed, in which the molecular aspects of these phenomena are taken into account. This is precisely the purpose of the theory of fluctuations in nonequilibrium systems, different aspects of which will be discussed in the present paper.

Essentially, fluctuation theory provides us with the microscopic counterpart of the phenomena of instability and bifurcation. We discuss the static aspects of this problem in section 3, after surveying in section 2 the problems related to the modelling of the fluctuations. Section 4 is devoted to the origin of coherent behavior in nonequilibrium systems. Specifically, we study the spatial correlation function and show that, as soon as a system deviates from thermodynamic equilibrium, it generates spatial correlations of macroscopic range. This phenomenon, which has no deterministic analog, is further accentuated near bifurcation by the fact that the correlation length tends to infinity and order encompasses the entire system. In section 5 we survey the time-dependent aspects of fluctuations with special emphasis on the kinetics of explosive reactions. A summary of the results and a short discussion of some open questions is given in the final section 6.

We want to emphasize that the need to study the molecular aspects of nonequilibrium transitions stems from a deep conceptual question which cannot be overlooked in the name of pragmatism. True, in most cases of practical interest the effect of thermodynamic fluctuations, whose size is necessarily small, is masked by the effect of external disturbances of macroscopic size. A completely analogous situation occurs in connection with ordinary phase-transitions where external fields, finite size effects, or even deliberate intervention like the seeding in crystal growth, turn out to be more efficient than the internally generated fluctuations. In both cases, however, one may regard these macroscopic effects as probes of what the system would be capable of doing anyway, albeit on a much longer time scale. We are therefore convinced that, having mastered to some extent the phenomenology of nonequilibrium transitions, physical chemists will discover growing evidence about the importance of the fluctuations. Sign posts in this direction are the recent intriguing results on the role of incomplete stirring [1], which reveal the importance of inhomogeneous disturbances even for transitions between homogeneous states. This is very much like the theoretical prediction concerning the role of the inhomogeneous fluctuations in the critical behavior around a bifurcation point as well as in the transition between the simultaneously stable states arising beyond bifurcation.

2. Some comments on the modelling of the fluctuations

Space and time-correlation functions of macrovariables around nonequilibrium steady states have recently been calculated from nonequilibrium statistical mechanics [2]. Despite this progress, fluctuation theory for nonequilibrium states remains based, in its essential aspects, on stochastic theory. Specifically, it is assumed that one can define an appropriate set of discrete variables $\underset{\sim}{X} = \{X_\alpha\}$, generally localized in space, which constitute a Markov process. The following gain-loss balance equation, usually referred to as multivariate master equation, can then be written [3] :

$$\frac{d}{dt} P(\underset{\sim}{X};t) = \sum_\rho [W_\rho(\underset{\sim}{X}-\underset{\sim}{\nu}_\rho \rightarrow \underset{\sim}{X}) P(\underset{\sim}{X}-\underset{\sim}{\nu}_\rho;t) - W_\rho(\underset{\sim}{X} \rightarrow \underset{\sim}{X}+\underset{\sim}{\nu}_\rho) P(\underset{\sim}{X};t)] \quad (1)$$

Here $W_\rho(\underset{\sim}{X} \rightarrow \underset{\sim}{X}+\underset{\sim}{\nu}_\rho)$ is the transition probability per unit time for the occurrence of an elementary process which changes the state of the system from $\underset{\sim}{X}$ to $\underset{\sim}{X}+\underset{\sim}{\nu}_\rho$. Its specific structure has to be derived from the available information on the nature of the processes that take place in the system and from the constraints imposed by equilibrium and other limiting laws. A chemical reaction is thus modelled as a birth or as a death process, whereas transfer of energy or matter across neighboring space cells is modelled as a random walk.

We have invoked so far Markov processes in discrete state space as the natural model of fluctuations, since the latter are the consequence of the discrete nature of the microscopic processes underlying the macroscopic evolution laws,

$$\frac{d}{dt} \underset{\sim}{X} = \underset{\sim}{F}(\underset{\sim}{X}, \lambda) \tag{2}$$

where $\underset{\sim}{F}$ is the (generally nonlinear and dissipative) evolution operator, and λ a set of control parameters. Nevertheless, considerable success has been achieved by describing fluctuations as a continuous Markov process [4,5]. In this approach it is assumed that a generalized Langevin equation descriptive of a diffusion process can be written,

$$\frac{d}{dt} \underset{\sim}{X} = F(\underset{\sim}{X}, \lambda) + \mathcal{F}(t) \tag{3}$$

where $\mathcal{F}$ is a random force obeying a fluctuation-dissipation like theorem, similar to the one established near the state of equilibrium. Some powerful theorems ensuring the equivalence of both descriptions for finite time are available [6,7]. However, in general, this equivalence does not extend for all times. We discuss this problem briefly again in sections 3 and 5.

3. Stochastic analog of instability and bifurcation

Let λ_c be a set of parameter values for which a solution $\underset{\sim}{X}_s$ of eqs. (2), referred to as reference state, loses its stability and gives rise to new branches of solutions by a bifurcation mechanism. We want to see how the solution of the master equation, eq. (1), behaves under these conditions, and how this behavior depends on small changes of the parameters λ around λ_c. The answer to this question depends on the kind of bifurcation considered, on the nature of the reference state, and on the number of variables involved in the dynamics. The simplest case is, by far, the pitchfork bifurcation occurring as a first transition from a previously stable spatially uniform stationary state. This transition is characterized by a remarkable universality. First, whatever the number of variables present initially, it is always possible to cast the stochastic dynamics in terms of a single, "critical" variable. This is the probabilistic analog of adiabatic elimination or, in more modern terms, of the center manifold theorem [4,8-10]. Second, the stationary probability distribution of the critical variable can be cast in the form (we set $\delta \underset{\sim}{X}_{\underset{\sim}{r}} = \underset{\sim}{X}_{\underset{\sim}{r}} - \underset{\sim}{X}_s$, $\underset{\sim}{r}$ stands for the spatial coordinates) :

$$P \approx \exp - U(\{\delta X_{\underset{\sim}{r}}\}) \tag{4}$$

where the stochastic potential U is an extensive quantity similar to the Landau-Ginzburg type of potential familar from equilibrium phase transitions [4,5,9,12]. For a cubic nonlinearity one obtains the following simple form, up to a mutiplicative constant (D is the Fick's diffusion coefficient) :

$$U \approx \int \left\{ - (\lambda - \lambda_c) \frac{(\delta X_{\underset{\sim}{r}})^2}{2} + \frac{(\delta X_{\underset{\sim}{r}})^4}{4} + \frac{D}{2} (\nabla . \delta X_{\underset{\sim}{r}})^2 \right\} d\underset{\sim}{r} \tag{5}$$

Many important properties, such as critical dimensionality and critical exponents describing the divergence of correlation length and other quantities can thus be obtained from renormalization group analysis. It is worth noting that for λ well below λ_c only the quadratic terms of the potential contribute to the asymptotic properties of P, which reduces therefore to a multigaussian distribution in accordance with the central limit theorem [13]. There exists, however, a (frequently very narrow) vicinity of λ_c

such that the quartic part of U plays a role, owing to the coupling between different spatial modes of fluctuations. We are therefore witnessing a breakdown of the central limit theorem in the present problem. This reflects the coherence associated with bifurcation : the system can no longer be partitioned into a collection of weakly correlated subsystems.

The above results can be derived from both the master equation and the generalized Langevin equation as starting point. As a matter of fact one can show that, in some well-defined asymptotic sense, the discrete Markov process described by the master equation reduces to a Markovian process with continuous realizations equivalent to eq. (3) wherein $\mathcal{F}(t)$ represents a state-independent white noise in time [12]. This implies, in turn, that the stochastic potential U is identical up to a multiplicative constant to the deterministic potential for the critical variable. The situation however changes considerably in the region of multiple simultaneously stable states beyond bifurcation. Fig. 1 represents schematically the probabilistic analog of multistability. The dominance of a particular state is measured by the peak of the probability distribution. Because the extensive character of the stochastic potential selection is very sharp, in the sense that the non-dominant state occurs with a very low probability. There exists, however, a set of values of the control parameters for which both states are equally dominant (Fig. 1e). We refer to it as the "coexistence" case : the two macroscopic states X_- and X_+ occur with an equal a priori probability. Now, the study of solvable models shows that the deterministic and the stochastic potential obtained from the master equation are different near the coexistence region [14-15]. As a consequence certain observable quantities such as the mean first passage time from X_- to X_+, differ by several orders of magnitude according to the description we adopt. In other words, we can no longer model the stochastic dynamics as a diffusion process in state space with a constant diffusion coefficient. On inspecting Fig. 1e one can convince oneself that this conclusion was to be expected. Indeed, we deal here with a composite stochastic process : a small scale motion around each of the maxima at X_- and X_+, which may well be a diffusion process albeit with a state-dependent diffusion coefficient ; and a large scale transition between X_+ and X_-, which should be similar to a random telegraph signal. There is thus no reason that the overall process be a Markovian process of the diffusion type. One may speculate that if the process was still to admit continuous realizations, it would have to be a non-Markovian process. We are witnessing here a breakdown of the universality that was characterizing the behavior of the probability distribution up to the bifurcation point.

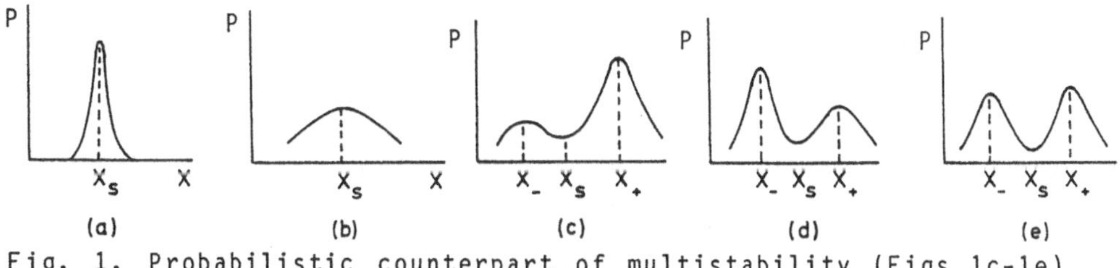

Fig. 1. Probabilistic counterpart of multistability (Figs 1c-1e) following a bifurcation from an initially stable unique state (Figs 1a-1b)

Symmetry-breaking bifurcations leading to spatially inhomogeneous steady states, and Hopf bifurcation leading to spatially uniform time-periodic states of the limit cycle type, are the next two simplest examples of transitions in nonequilibrium systems. The master equation can be solved asymptotically in both cases, and the corresponding stochastic potential can be constructed explicitly [17,18]. Agreement with the generalized Langevin description [5,19] is found in the vicinity of the bifurcation point. In both cases, a most remarkable phenomenon is the formation of defects of various kinds, and their propagation within the system. Phase fluctuations play here a very important role and tend, at least in two-dimensional systems, to desynchronize the homogeneous limit-cycle oscillation by initiating wave patterns. We refer to the lecture by D. Walgraef for a detailed presentation of this phenomenon, which may well turn out to provide the first satisfactory explanation of the intrinsic origin of propagating waves in chemical systems.

Implicit in the analysis of the bifurcations discussed so far was the assumption that they constituted the first transition from an initially stable reference state. In many instances a whole cascade of bifurcations is observed, especially when several control parameters are varied simultaneously [20]. For instance, the branch of solutions arising from one of the above mentioned primary bifurcations can in turn lose its stability and give rise to a secondary or even a higher order bifurcation. Quasi-periodic oscillations, homoclinic orbits and chaos are some of the typical states that can be reached through these transitions. Again, we can raise the question of the stochastic counterpart of these phenomena. One can suspect that fluctuations may play here an even more interesting role, in view of the sensitivity of these regimes to variations of parameter values and, especially, to initial conditions, as a result of which thermal noise should be amplified tremendously. We refer to the lecture by H. Lemarchand and A. Fraikin and to the poster by A. Fraikin for some results on this new class of stochastic problems.

4. Spatial correlations

In the preceding section it was tacitly assumed that macroscopic systems functioning away from phase transition points and dominated by short range intermolecular forces are, nevertheless, capable of presenting coherent behavior on a supermolecular space and time scale. In this section we comment on the origin of this coherence and, in particular, on the role of the nonlinearity and of the nonequilibrium constraints. Essentially, we want to understand how the system may deviate spontaneously from the completely random distribution of chemical species within the reaction volume described by Poissonian statistics. To this end we need to compute the spatial correlation function between two different spatial regions of volume ΔV centered on points $\mathfrak{r}$ and $\mathfrak{r}'$:

$$g_{\mathfrak{r},\mathfrak{r}'} = \langle \delta X_{\mathfrak{r}} \delta X_{\mathfrak{r}'} \rangle - \langle X_{\mathfrak{r}} \rangle \frac{\delta^{kr}_{\mathfrak{r},\mathfrak{r}'}}{\Delta V} \tag{6}$$

where $\langle \delta X_{\mathfrak{r}} \delta X_{\mathfrak{r}'} \rangle$ is the second moment of the multivariate probability distribution, eq. (4). We limit ourselves, for simplicity, to systems involving only one variable.

Let us first present some intuitive arguments. Suppose that the system is described by linear chemical kinetics. Particles will then be created ($A \to X$) or destroyed ($X \to A$) at random places in space. However, in the case of nonlinear chemical kinetics, particles will either be created in clusters (for instance $A \to 2X$ gives pairs of particles in the same spatial region) or particles close to each will be destroyed reactively (the reaction $2X \to A$ depletes the system from pairs of particles). In equilibrium, the property of detailed balance ensures that any reaction, such as $A \to 2X$, will be, in average, as frequent as its inverse, $2X \to A$, and we recover the Poissonian behavior. In non-equilibrium however, this balance is disturbed and deviations from the Poisson law are thus generated. How can one relate these deviations to the existence of spatial correlations ? Let us consider again the reaction step $A \to 2X$. When such a reaction occurs, the two X-particles are obviously correlated, being created together. Due to diffusion this correlation extends in space until one of the particles again undergoes a reaction. This will happen in average at a certain well-defined rate k. The resulting correlation length is thus given by :

$$l_c \approx \sqrt{D/k} \qquad (7)$$

where D is the Fickian diffusion coefficient of the chemical species considered.

To go beyond these qualitative arguments, we turn to the Multivariate Master Equation, eq. (1). We may easily derive an equation for the spatial correlation function, eq. (6) by neglecting third and higher order moments. This Gaussian type of approximation is expected to be valid well before bifurcation, as pointed out in the previous section. One obtains :

$$\frac{D}{2d} \sum_{\underline{l}} (g_{\underline{r}+\underline{l},\underline{r}'} - g_{\underline{r},\underline{r}'}) + F'(X_s)\, g_{\underline{r},\underline{r}'} + [\frac{1}{2} Q(X_s) + X_s F'(X_s)] \frac{1}{\Delta V} \delta^{kr}_{\underline{r},\underline{r}'} = 0 \qquad (8)$$

X_s is the homogeneous macroscopic steady state of the system, F and Q being the macroscopic chemical rate and chemical noise strength respectively, and the sum over $\underline{l}$ runs over all nearest neighbors. The last term in the l.h.s. of (8) is a source term for spatial correlations. Explicit calculations [21] show that it is proportional to the net material flux, J_s , through the system at the steady state that is, ultimately, to the nonequilibrium constraint acting on the system.

The equation for g in the continuum limit takes thus the form :

$$D \nabla^2_{\underline{r}-\underline{r}'}\, g(\underline{r}-\underline{r}') + F'(X_s)\, g(\underline{r}-\underline{r}') = -2\, J_s\, \delta(\underline{r}-\underline{r}') \qquad (9)$$

Its solution reads :

$$g(\underline{r}-\underline{r}') = \frac{J_s}{2\pi D} \frac{1}{|\underline{r}-\underline{r}'|} \exp\left\{ -\left[\frac{F'(X_s)}{D}\right]^{1/2} |\underline{r}-\underline{r}'| \right\} \qquad (10)$$

The exponential in the r.h.s. of (10) features a correlation length of the form of eq. (7) with $k = F'(X_s)$.

In short, nonlinear chemical systems are characterized by an intrinsic correlation length l_c of macroscopic range. This quantity is a smooth function of the distance from equilibrium (see

Fig. 2b). As expected from the discussion of section 3, in the vicinity of a bifurcation point it diverges ($F'(X_s) \to 0$), as correlations invade the entire system. On the other hand, the amplitude of the correlation function and, concomitantly, the deviation from the Poisson law, can be considered as an "order parameter" characterizing the "transition" from equilibrium to nonequilibrium (see fig. 2a). Although the correlation length is an intrinsic property of the chemical system, it will not be perceived in the equilibrium state where direct and inverse elementary processes cancel out exactly due to detailed balance. Therefore the transition to nonequilibrium witnesses the sudden arousal of long range spatial correlations.

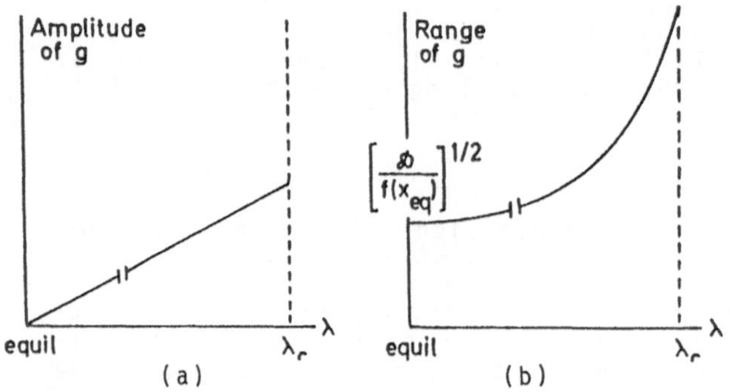

Fig. 2. Amplitude of spatial correlation function (a) and range (b) in a nonlinear chemical system, plotted against the nonequilibrium constraints

The chemical system discussed above is subject to a scalar nonequilibrium constant, namely the nonequilibrium concentrations of appropriate reservoir variables A, B, ... This allows to consider an infinite system, for which no parameter related to the system size enters into the problem. Let us now briefly discuss the case of a nonequilibrium constraint which breaks the translational symmetry of the system. An example is provided by the diffusion of heat through a one-dimensional system of length L, parallel to the z-axis, whose end points a and b are maintained at different temperatures T_a and T_b . If one supposes that heat conduction following Fourier's law is the only transport mechanism taking place in the system, one verifies that the correlation function

$$g(z,z') = \langle \delta T(z)\ \delta T(z') \rangle - \frac{k_B \bar{T}_z^2}{C_V}\ \delta(z-z') \tag{11}$$

obeys the following equation [21]

$$\kappa\left(\frac{\partial^2}{\partial z^2} + \frac{\partial^2}{\partial z'^2}\right) g(z,z') = -\ 2\ k_B\ \frac{\gamma^2}{C_V}\ \delta(z-z') \tag{12}$$

Here κ is the thermal diffusivity coefficient (supposed constant), C_V the specific heat of the system, $\gamma = T_b - T_a/(z_b - z_a)$ the temperature gradient and $T_z = T_a + \gamma(z - z_a)$ the average temperature at the position z. As in the case of the reaction-diffusion system, the source term of the spatial correlations is related to the nonequilibrium flux (in the present case the heat flux) through the system. However, no intrinsic length scale l_c appears in the present problem. Therefore, it is not surprising that the correlation length is determined by the total length $L = z_b - z_a$, i.e. it encompasses the entire system. Indeed, the solution of eq. (12) reads :

$$g(z,z') = g(z',z) = \frac{k_B \gamma^2}{L\ C_V} z'(L-z) \quad ; \quad z' \leqslant z \qquad (13)$$

to be compared with the result of eq. (10) for chemical systems.

Similar results can be obtained from the generalized Langevin equation approach [23]. Molecular dynamics simulations corroborate the main qualitative predictions [24]. For a detailed presentation of these aspects we refer to the lecture by M. Mareschal.

In short, the occurrence of spatial correlations of macroscopic range is a general feature for nonequilibrium systems, even independently of bifurcation phenomena. The correlation length may be either intrinsic or determined by the system size, but the amplitude of the correlation function is always related to the strength of the nonequilibrium constraint, e.g. the material or energy flux through the system. At equilibrium, this amplitude is zero masking the presence of the inherent long-range coherence in the system.

5. Time-dependent properties of the fluctuations

A transition between states involves, perforce, a stage during which the variables evolve in time. If the transition is to be attributed solely to the fluctuations, because of the localized character of the latter it will necessarily start in a volume element ΔV of the system, whose size is much smaller than the size of the system as a whole : $\Delta V/V \ll 1$. As time grows this localized fluctuation may decay or, on the contrary, may grow and invade the whole system. The details of the kinetics and, in particular, the time scales involved in the transition will depend crucially on the nature of the initial and final states but also, surprisingly, on intermediate unstable states that may exist between them. In this section we discuss certain characteristic features of this phenomenon.

A first approximation to the description of the full process is to assume that, by some mechanism independent of the fluctuations, the system remains uniform in space. In this "zero-dimensional" description, considerable attention was focussed on the decay from an initial unstable state, and on the passage times between simultaneously stable states [25-29]. Two characteristic scales emerge from this analysis. For the evolution around the unstable state, the time needed for the probability distribution to forget the initial condition and begin to develop peaks toward the stable attractors is

$$\tau_u \sim \ln \frac{1}{\varepsilon + \Delta} \qquad (14)$$

where $\varepsilon = V^{-1}$ is the inverse of the size of the system and Δ measures the width of the initial probability distribution around the unstable state. If on the other hand the initial condition is centered on a stable state in the range of parameter values for which this state coexists with other attractors, then the transition occurs on the much longer time scale

$$\tau_s \sim \exp \frac{1}{\varepsilon} \Delta U \tag{15}$$

The "potential barrier" ΔU is a positive number whose value depends on the properties of both the initial stable state and the (homogeneous) unstable steady states via which the transition is taking place. A very interesting question concerns the statistics of these "passage times", whose average values are given by eqs. (14) and (15). This problem requires the computation of "level crossing" probabilities and has been investigated recently by Arecchi and coworkers [28]. It is worth pointing out that the master equation and the generalized Langevin equation description give different predictions for this class of problems, for the reasons invoked already in Section 3.

A more satisfactory analysis of transitions between states is to incorporate in the description the spatially inhomogeneous character of the fluctuations. The full problem has not yet been solved in a consistent way, but interesting results have been obtained for a two-box model, both by analytic solution of the master equation and by Monte Carlo simulations [30-32]. Specifically, if the diffusion constant coupling the two boxes has an intermediate strength compared to the chemical rate constants, it is found that the transition between two homogeneous steady states occurs via an inhomogeneous transient state. This is reminiscent of nucleation, familiar from phase transitions. We refer to the lecture by M. Moreau and D. Borgis and to the poster by M. Frankowicz for a detailed description of this mechanism.

In the remaining part of this section we shall concentrate on a class of transient phenomena characterized by an initial induction regime, involving a very small rate of change of the pertinent variable, interrupted suddenly by a violent explosive behavior ignited at some characteristic time t_i. Eventually the system is stabilized at some state, which for simplicity is assumed to be the only stable state available in the range of interest of the parameter values (see fig. 3). A typical example in which these conditions are realized is explosion. In previous work [33-35] we analyzed the stochastic aspects of this phenomenon for adiabatic thermal explosion and for chemical explosion in an open system. A similar analysis was reported for the problem of switching in optical bistability [36], which bears many similarities with explosion. Here we describe some new results pertaining to thermal ex-

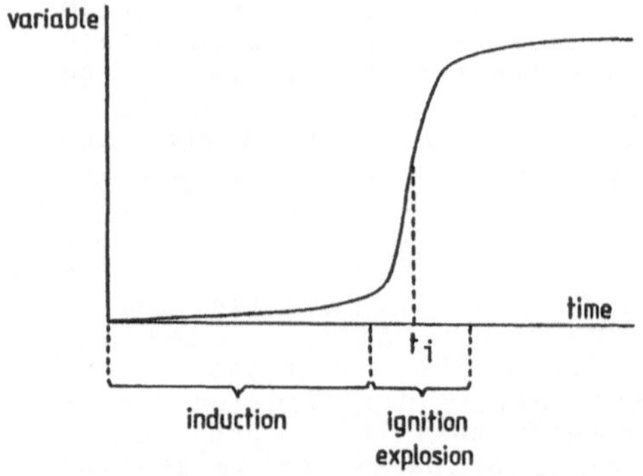

Fig. 3. Transient evolution toward a unique stable state exhibiting a long induction period and an abrupt explosion at $t=t_i$

plosion in a system which is in contact with a thermal reservoir. This problem is of central interest in the whole field of combustion [37]. Its modelling goes back to the classic work of Semenov, and has recently been reexamined by Lermant and Yip [38]. Let us outline the main steps of this analysis, in which fluctuations are neglected.

We consider a finite reaction vessel of volume V and surface area S which allows for thermal contact of the reactants with a reservoir at temperature T_0 at the boundary of the vessel but which does not allow for mass flow of the reactants across the boundaries. An exothermic reaction is supposed to occur in the vessel. We assume for simplicity that the constants are distributed homogeneously, for instance, by an efficient stirring mechanism. It should be realized, however, that the rate of stirring may have some highly nontrivial consequences on the observed behavior [1,39].

We denote by $\bar{c}$ the concentration of the reactant and by $\bar{T}$ the internal temperature. In the limit in which the thermal relaxation time is much faster than the chemical one, one may neglect the reactant consumption and identify $\bar{c}$ to its initial value, c_0. Assuming a single mth order reaction one obtains then the following equation of evolution for $\bar{T}$:

$$\rho\, C_p \frac{d}{dt}\bar{T} = Q\, c_0^m\, k_0\, e^{-E/R\bar{T}} - \gamma\,(\bar{T}-T_0) \qquad (16)$$

Here ρ and C_p are, respectively, the mass density and specific heat of the system ; Q is the heat of reaction ; E is the activation energy ; R the gas constant ; and $\gamma = \alpha S/V$, α being the Newton cooling coefficient.

It is well-known that there exists a range of parameter values for which eq. (16) admits two simultaneously stable states separated by an intermediate unstable state [37,38]. However, in the limit of high activation energy which is the usual situation in combustion,

$$\varepsilon' = \frac{RT_0}{E} \ll 1 \qquad (17a)$$

the high temperature "combustion" branch corresponds to unrealistically high values of temperature. On the other hand the low temperature "extinction" branch as well as the intermediate unstable branch are well described. Fig. 4 gives a plot of these latter steady states, T_s , as a function of the scaled parameter

$$\delta = \frac{\gamma}{Q\, c_0^m\, k_0} \qquad (17b)$$

Let now the system be started in the region of parameter values left to the ignition point δ_c , and at a value of temperature close to the value of $\bar{T}$ on the turning point of the lower branch of Fig. 4. One will observe then a thermal explosion, whereby after an induction period temperature increases suddenly and tends to the combustion branch of the steady state diagram, in a way similar to Fig. 3. Because of the unrealistically high value of the latter, we discard the part of the evolution referring to the final saturation, since for that stage the assumption of no consumption of the reactant breaks down. Fig. 5 describes the induction

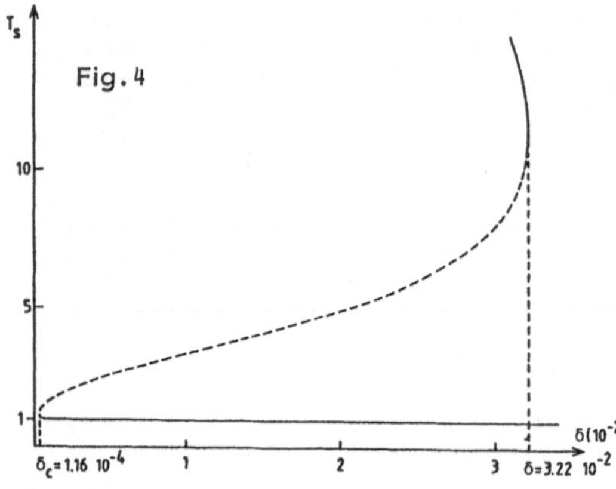

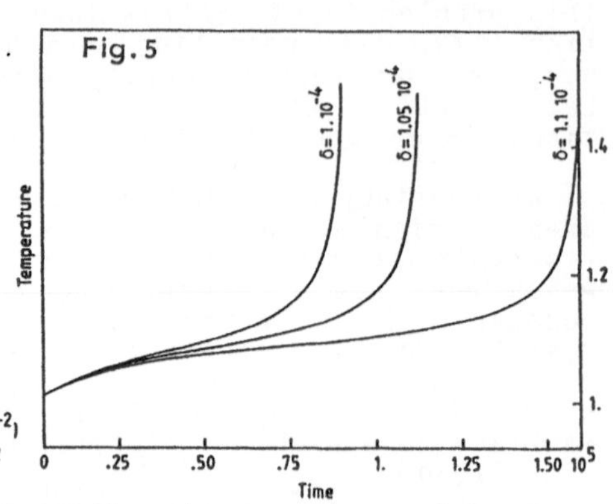

Fig. 4. Steady-state diagram for the exothermic reaction model of eq. (16). Parameter value : $\varepsilon' = 0.08$

Fig. 5. Time-dependent behavior of the solutions of eq. (16) in the domain of thermal explosion. Parameter value : $\varepsilon' = 0.08$. Time reduced in units of $\left[k_0 c_0^m \dfrac{Q}{\rho\, C_p\, T_0}\right]^{-1}$

period and the first stages of explosion for which the theory provides a satisfactory description. As δ comes closer to δ_c ignition becomes sharper in the sense that the initial plateau becomes increasingly longer or, in other words, ignition time t_i tends to infinity. A straightforward computation yields [40]

$$t_i \sim \frac{1}{\sqrt{1-\delta/\delta_c}}$$

which is analogous to the phenomenon of "critical slowing down" familiar from phase transitions.

We now turn to the stochastic aspects of the problem. Since we are dealing with the time-dependent properties that are independent of the combustion branch we expect, on the grounds of the discussion of section 3, that a diffusion process described by a generalized Langevin equation (eq. (3)) should provide a satisfactory description. We thus replace eq. (16) by the stochastic differential equation,

$$\frac{d}{dt} T = \frac{Q\, c_0^m\, k_0}{\rho\, C_p}\, e^{-E/RT} - \frac{\gamma}{C_p\, \rho} (T - T_0) + f(t) \tag{18}$$

where $f(t)$ is a Gaussian white noise whose strength is allowed to depend on the deterministic solution (cf. eq. (16)) :

$$\langle f(t) \rangle = 0$$

$$\langle f(t)\, f(t') \rangle = \varepsilon\, Q_{\bar{T}}\, \delta(t-t') \tag{19}$$

ε being the inverse of the system size (cf. eqs. (14)-(15)). The precise form of $Q_{\bar{T}}$ can be determined by requiring agreement with the equilibrium and detailed balance conditions. One thus finds [41]:

$$Q_{\overline{T}} = k_0 \, c_0^m \, \frac{Q^2}{\rho^2 \, C_p^2} \, e^{-E/R\overline{T}} + \frac{\gamma}{\rho \, C_p} \, \frac{k_B}{\rho \, C_p} \, (\overline{T}^2 + T_0^2) \qquad (20)$$

All desired properties of the fluctuations can be computed from eqs. (18)-(20) or from the Fokker-Planck equation for the underlying probability distribution :

$$\frac{\partial}{\partial t} P(T;t) = - \frac{\partial}{\partial T} \left(\frac{\partial U}{\partial T} P(T;t) \right) + \frac{\varepsilon}{2} Q_{\overline{T}} \frac{\partial^2}{\partial T^2} P(T;t) \qquad (21a)$$

The drift term of this equation features the deterministic potential, U,

$$U = - \int^{T} dT' \left[\frac{Q \, c_0^m \, k_0}{\rho \, C_p} \, e^{-E/RT'} - \frac{\gamma}{\rho \, C_p} (T' - T_0) \right] \qquad (21b)$$

which is represented in Fig. 6. for parameter values belonging to the explosion region.

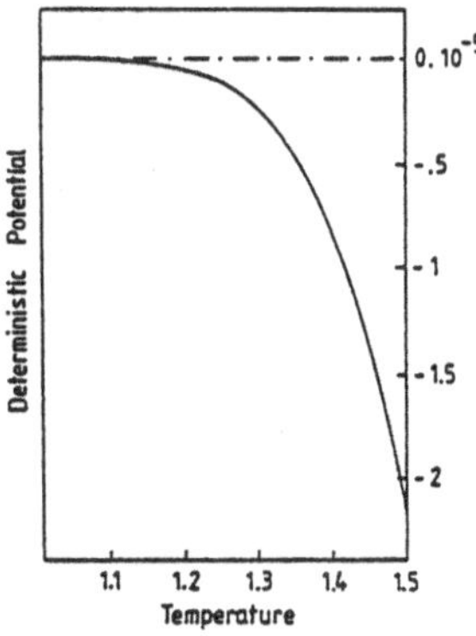

Fig. 6. Deterministic potential in the region of thermal explosion. Notice the occurrence of a flat part, associated with the induction region of Fig. 5. Parameter values : $\varepsilon' = 0.08$; $\delta = 1.1 \; 10^{-4}$

We first consider the region of parameter values between the ignition and extinction points δ_c and δ_c' (Fig. 4). We assume that the system is initially on the lower branch and inquire about the average time needed to undergo a spontaneous explosion in the form of a jump across the unstable state branch. A straightforward application of expression (15) in which the specific structure of the potential, eq. (21b), is taken into account, leads to the results summarized in Fig. 7. As expected, the transition time tends to zero as the system approaches the ignition point. Otherwise, the transition times are extraordinarily large, unless one deals with a small size system or, alternatively, with localized fluctuations in a sufficiently small volume. It seems therefore

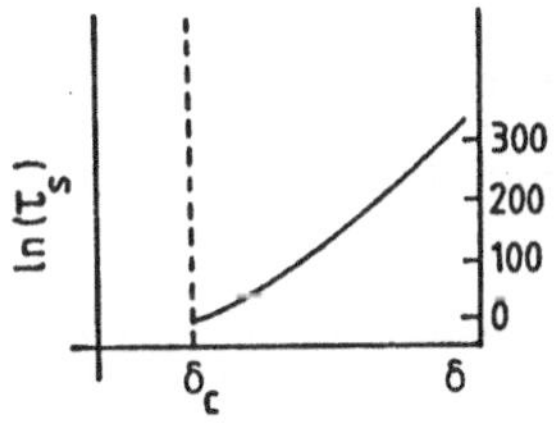

Fig. 7. Mean passage time for jumping across the unstable branch of Fig. 4 starting from the low-temperature steady-state, as a function of the control parameter δ. Parameter values : $\varepsilon' = 0.08$; $\varepsilon = 10^{-3}$;

that, in the region of bistability, stochastic effects can interfere with the behavior of the system only through a nucleation mechanism of the kind described earlier in this section. This requires however a finer analysis taking into account spatially inhomogeneous fluctuations.

The situation becomes very different when the system functions in the spontaneous explosion region, that is, for values of δ less than the ignition limit δ_c Fig. 8 reports the results of numerical solution of the Langevin equation (eqs. (18)-(20)) for different ε's and initial conditions. In each case, the stochastic trajectory is followed and the times at which it crosses a preassigned level of temperature value representative of explosion are determined. This allows one to compute the probability distribution of ignition times. We see that far from the ignition point this distribution is narrow, and its peak coincides with the deterministic explosion time. However, as δ approaches the ignition value δ_c the distribution becomes very broad, and the position of its maximum is substantially different from the deterministic time. The latter belongs, in fact, to the tail of the ignition time distribution. In other words, ignition becomes essentially a random event.

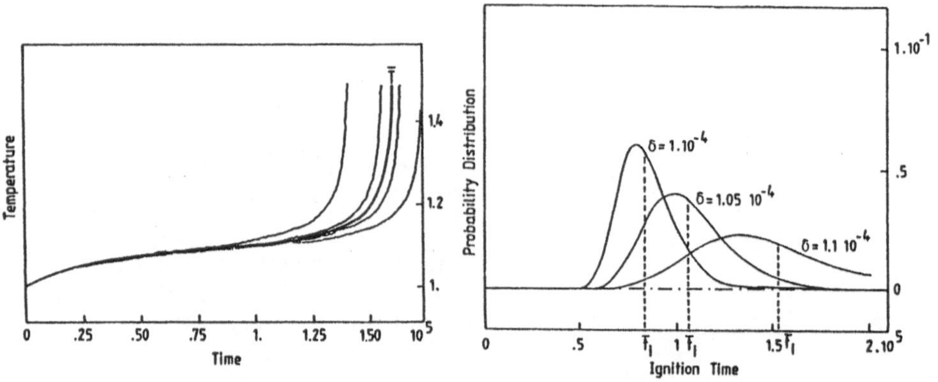

Fig. 8. Numerical solution of the Langevin equation (eq. (18)) in the region of thermal explosion. Left part : Individual realizations of the process of explosion featuring the considerable dispersion of ignition times. Right part : Probability distribution of ignition times, illustrating further the random character of explosion

The pronounced randomness suggested by the above results is to be related to the phenomenon of transient bimodality discovered earlier for adiabatic thermal explosion [33,35] as well as for chemical explosion [34] and for optical bistability [36]: initially the system is started by a sharply peaked distribution centered on a state belonging to the induction (flat) part of Fig. 5. One observes subsequently that, while one of the maxima of the probability distribution remains centered close to the initial state, a second maximum is formed near the state of explosion. Eventually the first peak disappears, and the system is driven to its unique stable state. We have therefore a "bifurcation behavior in time" during the transient evolution of the system.

Let us outline a qualitative explanation of these unexpected phenomena. Remember that we are dealing with a process involving two widely separated time scales, and that our system is initially prepared in a state in which the deterministic rate is very small. The maximum of the underlying probability distribution, whose motion roughly follows the deterministic one, will therefore move very slowly toward the region of higher values of T. Meanwhile, because of the fluctuations, the probability will develop a width proportional to the length of the induction period and inversely proportional to the square root of the size V. If the length of the induction period is large, the size effect will be counteracted and the width will be appreciable. As a result, a substantial part of the probability mass will reach the ignition point well before the maximum does so. At this moment it will be quickly entrained by the fast motion toward the region of high values of T, a phenomenon that will be interpreted by the observer as a precocious ignition. This leak of probability will go on continuously, but since the system cannot reach infinite values of temperature a "traffic jam" will arise as a result of which a new probability peak will emerge in the region of high T. Eventually the primary peak, by then considerably diminished, will reach the ignition point and this will mark the end of transient bimodality. The same argument suggests also that the deterministic ignition time should belong to the tail of the ignition probability distribution, in agreement with the numerical results of Fig. 8.

6. Concluding remarks

The theory of fluctuations allows us to understand how an instability and a transition between different macroscopic states is switched on at the molecular level. Many open questions subsist in this study. The microscopic foundations of the Markovian description are poorly understood. Progress in this direction is likely to occur through the analysis of simple, solvable models [42]. The incorporation of spatially inhomogeneous fluctuations in the description of the nucleation of an instability has so far been carried out either in an oversimplified or in a completely phenomenological way. In addition, most of the stochastic analyses of transient phenomena like thermal explosion are limited to the idealized model of perfectly well-stirred reactors. The understanding of the role of incomplete stirring and of inhomogeneous fluctuations is thus likely to bring new insights in this class of problems. For instance, it would allow one to understand how a "flame front" or a "hot spot" emerge spontaneously in a system. Molecular dynamics studies could provide a useful way of checking the validity of the various physical conjectures.

In this lecture we have not invoked the second general mechanism of intrinsic stochasticity in nature, namely chaotic dynamics. It has been shown [43] that the dynamics of certain chaotic maps can be cast in the form of a probabilistic description associated to a diffusion process. It would be very interesting to extend this analysis to wider classes of systems.

Experimental studies are, undoubtedly, what is needed mostly in the field of stochastic phenomena in Chemistry at the present time. Chemical systems constitute in this respect an ideal substratum, thanks to the possibility to control their characteristic parameters over wide ranges of values, and to excite a large number of closely-packed macroscopic modes (which is much more difficult in laser Physics where, nevertheless, there is ample evidence about the importance of fluctuations). We are convinced that in the coming years we will witness important developments in this field.

Acknowledgments

We are grateful to J.W. Turner for pertinent comments and fruitful discussions. This work is supported in part by U.S. Department of Energy under Contract No. DE-AS05-81 ER10947 and by the Belgian Government : A.R.C., Convention No. 76/81.II.3.

References

1. J.C. Roux, P. De Kepper and J. Boissonnade : Phys. Letters 97A, 168 (1983)
2. T.R. Kirpatrick, E.G.D. Cohen and J.R. Dorfman : Phys. Rev. Lett. 42, 862 (1979); 44, 472 (1980); Phys. Rev. A 26, 972 (1982); 26, 995 (1982)
3. G. Nicolis and I. Prigogine : Self Organization in Nonequilibrium Systems (Wiley, New York, 1977)
4. H. Haken : Advanced Synergetics : Instability Hierarchies of Self-Organizing Systems and Devices (Springer-Verlag, Berlin, 1983)
5. D. Walgraef, G. Dewel and P. Borckmans : Phys. Rev. A 21, 397 (1980); Adv. Chem. Phys. 49, 311 (1982)
6. T. Kurtz : Stoch. Proc. Applic. 6, 223 (1978)
7. P. Kotelenez : Report No. 81 (Universität Bremen, December 1982)
8. C. Van den Broeck, M. Malek Mansour and F. Baras : J. Stat. Phys. 28, 557 (1982)
9. F. Baras, M. Malek Mansour and C. Van den Broeck : J. Stat. Phys. 28, 577 (1982)
10. H. Lemarchand : Bull. Acad. R. Belgique 67, 343 (1981)
11. E. Knobloch and K.A. Wiesenfeld : J. Stat. Phys. 33, 611 (1983)
12. M. Malek Mansour, C. Van den Broeck, G. Nicolis and J. W. Turner : Ann. Phys. (New York) 131, 283 (1981)
13. W. Feller : An Introduction to Probability Theory and Its Applications (Wiley, New York, 1967)
14. G. Nicolis and R. Lefever : Phys. Lett. 62A, 469 (1977)
15. G. Nicolis and J.W. Turner : Ann. N.Y. Acad. Sci. 316, 251 (1979)
16. H. Lemarchand : Physica 101A, 518 (1980)
17. H. Lemarchand and G. Nicolis : J. Stat. Phys. (in press, 1984)
18. A. Fraikin and H. Lemarchand : J. Stat. Phys. (submitted, 1984)
19. D. Walgraef, G. Dewel and P. Borckmans : J. Chem. Phys. 78, 3043 (1983)
20. See, for instance, J. Guckenheimer and P. Holmes : Nonlinear Oscillations, Dynamical Systems and Bifurcations of Vector Fields (Springer-Verlag, Berlin 1983)
21. G. Nicolis and M. Malek Mansour : Phys. Rev. A 29, 2845 (1984)
22. C.W. Gardiner, K.J. McNeil, D.F. Walls and I.S. Matheson : J. Stat. Phys. 14, 307 (1976)
23. A.M. Tremblay, M. Arai and E.D. Siggia : Phys. Rev. A 23, 1451 (1981)
24. M. Mareschal and E. Kestemont : Phys. Rev. A 30, 1158 (1984)
25. M. Suzuki : in Proc. XVIIth Solvay Conf. Phys. (Wiley, New York, 1981)
26. B. Caroli, C. Caroli and B. Roulet : J. Stat. Phys. 21, 415 (1979); Physica 101A, 581 (1980)
27. N.G. Van Kampen : J. Stat. Phys. 17, 71 (1977)
28. F.T. Arecchi, A. Politi and L. Ulivi : Il Nuovo Cim. 71B, 119 (1982)

29. B. Matkowsky, Z. Schuss, C. Knessel, C. Tier and M. Mangel, Techn. Rep. No. 8307, Applied Math. Department (Northwestern University, December 1983)
30. W. Ebeling and L. Shimansky-Geier : Physica 98A, 587 (1979)
31. M. Frankowicz and E. Gudowska-Nowak : Physica 116A, 331 (1982)
32. D. Borgis and M. Moreau : J. Stat. Phys. (in press, 1984)
33. F. Baras, G. Nicolis, M. Malek Mansour and J.W. Turner : J. Stat. Phys. 32, 1 (1983)
34. M. Frankowicz and G. Nicolis : J. Stat. Phys. 33, 595 (1983)
35. G. Nicolis, F. Baras and M. Malek Mansour : in Chemical Instabilities : Applications in Chemistry, Engineering, Geology and Material Science (G. Nicolis and F. Baras Eds., Reidel, Dordrecht, 1984)
36. G. Broggi and L.A. Lugiato : preprint (1984)
37. D.A. Frank-Kamenetskii : Diffusion and heat Transfer in Chemical Kinetics (Plenum, New York 1969)
38. J.C. Lermant and S. Yip : Combustion and Flame (submitted, 1984)
39. G. Nicolis and H. Frisch : Phys. Rev. A (in press, 1984)
40. T. Boddington, P. Gray and S.K. Scott, J. Chem. Soc. Faraday Trans. 2 78, 1721 and 1731 (1982)
41. F. Baras : (to be published, 1984)
42. Y. Elskens, H. Frisch and G. Nicolis : J. Stat. Phys. 33, 317 (1983)
43. T. Geisel : Phys. Rev. Lett. 48, 7 (1982)

Stochastic Theory of a Bistable Chemical Reaction in Non-Homogeneous Liquids

M. Moreau and D. Borgis

Laboratoire de Physique Théorique des Liquides, T. 16, Université Pierre et Marie Curie, 4, Place Jussieu, F-75230 Paris Cédex 05, France

1. INTRODUCTION

Bistable chemical reactions are the object of increasing interest from the experimental and theoretical points of view. The simplest abstract example is the Schlögl model ; well known experimental cases of bistability are the chlorite-iodide reaction or the iodate oxydation of the arsenous acid.

The theoretical study of a bistable reaction in an inhomogeneous medium is a complex but exciting problem because of the large variety of possible behaviours, leading for instance to nucleation processes. The macroscopic analysis is not quite easy, and it does not account for the chemical relaxation between the stable sates. As for the stochastic kinetics of the reaction, it is described by a multivariate Master Equation which cannot be solved exactly : even its stationary solution is unknown. The continuous approximations only yield local information on the probability distribution, which are not sufficient to study the relaxation between the macroscopic stationary states, at least far from the threshold of bistability.

However we will see that the multivariate M.E. may be studied approximately if the diffusion coefficient is large or small. In the first case, quasihomogeneity is rapidly established, and a Chapman-Enskog method applies. In the case of a small diffusion coefficient a quite different approximation is obtained by supposing that the probability distribution is quasistationary in the neighbourhood of the macroscopic stable states : when applied to the simplest example, a Schlögl reaction in two homogeneous cells coupled by diffusion, this method leads to non-intuitive results which agree with Monte-Carlo simulations. They are expected to hold for reactions in two macroscopic cells which could perhaps be realized experimentally. On a smaller scale it seems possible to extend the method to a succession of homogeneous cells and to treat simple models of nucleation.

2. HOMOGENEOUS SYSTEMS

2.1. Deterministic description of a closed system

Let us take as an example the Schlögl model [1] :

$$A + 2X \underset{k_2}{\overset{k_1}{\rightleftarrows}} 3X \quad , \quad B \underset{k_4}{\overset{k_3}{\rightleftarrows}} X \tag{1}$$

We suppose that these reactions take place in a box of volume V which is closed for the molecules X, but not necessarily for the other ones, so that the concentrations c_A, c_B of A and B remain constant. If n(t) is the number of molecules X at time t, the kinetics of this system is given by the deterministic equations :

$$\frac{d}{dt} n = w(n) - \overline{w}(n) \tag{2}$$

$$w(n) = k_1 c_A n^2/V + k_3 c_B V \quad ; \quad \overline{w}(n) = n(k_2 n^2/V^2 + k_4) \tag{3}$$

It is easily seen that in some conditions (2) admits three stationary solutions $\alpha < \beta < \gamma$: α and γ are stable whereas β is unstable. (Clearly the same analysis applies to other systems with one free variable concentration implied in trimolecular reactions). It is well known that a deterministic equation such as (2) does not allow the system to pass from one stable state to the other. To take this effect into account a stochastic description is necessary.

2.2 Stochastic theory

In the birth and death formalism the probability p(n,t) to have n particles at time t in the homogeneous reactive system obeys the Master Equation :

$$\frac{d}{dt} p(n) = W_{n-1}\, p(n-1) - (W_n + \bar{W}_n)\, p(n) + \bar{W}_{n+1}\, p(n+1) = R\,(\dot{p}(n)) \tag{4}$$

where the birth and death rates W_n and $\bar{W}_n$ are, with the notations of (3) :

$$W_n = k_1 c_A\, n(n-1)/V + k_3\, c_B V \;; \quad \bar{W}_n = k_2\, n(n-1)\,(n-2)/V^2 + k_4\, n \tag{5}$$

The stationary solution $p(n,\infty) = q_n$ satisfies the detailed balance condition $W_n q_n = \bar{W}_{n+1}\, q_{n+1}$, so that

$$q_n = g_n / \sum_k g_k \tag{6}$$

with

$$g_n = W_o W_1 \ldots W_{n-1} \,/\, (\bar{W}_1 \bar{W}_2 \ldots \bar{W}_n) \tag{6'}$$

It may be shown that q_n has two sharp maxima at the deterministic stationary states α and γ and a minimum at β. The spectral analysis of (5) shows that, satrting from an initial distribution concentrated around some state $n_o \neq \beta$, p(n,t) at first evolves towards a quasistationary distribution with one peak on the nearest stationary state, α or γ ; at long times a second peak appears on the other stationary state. The final evolution does not change the shape of these two peaks, but only their probability weight, until the stationary distribution q_n is realized; if p_α and p_γ are the total probabilities of the peaks α and γ respectively, they satisfy the very simple approximate equations [2] :

$$\frac{d}{dt} p_\alpha = - \frac{d}{dt} p_\gamma = - k^\alpha\, p_\alpha + k^\gamma\, p_\gamma \tag{7}$$

with, if $x = \alpha$ or γ

$$k^x = \{q^x \sum_{k=m-1}^{M} 1/W_k q_k\}^{-1} \tag{8}$$

Here $q^\alpha = \sum_{k<\beta} q_k$ and $q^\gamma = \sum_{k>\gamma} q_k$ are the stationary values of p_α and p_γ respectively, m and M are the limits of an intermediary region around the unstable state γ ($a<m<\beta<m<\gamma$). On an other hand, continuous approximations of (4) are obtained by taking as a variable the concentration n/V, and expanding the equations in powers of 1/V [3]. The method of stochastic potentials [4] for instance, consists in supposing p(n,t) to be of the form

$$p(n,t) \propto \exp\{- V\, \psi\, (n/V,t)\}$$

and in studying the successive approximations of ψ. In this way it is possible to find the stationary probability, and time-dependent evolution of p near its extrema.

2.3. Homogeneous cell with diffusion through the walls

If now the same system is allowed to exchange molecules X with the exterior by diffusion through the walls, we have to add a diffusion term to (2) and (4). Assuming that the X molecules can freely cross the walls of the cell and that their external concentration c_o is kept constant, this diffusion term id $D(c_oV - n)$ in the deterministic equation (2) and

$$Dc_oV\ p(n-1) - D(c_oV + n)\ p(n) + D(n+1)\ p(n+1) \tag{7}$$

in the stochastic formalism of (4). The diffusion coefficient D is related to Fick's coefficient d by

$$D = d/(\lambda\ell \tag{8}$$

where λ is the mean free path of the X molecules and ℓ the cell length. Then it is easily seen that all the results of 2.2. hold if W_k and $\overline{W}_k$ are replaced by $W_k + Dc_oV$ and $\overline{W}_k + Dk$ respectively. It turns out that bistability is lost if D is superior to a critical value D_c ; if $D < D_c$ and if C_o corresponds to one of the steady states α or γ, the diffusion increases the stability of this steady state in the cell.

3. TWO-CELLS MODEL

3.1. Deterministic analysis

We now consider a system consisting of two identical, homogeneous cells coupled by diffusion. In each cell takes place the Schlögl reaction (1), with the same constant values of c_A and c_B. The equations for the numbers n_1 and n_2 of molecules X in each cell are

$$\frac{d}{dt} n_1 = w(n_1) - \overline{w}(n_1) + D\ (n_2 - n_1) \quad ; \quad \frac{d}{dt} n_2 = w(n_2) - \overline{w}(n_2) + D(n_1 - n_2) \tag{9}$$

where w and $\overline{w}$ are given by (3) and D by (8). The stationary solutions of (9) can be found exactly for the Schlögl reaction [5] or any trimolecular reactive scheme. When the chemical equation (2) has three possible solutions α, β, γ, the system (9) can have up to 9 stationary solutions. Two critical values D_{c1} and D_{c2} of D exist. For $D < D_{c1}$, both homogeneous and inhomogeneous solutions occur : among the homogeneous ones, (α,α) and (γ,γ) are stable nodes and (β,β) is an unstable point ; the inhomogeneous solutions include two stable nodes near (α,γ) and (γ,α) and four saddle points. For $D_{c1} < D < D_{c2}$, the two inhomogeneous stable nodes and two of the inhomogeneous saddle points disappear. For $D > D_{c2}$, only the homogeneous solutions exist ; (α,α) and (γ,γ) are stable nodes, (β,β) is a saddle point.

3.2. Stochastic description : the multivariant Master Equation

According to the multivariate Master Equation formalism the probability $p(n_1,n_2;t)$ of having n_1 molecules in cell 1 and n_2 molecules in cell 2 at time t satisfies the equation

$$\frac{d}{dt} p(n_1,n_2) = (\mathbf{R}_1 + \mathbf{R}_2)\ (p(n_1,n_2)) + D(n_1+1)\ p(n_1+1,n_2-1) - D(n_1+n_2)\ p(n_1,n_2) + D(n_2+1)\ p(n_1-1,n_2+1) \tag{10}$$

where $\mathbf{R}_i$ is the reaction operator defined by (4), operating on n_i. No detailed balance prevails in (10), and the stationary solution cannot be found. As for the method of probability potentials, it yields a non-linear partial differential equation which cannot be solved exactly even in the stationary case ; the time-dependent evolution may only be known locally, and the long time evolution cannot be found in this way beyond the threshold of bistability. However, approximations of the discrete Master Equation (10) may be used if D is large or small. We will only study the last case, since for large D the system becomes rapidly homogeneous.

3.3. Small diffusion coefficient ($D < D_{c1}$)

It has been shown by Monte-Carlo simulation [6] that the stationary distribution presents two peaks on the homogeneous steady states (α,α), (γ,γ) and two peaks on the inhomogeneous steady states (α',γ') and (γ',α'). If $D = 0$, $\alpha' = \alpha$ and $\gamma' = \gamma$; if D increases, α' and γ' are slightly shifted towards γ and α respectively. Thus in the plane (n_1,n_2) there are only four regions, around these peaks, where the probability differs significantly from 0 at long times ; they are denoted $(\alpha\alpha)$, $(\alpha\gamma)$, $(\gamma\alpha)$, $(\gamma\gamma)$, for instance $(n_1,n_2) \in (\alpha\gamma)$ if $n_1 < m$ and $n_2 > M$, where m and M are the limits of an intermediary region around the unstable homogeneous state β. We assume that during the final evolution the shape of $p(n_1,n_2;t)$ does not change appreciably in each of these four regions, whereas their total probabilities $p_{\alpha\alpha}$, $p_{\alpha\gamma}$, $p_{\gamma\alpha}$, $p_{\gamma\gamma}$, evolve towards their stationary (but unknown) values ; this hypothesis is completely justified if $D = 0$ (uncoupled cells) and presumably applies if D is small.

Then it may be shown [7] that an approximate Master Equation holds for the probabilities p_{xy} (where $x,y = \alpha$ or β) :

$$\frac{d}{dt} p_{\alpha\alpha} = -2 k_\alpha^\alpha p_{\alpha\alpha} + k_\alpha^\gamma (p_{\alpha\gamma} + p_{\gamma\alpha}) \qquad (11)$$

$$\frac{d}{dt} p_{\alpha\gamma} = k_\alpha^\alpha p_{\alpha\alpha} - (k_\gamma^\alpha + k_\alpha^\gamma) p_{\alpha\gamma} + k_\gamma^\gamma p_{\gamma\gamma}$$

($\frac{d}{dt} p_{\gamma\gamma}$ and $\frac{d}{dt} p_{\gamma\alpha}$ being obtained by interchanging α and γ), with

$$k_y^x = \{q_y^x \sum_{k=m-1}^{M} 1/\, \mu_k^x\, g_k^x\}^{-1} \qquad (12)$$

$$g_k^x = \mu_o^x\, \mu_1^x \ldots \mu_{k-1}^x \,/\, (\bar{\mu}_1^x\, \bar{\mu}_2^x \ldots \bar{\mu}_k^x) \quad ; \quad q_y^x = \sum_{k \in y} g_k^x \qquad (13)$$

and finally

$$\mu_k^x = W_j + D < n/j >_x \, , \quad \bar{\mu}_k^x = \bar{W}_j + D\, j \qquad (14)$$

where $< n/j >_x$ is the average number of particles in one cell, knowing that this cell is in region (x) and that the other cell is in state j. The last difficulty implied by the coupling between the cells consists in evaluating $< n/j >_x$.

The simplest approximation is to take $< n/j >_x = x$, which in fact amounts to neglecting the correlation between the cells. Then the eigenvalues and eigenvectors of (11) are easily found, and the long time evolution is completely determined and quatitatively agrees with the results of Monte-Carlo simulations [6] for small D. In particular it is shown that the system passes from $(\alpha\alpha)$ to $(\gamma\gamma)$ via the inhomogeneous states $(\alpha\gamma)$ and $(\gamma\alpha)$.

However, the stationary distribution obtained in this way does not reflect the deterministic results reported in 3.1. When D is not very small, since the correlation between cells cannot been neglected then. But it is possible to show that the conditional average $i = < n/j >_x$ approximately satisfies the equation :

$$W_i - \bar{W}_i + D(j+1) \frac{W_j + Di}{\bar{W}_{j+1} + D(j+1)} - D\, i\, \frac{\bar{W}_j + Dj}{W_{j-1} + Di} = 0 \qquad (15)$$

For a given value of j a solution of (15) is also an extremum of $p(n_1,j)$ for variable n_1 ; thus the extrema of $p(n_1,n_2)$ are the solutions of (15) and of the symmetrical equation obtained by exchanging i and j. These equations may be treated numerically, and the results agree completely with the conclusions of the deterministic analysis (Fig.1). As for the mean passage times, computed from (11) they do not differ very much from those of the previous simple approximation.

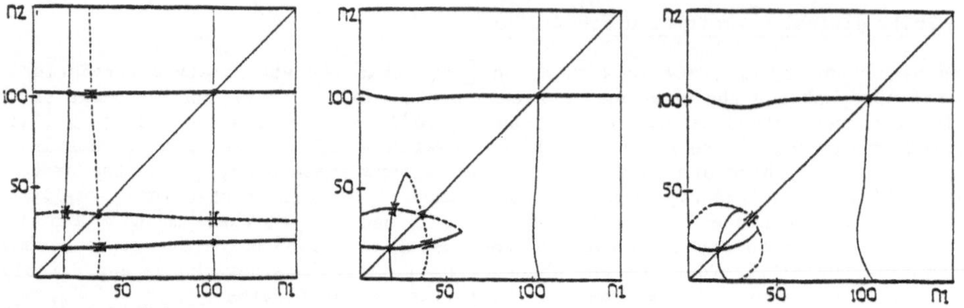

(a) (b) (c)

Fig. 1. Maximum (——) and minimum (---) of the conditional probabilities, in both directions, and for different values of D : (a) $D < D_{c1}$ (b) $D_{c1} < D < D_{c2}$ (c) $D_{c2} < D$; the intersection of two solid lines gives a stable node (o) ; the intersection of one solid line and one dashed line gives a saddle point (⤫) ; the intersection of two dashed lines gives an unstable point

4. CONCLUSION : THE NUCLEATION PROBLEM

The previous results as well as numerical simulations [8] suggest that a non-equilibrium bistable reactive system may evolve by a nucleation process, in agreement with the theory of nucleation-induced transitions [9] and with experimental observations [10]. If the whole system is initially in the metastable state α, in such conditions that the lifetime of α is of the order of the observation time, spontaneous local fluctuations may form a microdroplet of the stable state γ ; then the chemical reaction tends to extend its volume, whereas the diffusion tends to reduce it. The deterministic theory shows [11] that if the radius of the droplet happens to exceed a critical value it will grow until the whole system switch to state γ ; but a droplet of smaller radius should disappear. Thus a stochastic theory is necessary to describe the formation of a droplet of critical size.

A way to do it is to modelize the possible states of the droplet as a succession of spheres of increasing radii R_ν , $\nu = 0,1,2,...$ The system is in state ν if the droplet of (γ) has a radius R_ν, and it can jump from ν to $\nu \pm 1$, with forward and backward rates of type (14). Only the simplest cases have been studied [11]. It would be interesting to improve these treatments, and in particular to account for the correlation between the different zones by extending the method exposed for the two cells model.

REFERENCES

1. F. Schlögl, Z. Phys. 253, 147 (1972)
2. D. Borgis and M. Moreau, Physica 113A, 109 (1983)
3. M. Malek-Mansour, C. Van den Broeck, G. Nicolis and J.W. Turner, Ann. Phys. (N.Y.) 131, 282 (1981)
4. H. Lemarchand, Physica 101A, 518 (1980)
5. W. Ebeling and H. Malchow, Ann. Physik 36, 121 (1979)
6. M. Frankowicz and E. Gudowska-Novak, Physica, 116A, 331 (1982)
7. D. Borgis and M. Moreau, to be published
8. J. Boissonade, Physica, 113A, 607 (1982)
9. I. Prigogine and G. Nicolis, Proc. 3rd Intern. Conf. : From Theoretical Physics to Biology, p. 89 (Karger, 1973)
10. J.C. Roux, P. de Kepper and J. Boissonade, Phys. Letters 97A, 168 (1983)
11. L. Schimansky-Geier and W. Ebeling, Ann. der Phys. 40, 10 (1983).

Stochastic Description of Various Types of Bifurcations in Chemical Systems

H. Lemarchand and A. Fraikin

Laboratoire de Chimie Générale, Université P. et M. Curie, Tour 55
F-75230 Paris Cédex 05, France

When a nonlinear system evolves under far-from-equilibrium conditions in the vicinity of a bifurcation point, a purely deterministic description often proved to be incomplete. The fluctuations of the dynamical variables can play an essential role and obstruct the observation of a transition expected by a deterministic analysis. In the framework of the deterministic approach, the stability of the different states according to the values of the control parameters is studied through a mathematical analysis of the velocity field. In particular, the theory of normal forms leads to the determination of the various kinds of attractors [1,2]. As far as we are concerned with the stochastic approach, the *master equation* has been widely used to analyze bifurcations of homogeneous or spatially ordered steady states or of limit cycles [3,4]. Our aim in the present contribution is to insist on the generality of the method to analyze various kinds of bifurcations in nonlinear nonequilibrium systems. The general procedure proposed to obtain a local description of the probability, which allows us to determine the system's attractors, turns out to display marked analogies with the theory of normal forms.

In section 1, we recall the general formalism leading to the local description around a reference steady state of the stationary solution of the master equation. The result is expressed in terms of the first and the second moments of the transition probabilities. Section 2 is devoted to the application of the general formalism to complex bifurcations which are characterized by the existence, in the spectrum of the linearized operator around the reference state, of two kinds of singularities controlled by two parameters. The first example refers to a reaction-diffusion system in conditions close to those giving to the linearized operator both a pair of pure imaginary eigenvalues and a zero eigenvalue. As a second application we refer to the poster presented by A. FRAIKIN the abstract of which is included in this volume. It is devoted to the stochastic analysis of a normal form of two variables which may display a homoclinic trajectory.

1. General Formalism

Details of the method outlined below can be found in [5]. A stochastic system of several extensive variables X_j is supposed to be described by a master equation which can be explicitly written when the transition probabilities per unit time $W(\{X_j\}\rightarrow\{X'_j\})$ are known. In a reaction diffusion system, X_j may be the number of chemical species α in a cell located by the vector r and is denoted by $X_j = X_{r\alpha}$. Introducing the *stochastic potential* U defined by $P = \exp(-S - \mathcal{N}\, U)$, where P is the probability, $\mathcal{N}$ is proportional to the total volume of the system and S stands for the normalization factor, we switch to the quasicontinuous intensive variables $x_j = X_j/N$, where N may be the mean number of particles in one cell of a reaction-diffusion system. If we assume that for all states for which $W(\{X_j\}\rightarrow\{X'_j\})$ are nonnegligible, $x_j - x'_j$ is much smaller than 1, the equation for U can be expressed, at the zeroth order in 1/N, in terms of $\{x_j\}$ and $\{\partial U/\partial x_j\}$. We thus obtain a Hamilton Jacobi type of equation :

$$-\frac{1}{\mathcal{N}}\frac{dS}{dt} - \frac{\partial U}{\partial t} = H(\{x_j\},\{\frac{\partial U}{\partial x_j}\}) \tag{1}$$

The hamiltonian H is completly determined when the transition probabilities are known. Moreover, its derivatives with respect to $\partial U/\partial x_j$, $\partial U/\partial x_{j'}$, ... at a point where $\{\partial U/\partial x_j\} = 0$,are shown to represent the successive moments of the transition probabilities. They are denoted by H_j, $H_{jj'}$, ...

Looking for a local description of the stationary solution of (1) around a state $\{\bar{x}_j\}$ for which $\{\partial U/\partial x_j\} = 0$, we write U in the form of a Taylor expansion in $\xi_j = x_j - \bar{x}_j$ as

$$U = \frac{1}{2}\, U^{j^1 j^2}\, \xi_{j^1}\xi_{j^2} + \frac{1}{3!}\, U^{j^1 j^2 j^3}\, \xi_{j^1}\xi_{j^2}\xi_{j^3} + \frac{1}{4!}\, U^{j^1 j^2 j^3 j^4}\, \xi_{j^1}\xi_{j^2}\xi_{j^3}\xi_{j^4} + \ldots \tag{2}$$

where all indices with numerical exponent are summed over. The equations for the coefficients of (2) are easily deduced from (1) by successive derivations. They are expressed in terms of the moments H_j, $H_{jj'}$, ... and their derivatives with respect to $x_j, x_{j'}$, ..., denoted by upper indices: $H_j^{j'}$,$H_j^{j'j''}$,.... It is easily shown that $H_j^{j'}$ is the linearized operator around the reference state $\{\bar{x}_j\}$ which is taken to be a uniform steady state. The point is that all the equations for the coefficients of (2) can be explicitly solved in the representation which diagonalizes the linear operator $H_j^{j'}$. Switching to this representation and denoting by ξ_l the new variables (the new indices l label the eigenvectors of $H_j^{j'}$), we obtain a local expansion of U in the same form as (2) by replacing all the indices j by l. The expression of U is then reduced by taking advantage of the simplifications arising in the vicinity of a bifurcation point. Note that far from such conditions, (2) can be truncated to its quadratic terms. Near a bifurcation, some eigenvalues of the linear operator, $\omega_{1_o} = \eta_{1_o} + i\,\theta_{1_o}$, are marked by a vanishing real part η_{1_o}. As a result, the coefficients of the quadratic terms $U^{11'}$ are shown to take well-separated orders of magnitude, depending on whether the eigendirections 1,1' are associated to singular eigenvalues 1_o or nonsingular ones $1_\emptyset$. We find :

$$U^{1_o 1'_o} = O(\varepsilon^2_{1_o}) + \varepsilon_{1_o}\,\delta_{1'_o \bar{1}_o} \quad ; \quad U^{1_o 1_\emptyset} = O(\varepsilon_{1_o}) \quad ; \quad U^{1_\emptyset 1'_\emptyset} = O(1)$$

where $\bar{1}$ labels the eigenvector complex conjugate of the 1th and $\varepsilon_{1_o} = 1/U^{-1}_{1_o \bar{1}_o} = -2\,\eta_{1_o}/H_{1_o \bar{1}_o}$ vanishes at the bifurcation point. This allow us to separate the stochastic potential into two parts: a quadratic contribution relating to the "noncritical" variables $\xi_{1_\emptyset}$ and a critical contribution including terms of third and fourth order in the variables ξ_{1_o}. We give below the resulting fourth order poly nomial U_{cr} :

$$U_{cr} = \frac{1}{2}\,\varepsilon_{1^1_o}\,\xi_{1^1_o}\,\xi_{\bar{1}^1_o} + \frac{1}{3!}\,U^{1^1_o 1^2_o 1^3_o}\,\xi_{1^1_o}\,\xi_{1^2_o}\,\xi_{1^3_o} + \frac{1}{4!}\Big(U^{1^1_o 1^2_o 1^3_o 1^4_o} - 3\,U^{1^1_o 1^2_o 1^5_\emptyset}\,U^{-1}_{1^5_\emptyset 1^6_\emptyset}\,U^{1^6_\emptyset 1^3_o 1^4_o}\Big)\,\xi_{1^1_o}\,\xi_{1^2_o}\,\xi_{1^3_o}\,\xi_{1^4_o} \qquad (3)$$

where the coefficients are given, at the dominant order, by

$$U^{1_o 1'_o 1''_o} = \frac{S\{\varepsilon_{1_o}\,H^{1_o 1'_o 1''_o}_{1_o}\}}{\omega_{\bar{1}_o} + \omega_{\bar{1}'_o} + \omega_{\bar{1}''_o}} \qquad (4)$$

$$U^{1_o 1'_o 1''_o 1'''_o} - S\{U^{1_o 1'_o 1^1}\,U^{-1}_{1^1 1^2}\,U^{1^2 1''_o 1'''_o}\} = \frac{S\{\varepsilon_{1_o}\,H^{1_o 1'_o 1''_o 1'''_o}_{1_o} - \varepsilon_{1_o}\,H^{1_o 1'_o 1^1}_{1_o}\,U^{-1}_{1^1 1^2}\,U^{1^2 1''_o 1'''_o}\}}{\omega_{\bar{1}_o} + \omega_{\bar{1}'_o} + \omega_{\bar{1}''_o} + \omega_{\bar{1}'''_o}} \qquad (5)$$

The symmetrization symbol S, applied to a function of the indices 1,1'.. represents the sum of all distinct terms obtained by permutation of these indices.

It is worth pointing out that (4) and (5) show that the coefficients of the cubic and quartic terms of U_{cr} are proportional to ε_{1_o} and will thus vanish at the bifurcation point unless the denominators, $\omega_{\bar{1}_o} + \omega_{\bar{1}'_o} + \omega_{\bar{1}''_o}$ or $\omega_{\bar{1}_o} + \omega_{\bar{1}'_o} + \omega_{\bar{1}''_o} + \omega_{\bar{1}'''_o}$, are also of order ε_{1_o}. These last conditions are strictly equivalent to the resonance conditions which lead to the retention of the dominant nonlinear terms in the theory of normal forms.

This general result, applied to a bifurcation of spatially ordered steady states [4] or to a Hopf bifurcation in a reaction-diffusion system exhibiting inhomogeneous fluctuations [8], proves to be in agreement with the critical dynamics analysis carried out by WALGRAEF et al. [6,7].

2. Application to Complex Bifurcations

As an example of application of the general formalism, let us consider a two component reaction-diffusion system of large size described by the variables $\xi_j = \xi_{r\alpha} = x_{r\alpha} - \overline{x}_\alpha$ where $\{\overline{x}_\alpha\}$ determines the reference uniform steady state. In such a system, the eigenvectors of the linear operator are labelled with the indices $l = m\beta$ where m plays the role of a wave-vector and β takes two values 1 and $-1 = \overline{1}$.

We consider the complex bifurcations characterized by a set of complex conjugate eigenvalues with small real parts, $\omega_{l_0} = \omega_{q\beta}$ and a set of small eigenvalues $\omega_{l_0'} = \omega_{k1}$. The characteristic wave-vectors q associated with the first type of bifurcation have a small modulus leading to a quasi-uniform time periodic structure. The wave-vectors k characteristic of the second bifurcation have a modulus close to k_c, specifying the corresponding spatially ordered stationary state.

Following the general results of section 1, we easily select the resonant terms of U_{cr}. At the third order, a resonance appears for $(l_0\, l_0'\, l_0'') = (k\,1\; k'1\; k''1)$ or $(k\,1\; q\,1\; q'\overline{1})$. The latter makes no contribution to U_{cr} because of the Kronecker delta, $\delta(q + q' + k) = 0$, multiplying the corresponding term in the case of periodic boundary conditions. At the fourth order the resonance conditions lead to the retention of three kinds of terms: $(l_0\, l_0'\, l_0''\, l_0''') = (k\,1\; k'1\; k''1\; k'''1)$, $(k\,1\; k'1\; q\,1\; q'\overline{1})$ and $(q\,1\; q'1\; q''\overline{1}\; q'''\overline{1})$. The resulting critical part of the stochastic potential U_{cr} takes thus the following form:

$$
\begin{aligned}
U_{cr} = {} & \varepsilon_{q^1}\, \xi_{q^1 1}\, \xi_{\overline{q^1}\,\overline{1}} + u(q^1 q^2 q^3 q^4)\, \delta(q^1 + q^2 + q^3 + q^4)\, \xi_{q^1 1}\, \xi_{q^2 1}\, \xi_{q^3 \overline{1}}\, \xi_{q^4 \overline{1}} \\
& + \Big[\varepsilon_{k^1}\, \xi_{k^1 1}\, \xi_{\overline{k^1} 1} + \gamma\, \delta(k^1 + k^2 + k^3)\, \xi_{k^1 1}\, \xi_{k^2 1}\, \xi_{k^3 1} \\
& \quad + v(k^1 k^2 k^3 k^4)\, \delta(k^1 + k^2 + k^3 + k^4)\, \xi_{k^1 1}\, \xi_{k^2 1}\, \xi_{k^3 1}\, \xi_{k^4 1} \Big] \\
& + w(q^1 q^2 k^1 k^2)\, \delta(q^1 + q^2 + k^1 + k^2)\, \xi_{q^1 1}\, \xi_{q^2 \overline{1}}\, \xi_{k^1 1}\, \xi_{k^2 1}
\end{aligned} \tag{6}
$$

where the coefficients γ, u, v, w, are calculable in terms of the transition probability moments. This expression contains three distinct parts. The first one, relating to the variables $(\xi_{q1}, \xi_{q\overline{1}})$, describes the potential obtained in the vicinity of a Hopf bifurcation. The second one, depending on the variables ξ_{k1}, results from the vicinity of the bifurcation of a spatially ordered structure. Note that the coefficient v depends on the imaginary part of the eigenvalues characterizing the Hopf bifurcation and makes the corresponding quartic terms slightly different from those obtained in the case of a simple bifurcation of a spatially ordered structure. Finally, the third part of U_{cr}, containing both the variables $(\xi_{q1}, \xi_{q\overline{1}})$ and ξ_{k1} expresses the interaction between the two kinds of bifurcations. We easily deduce from (6) the equation of the extrema of U_{cr} which admits a solution of the following form:

$$\xi_{q1} = \rho\, e^{i\phi}\, \delta_{q0} \;;\; \xi_{q\bar{1}} = \rho\, e^{-i\phi}\, \delta_{q0} \;;\; \xi_{k1} = \sigma\, (e^{i\psi}\, \delta_{kk_0} + e^{-i\psi}\, \delta_{k\bar{k}_0}) \quad (7)$$

where k_0 is one particular wave-vector of modulus k_c and the amplitudes ρ and σ are imposed by (6). This solution, depending on the phase parameter ϕ, represents the set of states described during the time oscillation of a spatially-ordered structure. The stability of this structure can be analyzed through the technique already used for the simple bifurcation of a limit cycle [8]. It consists in determining the matrix of the second derivatives of U_{cr} at one state of the set given by (7) and in analyzing the sign of its eigenvalues. We expect that this method will allow us to evaluate the spatial correlation of the phase of the oscillation and to determine the conditions of destruction of long range order.

References

1. V. ARNOLD, *Chapitres Supplémentaires de la Théorie des Equations Différentielles Ordinaires* (Mir, Moscou 1980)
2. S.N. CHOW and J.K. HALE, *Methods of Bifurcation Theory* (Berlin, Heidelberg, New York, 1982)
3. M. MALEK MANSOUR, C. VAN DEN BROECK, G. NICOLIS, and J.W. TURNER, Ann. Phys. (New York) 131, 283 (1981)
4. H. LEMARCHAND and G. NICOLIS, J.Stat. Phys. (Submitted)
5. H. LEMARCHAND , Bull. Acad. Roy. Belg. (in press)
6. D. WALGRAEF, G. DEWEL and P. BORCKMANS, Adv. Chem. Phys. 49,311(1982)
7. D. WALGRAEF, G. DEWEL and P. BORCKMANS, J.Chem. Phys. 78 3043 (1983)
8. A. FRAIKIN and H. LEMARCHAND, J. Stat. Phys. (Submitted)

Long Range Correlations in a Non Equilibrium System

M. Mareschal

Faculté des Sciences de l'Université Libre de Bruxelles, Campus Plaine, CP 231
B-1050 Bruxelles, Belgium

1. Introduction

A very **remarkable** property exhibited by non equilibrium systems is the appearance of coherence and long range order [1]. Once subjected to a constraint which **maintains** a system in a nonequilibrium steady state, fluctuations which occur in its different parts, are not **independent.** The correlation extends over the size of the sample [2-5]. We shall here study the origin of this behaviour in a simple example and describe a simulation set up in order to demonstrate this property.

The model we present here is very simple, although it contains all the ingredients needed for the description **of long** range order. It consists of a system obeying the Fourier law of conduction of heat and maintained out of equilibrium by appropriate boundary conditions. We shall use a Langevin approach in order to calculate the temperature fluctuations in the steady state. These calculated fluctuations differ from those of, even local, equilibrium. However small the constraint is, there will appear a correlation between fluctuations taking place everywhere in the system[2c,5].

The first part of this article deals with the Langevin approach; it is very intuitive and it is, in fact, difficult to justify. One needs either experiments which confirm theoretical pre**dictions, or** simulations which validate the hypotheses that feed the theories. The problem is difficult, as the expected effects are very small [6]. In the second part we describe a nonequilibrium simulation of a dynamical system submitted to an external constraint we made in order to check on the system itself the validity of the Langevin hypotheses [6b] .

2. The Fourier-Langevin approach

We consider a system where the temperature obeys the Fourier equation, namely

$$\frac{\partial T(\vec{r},t)}{\partial t} = \frac{-1}{nc_p} \vec{\nabla}\cdot\vec{J}_q(\vec{r},t) \qquad (1)$$

$$\vec{J}_q(\vec{r},t) = -\kappa\, nc_p \vec{\nabla} T(\vec{r},t) \qquad (2)$$

where κ is the thermal diffusivity, supposed to be **independent** of the temperature and $\vec{J}_q(\vec{r},t)$ is the heat flux in the system.

The study of fluctuations at equilibrium is based on the following scheme[7]. The temperature fluctuations, $\delta T(\vec{r},t)$,are supposed to obey the same Fourier eq. (1), with the the **addition** of a fluctuating source term. This fluctuating heat flux, $\vec{J}_q^{\,f}(\vec{r},t)$, is varying on a time scale of the order of the relaxation time, that is small compared to the hydrodynamical time scale, and the correlation in space is also restricted to a molecular scale. This suggests the following assumptions for the mean values of the fluctuating heat flux

$$\langle \vec{J}_q^{\,f}(\vec{r},t)\rangle = 0 \tag{3}$$

$$\langle J_{q\alpha}^{f}(\vec{r},t) J_{q\beta}^{f}(\vec{r}',t')\rangle = A\,\delta(\vec{r}-\vec{r}')\,\delta(t-t')\,\delta^{kr}_{\alpha,\beta} \quad (\alpha,\beta=x,y,z) \tag{4}$$

The amplitude A is determined by the requirement that one should recover the thermodynamic result, derived from Einstein's formula, when solving the fluctuating Fourier equation, namely

$$\langle \delta T(\vec{r},t)\,\delta T(\vec{r}',t)\rangle = \frac{k\ T^2}{nc_V}\,\delta(\vec{r}-\vec{r}') \tag{5}$$

where n is the particle number density, c_V is the heat capacity per particle, k is the Boltzmann's constant and T the average temperature. One finds that $A=2\kappa kT^2$, and with this value of A, eq.(4) is often referred to as a " Landau Lifschitz " formula [7].

The basic idea in the extension of this theory to a nonequilibrium situation is to assume that eq.(4) remains valid with the amplitude A now becoming space-dependent through the nonuniform temperature. The procedure to calculate the averages of temperature fluctuations is then the following:

(i)find the stationary state's solution , $T(\vec{r})$, of eq.(1)

(ii)assume a nonequilibrium " Landau Lifschitz " formula

$$\langle J_{q_\alpha}^{f}(\vec{r},t) J_{q_\beta}^{f}(\vec{r}',t')\rangle = 2\ kT^2(\vec{r})\,\delta(\vec{r}-\vec{r}')\,\delta(t-t')\,\delta^{kr}_{\alpha,\beta} \tag{6}$$

(iii)solve then the equation for the temperature fluctuations

$$\frac{\partial\,\delta T}{\partial t}(\vec{r},t) = \kappa\nabla^2\,\delta T(\vec{r},t) - \frac{1}{nc_p}\,\vec{\nabla}\vec{J}_q^{\,f}(\vec{r},t) \tag{7}$$

This procedure leads to the following result for the equal-time temperature fluctuations,

$$\langle \delta T(\vec{r},t)\,\delta T(\vec{r}',t)\rangle = \frac{kT^2(\vec{r})}{nc_V}\,\delta(\vec{r}-\vec{r}') + g(\vec{r},\vec{r}') \tag{8}$$

In three dimensions the correlation function $g(\vec{r},\vec{r}')$ reads,

$$g(\vec{r},\vec{r}') = \frac{k|\vec{\nabla} T|^2}{nc_V} \int \frac{d\vec{q}}{(2\pi)^3}\, e^{-i\vec{q}\cdot(\vec{r}-\vec{r}')}\,\frac{1}{q^2} \tag{9}$$

Actually, $g(\vec{r},\vec{r}')$ is the solution of the Poisson equation which determines the potential due to a charge located at $\vec{r}=\vec{r}'$. So it diverges in an infinite one or two-dimensional system. In a finite rectangle ,for example , if we look for solutions which vanish at the boundary, $g(\vec{r},\vec{r}')$ is the logarithm of the elliptic function; it behaves like $-\log(\vec{r}-\vec{r}')$ near $\vec{r}=\vec{r}'$ [10] . In the one-dimensional line of length L , the correlation function simply reads,

$$g(x,x') = \frac{k|\vec{\nabla}T|^2}{nc_V} \begin{cases} x(L-x') & x > x' \\ x'(L-x) & x < x' \end{cases} \tag{10}$$

Let us also mention that:

a) in the chemical case [1] ,the same long range correlation function appears multiplied by a decreasing exponential which screens the effect. Only near an instability does this screening disappear.

b) the same result was obtained by Nicolis and Malek Mansour using the Master Equation. This is a more deductive approach where energy-transition probabilities are deduced from the requirement of detailed balance at equilibrium. The advantage of the Master Equation with respect to the Langevin formulation is that it permits to study situations where the nonequilibrium " Landau Lifschitz " formulas cannot be guessed from a intuitive argument : this is the case, for example, of a fluid submitted to a shear [8] .

c) a more rigourous derivation of the form of the correlation function is possible through the use of kinetic theory [3,4] . However, some of the hypotheses made in kinetic theory are not under control, like, for example the convergence of the expansion necessary to find the microscopic stationary state's solution.

3.The Molecular dynamic simulation.

The quantities measured in a physical system are in fact averaged over a small volume, say $\Delta V=\Delta L^d$,

$$\Delta T=\frac{1}{\Delta V}\int_{\Delta V} d\vec{r}\,\delta T(\vec{r},t) \tag{11}$$

and if we compare equilibrium and nonequilibrium contributions to the fluctuations, we obtain the following ratio

$$\frac{\langle\Delta T^2\rangle_{eq}}{\langle\Delta T^2\rangle_{noneq}}=\frac{T^2\,L^2}{\Delta T^2\Delta L^2} \tag{12}$$

where $\Delta T/L$ is the imposed gradient. It is therefore not easy to measure directly in an experiment. Nonequilibrium effects have been measured in fluids but

(i)they are of first order in the gradient, like the product at different times of the density and velocity fluctuations [6a].

(ii)these effects are very sensitive to boundary conditions which are not really known[9].

On the contrary nonequilibrium molecular dynamic simulations deal with systems where $T/\Delta T$ may be less than 1 and $L/\Delta L$ may be chosen so that the nonequilibrium effects are not small. The system that we have chosen is an assembly of a few hundreds of hard disks (up to 3000) enclosed in a rectangle of sides L_x and L_y; L_x is three times larger than L_y , periodic boundary conditions are imposed in the y direction and thermal reflection **occurs in the** other direction: each time a disk hits a thermal boundary, it is reinjected in the rectangle with a velocity chosen in an equilibrium velocity distribution, corresponding to temperatures T_1 and $T_1+\Delta T$ at, respectively left and **right.** The choice of a two-dimensional system is due to the fact that we need to have a system with a large hydrodynamic distance integrated during a large hydrodynamical time and this is easier to achieve in lower dimensionality. The averaged density is that of a mean free path of approximately one disk diameter and the largest L_x were of the order of two hundred molecular diameters.

We divide the rectangle in cells of approximately 50 particles and measure particle number density, velocity and temperature in each cell, the temperature being defined as the local time averaged kinetic energy per particle. We also measure averages of products of fluctuations in the same or different cells. As all averages are time averages the precision of the measurement depends strongly on how long the trajectory of the system is calculated. This time may conveniently be expressed in collisions number. For example, after a run of two million collisions for a system of 3000 hard disks at equilibrium, the dispersion of the temperature over **different** cells was less than one percent and that of temperature fluctuations less than seven percent. That number of collisions corresponds approximately to 1500 relaxation times and twenty times the crossing time of a sound wave.

(i) The stationary state reached by the system is that of local equilibrium. Indeed the temperature profile that is shown in fig.2 fits very well the theoretical **straight** line. The pressure measured inside the system is constant everywhere and the density obtained through the hard disks equation of state agrees very well with the measured density (fig.1) . The velocity distribution function inside each cell is much like a gaussian, as can be checked by measuring the moments of this function.

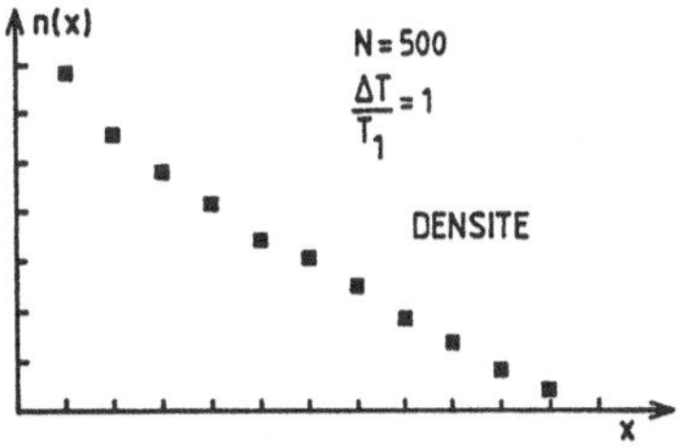

Figure 1 : density profile measured in a run of 1 million collisions.

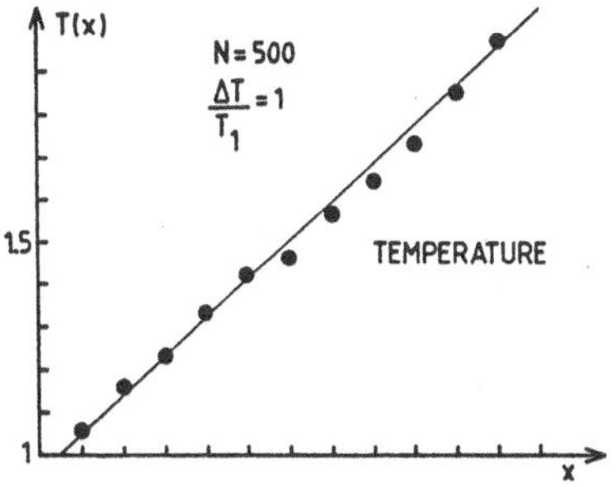

Figure 2 : temperature profile in the same run.

(ii) the validity of eq.(6), that is the nonequilibrium Landau Lifschitz formula, has been checked in the stationary state. Figure 3 shows the evolution of the fluctuating heat flux at equilibrium in two neighbouring cells. The relaxation time in the system's units is 2, and that should be the typical time over which the fluctuating heat flux is self correlated. This is indeed the case as it can be seen. Figure 4 **confirming** the absence of spatial correlations in the system. Measuring the time-averaged product of fluxes **gives** a negligible contribution unless **it refers** to the same cell. These properties remain the same in a nonequilibrium steady state. Beside, plotting $J_{qx}(i)^2$ as a function of the cell temperature ,see fig.5, is a direct check of eq.(6). The slope of the log-log curve is 2.1 and the extrapolated value of the thermal diffusivity agrees quite well with the value measured directly on the system.

(iii) looking at the long range correlations that are built into the system, we have plotted in fig.6 the points obtained for three different values of T/T_1. They refer to a system of 3000 hard disks and they are both time and space averages, as we averaged over all cells of a same line with same cell distance.

The points measured seem to obey the square of the gradient dependance of the correlation function and also they decrease linearly with the distance.

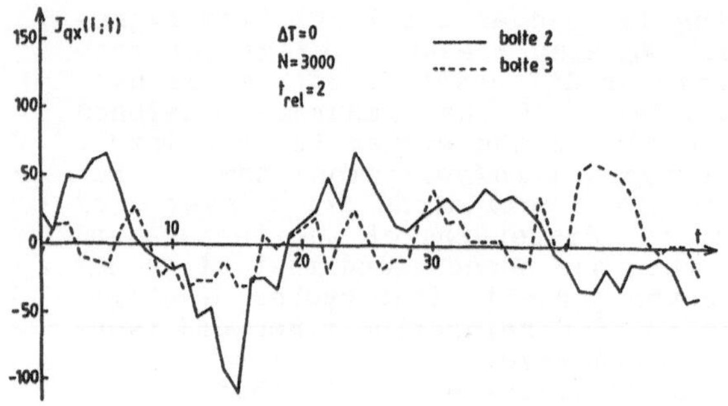

Figure 3 : fluctuating heat flux measured in 2 neighbouring cells.

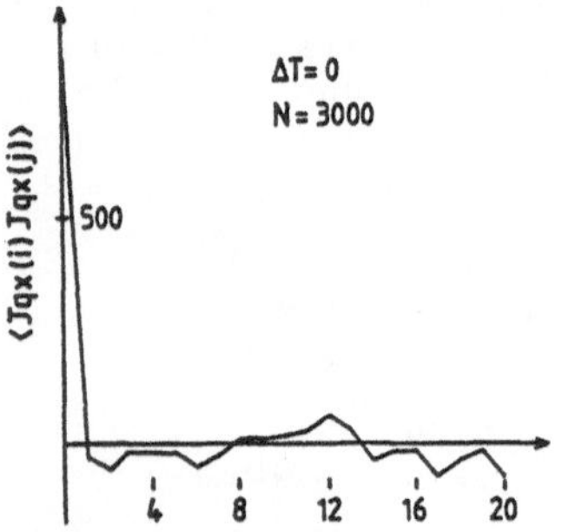

Figure 4 : spatial correlations of the fluctuating heat flux.

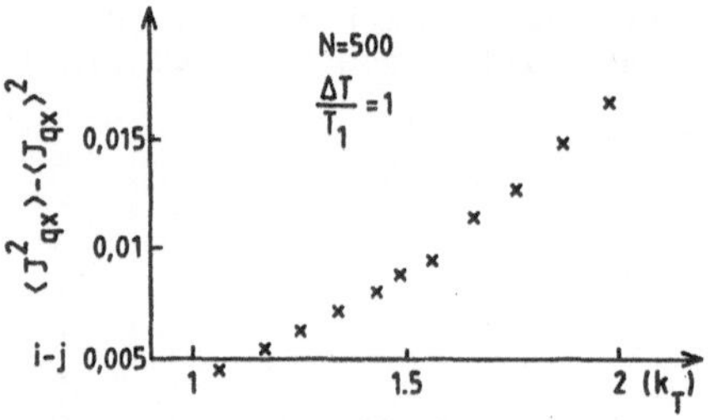

Figure 5 : fluctuating heat flux amplitude as a function of the temperature.

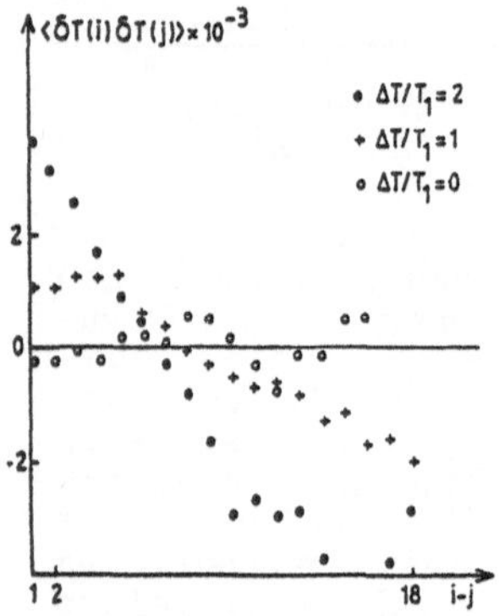

Figure 6 : temperature fluctuations as a function of cell distance.

4. Conclusions.

The coherent behaviour of nonequilibrium systems can be described in our model by a simple generalisation to nonequilibrium state of the Landau Lifschitz formula. On the other hand nonequilibrium molecular dynamics can be used to test the range of validity of the hypotheses that feed the theory. A direct check on the fluctuating fluxes is easier to perform from the point of view of computation time but it is also of a qualitative nature. Measurements on the correlations are feasable but the computation time is too large for three-dimensional systems. Simulations of one-dimensional systems may prove to be very useful in this respect.

Acknowledgements.
Many thanks to Eddy Kestemont whose collaboration was essential, and to Prof.I.Prigogine for his constant support. Much of the reported work **benefited** from discussions with Prof. G. Nicolis, Drs. M. Malek Mansour and F. Baras. Financial support of the "Actions de Recherches Concertées" of the Belgian Government is gratefully ackowledged.

References.

1.See the paper by G.Nicolis,F.Baras and M.Malek Mansour in this volume.
2.a.D.Ronis,I.Procaccia,I.Oppenheim,Phys.Rev.A19,1324(1979)
b.G.Vanderzwan,D.Bedeaux,P.Mazur,Physica 107A,491,(1981)
c.H.Spohn,J.Phys.A16,4275(1983)
3.A.Onuki,J.Stat.Phys.,18,475(1978)
4.T.R.Kirkpatrick,E.G.D.Cohen,J.R.Dorfman,Phys.Rev.A26,972(1982)
5.See the review article by A.M.S.Tremblay in Recent Developments in nonequilibrium thermodynamics, Springer Verlag (1984)
6.a.D.Beysens, Physica 118A ,250(1983)
b.M.Mareschal,E.Kestemont,Phys.Rev.A(August 1984)
7.L.D.Landau,E.F.Lifschitz, Fluids Mechanics, Pergamon,Oxford (1978)
8.G.Nicolis,M.Malek Mansour, Phys.Rev.A29,2845(1984)
9.G.Satten,D.Ronis,Phys.Rev.A26,940(1982)
10.R.Courant,D.Hilbert, Methods of Mathematical Physics,Interscience, New York (1953)

Part VII

Posters

Analysis of Chemical Network Dynamics Using Computer Programmed Theory

Bruce L. Clarke

Department of Chemistry, University of Alberta
Edmonton, Alberta, Canada T6G 2G2

Code for a package of computer programs in APL that can analyze chemical networks on an IBM Personal Computer has appeared [1]. This poster presents enhancements to this package. The package is not a simulation package but a package of computer coded theorems. It contains many new theoretical results in chemical network theory.

The package is intended to allow someone to play with a mechanism and discover its dynamics interactively. Once the reactions are entered in the computer any theorem can be immediately tested to see if it applies to the chemical network.

The initial setup of a reaction network is extremely simple and flexible. Ultimately one needs to have a chemical network with letters representing the species and with only those species whose dynamics is important appearing in the reactions. To get to this ultimate network from a mechanism, one can enter the whole mechanism using formulas and have the computer substitute letters for molecules, pick out the dynamically important species, and find the network.

Most of the package consists of programs that determine the dynamics of the intermediates when the overall reaction is negligibly slow. Of first importance is the nature of the manifold of steady states of the intermediates for all rate constants. One program finds the complete set of steady state manifolds on the closed nonnegative orthant of concentrations dynamics. Several other programs determine if a steady state exists for all rate constants, if such a steady state is unique or if there are multiple steady states for some rate constants.

Dynamical questions that can be answered concern stability and whether a steady state or region of concentration space is globally attracting. There is quite a powerful technique used for finding globally attracting steady states, even for networks where other steady states are unstable.

New programs in this package enable one to prove that a chemical network has a globally attracting region. This is a bounded region of concentration space that would have to contain one or more attractors, either steady states, periodic, quasiperiodic, or chaotic attractors. The program can be used when the steady states are unstable, to prove that some kind of interesting non-steady state attractor must exist. It works with chaotic networks.

The package does not do Hopf bifurcation analysis nor have any direct way to distinguishing between limit cycle and chaotic attractors. The package contains the Zero Deficiency Theorem, the "knot tree network theorem" as well as some older theorems that identify stable networks. The package solves the general reaction-balancing problem whose solution is a convex polyhedral cone of extreme reactions. It handles thermodynamic properties of reactions assuming ideality.

1. B.L. Clarke, "Qualitative Dynamics and Stability of Chemical Reactions Networks", in Chemical Applications of Graph Theory and Topology, Ed. R.B. King, (Elsevier, Amsterdam 1983) pp 322-357.

Heterogeneous Fluctuations in Chemical Oscillators

Michael Menzinger and Peter Jankowski
Department of Chemistry, University of Toronto, Toronto, Canada M5S 1A1

Experimental evidence is given for large-scale concentration fluctuations (of the order of ± 2-100%) in the ferroin-catalyzed BZ reaction, observed in a stirred batch-reactor. Their essentially heterogeneous, spatially distributed nature is established by monitoring simultaneously the redox potential in a micro-volume and a macro-volume by using electrodes of different size. A typical record is shown in Fig.1. Fluctuations are absent from the large-electrode record due to averaging over spatially distributed fluctuations. This represents a direct confirmation of the nucleation model and demonstrates a ubiquitous tendency of the system to differentiate spatially on a mesoscopic scale. We believe that fluctuations play a fundamental role in the global self-organization of unstirred systems.

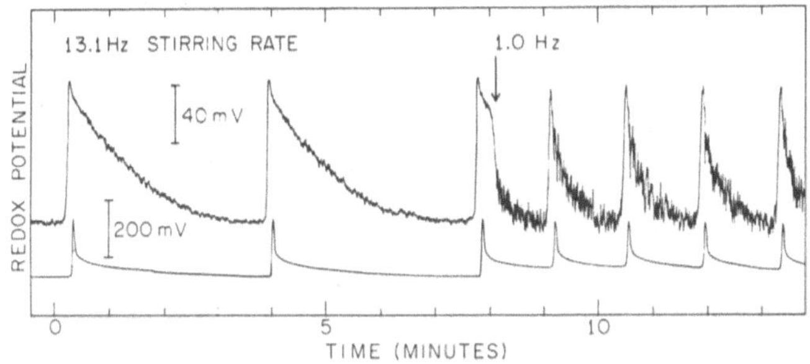

Fig.1. Oscillation of redox potential in the BZ reaction monitored by Pt-micro-electrode (upper trace) and a Pt-macro-electrode (lower trace).

The observed fluctuations are very sensitive to stirring. The pronounced decrease of the oscillation period with stirring rate is accompanied by a marked increase of the fluctuation amplitude. Figure 2 illustrates our interpretation in terms of the contraction of the limit cycle from its deterministic limit 1-2-3-4 to the stochastic limit cycle 1'-2'-3'-4'. The latter is dominated by noise-induced transitions. The interaction between local (stochastic) and global dynamics is seen to be profound.

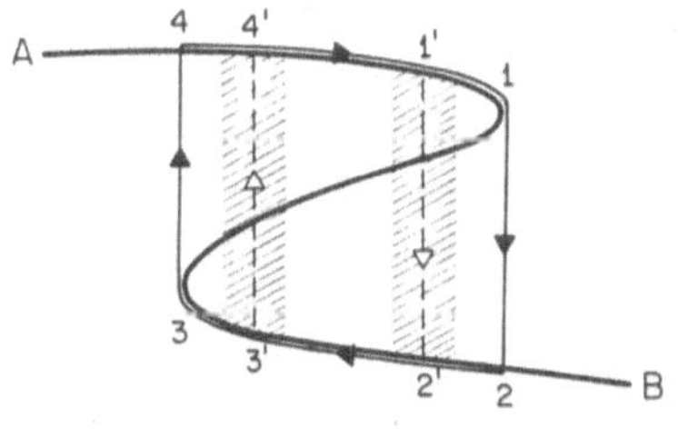

Fig.2. Schematic representation of noise-induced contraction of a limit cycle 1'-2'-3'-4' from its deterministic limit 1-2-3-4.

Basins of Attraction to Oscillatory Modes in a Forced Glycolytic Oscillator

Mario Markus and Benno Hess

Max-Planck-Institut für Ernährungsphysiologie, Rheinlanddamm 201
D-4600 Dortmund 1, Fed. Rep. of Germany

A glycolytic model containing two-substrate rate laws of the enzymes phosphofructokinase and pyruvate kinase was analyzed numerically for periodic substrate input flux [1-3]. Up to four attractors - corresponding to different time patterns - were found to coexist in phase space for the same values of the bifurcation parameters. The system can be switched from one of the coexisting attractors into the other by pulses of small metabolite additions or substractions. By choosing the right phase and the right metabolite amount, this switching may drive the system directly from one attractor to the other, i.e. without transients. However, switching can also be accomplished under less strict conditions by driving the system from one attractor into any phase space point inside the basin of attraction of the other attractor. In this case, transients occur before the new attractor is reached by the system.

We computed cross-sections (Ξ) of the basins of attraction with given planes in phase space. The boundary curves of the Ξ were computed by alternating between the inside and the outside of the Ξ in a zigzag motion defined by equilateral triangles. Often we found that the Ξ had long and narrow appendages looking like flagella. These appendages were computed by proceeding within a narrow region using linear segments of given length and having a slope determined by trial and error ("blindman's stick" method).

The knowledge of the Ξ and thus of the basins easily permits to determine conditions for switching between time patterns.

References

1 B. Hess and M. Markus, in: Synergetics - from Microscopic to Macroscopic Order, ed. by E. Frehland (Springer-Verlag, Berlin, 1984)pp.6-16

2 M. Markus and B. Hess: Proc.Natl.Acad.Sci. USA 81(1984)4394-4398

3 M. Markus and B. Hess: Arch.Biol.Med.Exp. (1984), in the press

Quasiperiodic, Entrained and Chaotic Responses of Yeast Extracts Under Periodic Substrate Input Flux: Model and Experiments

Mario Markus, Stefan C. Müller, and Benno Hess

Max-Planck-Institut für Ernährungsphysiologie, Rheinlanddamm 201
D-4600 Dortmund 1, Fed. Rep. of Germany

The numerical analysis of a two-enzyme model of glycolysis under an input flux given by $V = \bar{V} + A \sin(\omega t)$ reveals a profusion of oscillatory time patterns [1,2]. Scanning of the A-ω-plane leads to tongue-shaped regions, between which exist quasiperiodic regions for small A, and chaotic regions for large A and $2\omega_o < \omega < 3\omega_o$, where ω_o is the oscillation frequency of the system at constant input, i.e. at $V = \bar{V}$.

NADH fluorescence, as indicator of glycolysis, in extracts of baker's yeast (S. cerevisiae) was measured under continuous and periodic glucose injection. Previous investigations had been restricted to $A = \bar{V}$ [3]. We extended the measurement range by scanning the A-ω-plane and obtained entrainment with periods 1,2,3,4,5,7 and 9 times the input flux period, as well as a rich variety of quasiperiodic and chaotic oscillations. The regions on the A-ω-plane corresponding to the different types of responses agree well with the model predictions. On approaching some of the boundaries of these regions, we observed a lengthening of the transients before obtaining a final time pattern. This is expected from the weak stability of the cycles corresponding to maximum Liapunov exponents which become zero at the bifurcations [4]. Furthermore, we found boundaries, the positions of which depend on the direction in which they are crossed, i.e. hysteresis effects, as had been predicted in [1].

References

1 M. Markus and B. Hess: Proc.Natl.Acad.Sci.USA 81(1984)4394-4398
2 B. Hess, D. Kuschmitz and M. Markus, in: Dynamics of Biochemical Systems, ed. by J. Ricard and A. Cornish-Bowden (Plenum Press, N.Y., 1984), in press
3 M. Markus, D. Kuschmitz and B. Hess: FEBS Lett. 172(1984)235-238
4 B. Hess and M. Markus, in: Synergetics - from Microscopic to Macroscopic Order, ed. by E. Frehland (Springer, Berlin, 1984) pp.6-16

Critical Thresholds in Chemical Composition for Convective Pattern Formation in Protein Solutions and Cytoplasmic Media

Stefan C. Müller, Theo Plesser, and Benno Hess

Max-Planck-Institut für Ernährungsphysiologie, Rheinlanddamm 201
D-4600 Dortmund 1, Fed. Rep. of Germany

The evolution of convective patterns is studied in a protein solution (albumin), in cytoplasma extracted from yeast cells (S.cerevisiae), both containing up to 60 mg/ml protein, and in related model solutions. The samples are placed in a petri dish (diameter 3.2 cm) with a layer depth of 0.18 cm. Rod-like patterns form after about 2 minutes and are later transformed into a complex network of polygons. This is observed with a dark-field method for detection of refractive index gradients which reflect spatial inhomogeneities in temperature and/or chemical composition. Simultaneously, the appearence of horizontal temperature gradients up to 0.5°C/cm is measured. If a small amount of absorbent is mixed into the solutions before the experiment starts, the recording of directly transmitted light with a two-dimensional spectrophotometer can be used efficiently for pattern observation [1]. The method turns out to be very sensitive for the detection of weakly-pronounced structures. These form when the concentrations are close to a limiting value below which no convection flow occurs. Such threshold concentrations are expected to exist since convective patterns fail to develop in pure water if the layer depth is below 1 cm [2,3]. The measured upper and lower bounds for threshold concentrations are given in the following table.

Solution	Absorbing Substance	Threshold
Yeast extract	Methyl Orange	0.45 - 0.54 mg/ml
Albumin	(0.005 wt.%)	0.5 - 0.75 mg/ml
K-Phosphate buffer		20 - 30 mM
Na-Carbonate buffer	NADH (0.5 mM)	< 120 mM

The hydrodynamic instability leading to convective flow in the biochemical systems is driven by unbalanced forces (surface tension) at the liquid/gas interface, mainly caused by temperature gradients due to evaporative cooling [1,4]. Our experiments show that the chemical composition of the solution has to be accounted for as well. The significant parameters for the onset of pattern formation are the thermal and the solutal Marangoni numbers. Both are also important for spatial patterning in biochemically reactive liquid layers.

References

1 S.C. Müller and Th. Plesser, in: Modelling of Patterns in Space and Time (eds. W. Jäger and J. Murray) Lect.Notes in Biomath., Springer, Berlin (1984)
2 J.C. Berg, M. Boudart, and A. Acrivos, J.Fluid Mech. 24,721 (1966)
3 H.K. Cammenga, D. Schreiber, G.T. Barnes, and D.S. Hunter, J.Colloid Interface Sci. 98,585 (1984)
4 S.C. Müller, Th. Plesser, and B. Hess, Naturwissenschaften (1984), in press

Image Analysis in the Study of Dissipative Spatial Patterns

Michael L. Kagan[1], Shmuel Peleg[2], A. Tchiprout[2], and David Avnir[1]

Departments of Organic Chemistry[1] and Computer Sciences[2]
The Hebrew University of Jerusalem, Jerusalem 91904, Israel

The rapidly growing science of image analysis (IA) by artificial intelligence methods has been applied to many diverse fields such as: biology, geology, earth scanning by satellites, photography, and a variety of military uses. In chemistry IA has been used mainly for rapid analysis of spectral output [1] in, for instance mass spectroscopy. Recently we have applied this powerful tool for the analysis of real spatial structures in chemistry. One example is in fractal surface analysis [2] and the other is in dissipative spatial patterns. In the latter case, the discovery of the general phenomenon of spatial structure formation by chemical reactions at liquid interfaces [3] has led to questions previously unasked by chemists, for instance, 'How to measure the kinetic growth of a product not distributed evenly in space ?'; 'What is the change in entropy as a pattern develops ?'; 'How to qualitatively differentiate one pattern from another ?'.

Using both standard and original software we have been able to tackle these problems. The various steps are shown in the following scheme.

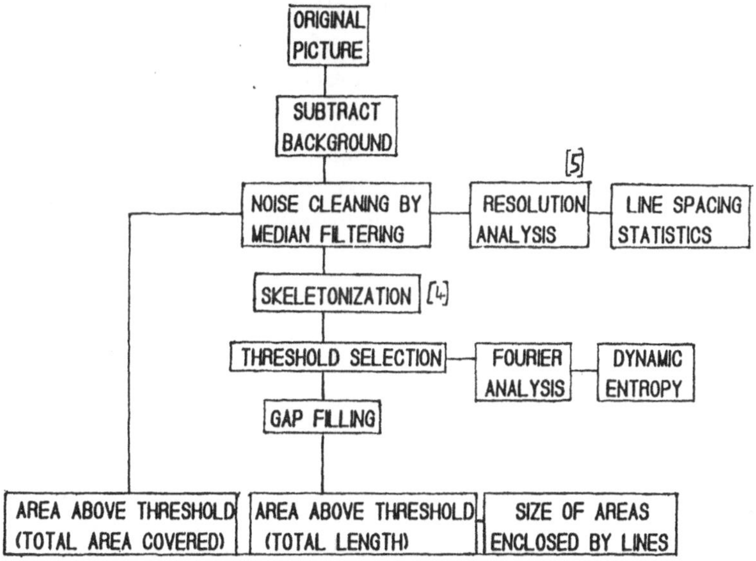

1) K. Varmuza, "Pattern Recognition in Chemistry", Springer, 1980.
P. Jurs, T. Isenhour; "Chemical Applications of Pattern Recognition", Wiley, 1975.
2) D. Avnir, D. Farin, S. Peleg, D. Yavin, in preparation.
3) See D. Avnir and M.L. Kagan - this volume.
4) S. Peleg, A. Rosenfeld; IEEE Trans. PAMI-3 , 208 (1981).
5) S. Peleg, J. Naor, R. Hartly, D. Avnir: IEEE Trans. Pattern Anal. Machine Intel. PAMI-6, 518 (1984).

Dissipative Structures in a Two Cells System

J. Breton, D. Thomas, and J.F. Hervagault
Laboratoire de Technologie Enzymatique, ERA n° 338 du CNRS
Université de Compiègne, B.P. 233, F-60206 Compiegne, France

Chloroplast thylakoids are immobilized within a calcium alginate gel layer. The photoassisted activity of the preparation is measured by following the reduction of DCPIPox, an artificial electron acceptor. Under illumination wavelengths ranging between 600 and 700 nm, the DCPIPox absorbs strongly. Thus, when traversing the acceptor solution before impinging upon the membrane, a part of the incident light intensity is absorbed according to a (decreasing) exponential law. The coupling between the light absorbance and the photoassisted reaction leads to an autocatalytic effect: An inhibition by excess DCPIPox is observed. When working under open conditions (CSTR), multiple steady-states occur with respect to either the DCPIPox concentration or the incident light intensity, as demonstrated experimentally.

In the present note we are dealing more precisely with the numerical study and the experimental evidences on the occurence of dissipative structures in an arrangment of two CSTRs with mutual mass exchange of DCPIPox through an inert membrane. The stable spatial structures are generated by creation of transient internal(1) or external(2) asymmetries:

(1) Identical boundary conditions for both the inlet DCPIPox concentration and the incident light intensity, and non-homogeneous (unstable) initial internal DCPIPox concentrations.

Identical initial DCPIPox concentrations within the reactor and transient asymmetrical boundary conditions (inlet DCPIPox concentration or incident light intensities).

In the latter situation, a non-trivial hysteresis effect is observed.

Photochemical Reactions and Hydrodynamic Instabilities

M. Gimenez, G. Dewel, and P. Borckmans

Laboratoire de Chimi-Physique II, Campus Plaine, Université Libre de Bruxelles
B-1050 Bruxelles, Belgium

J.C. Micheau

Laboratoire des IMRCP, ERA au CNRS n° 264, Université Paul Sabatier
F-31062 Toulouse, France

Some authors have reported the appearance of spatial structures during the irradiation of an homogeneous medium [1],[2],[3]. They appear when a 1-7 mm layer of solution is irradiated in a Petri dish. Using various photochemical systems of different mechanistic and structural features, we have suggested that their origin seems to be quite independent of the mechanisms involved during the photochemical reactions.

Using the schlieren technique we have shown that patterns were present in the system before irradiation, and that after irradiation the coloured structures appeared where the pre-patterns had already been localized. The photochemical reactions merely serve to reveal the convective pre-patterns induced by evaporative cooling. There exists a linear relationship between the average wavelength of the structures and the depth of the layer.

Having characterized similar pre-patterns in the solutions used in most of the photochemical reactions described by MOCKEL [1] and AVNIR [2] we think that the same phenomenon is responsible for the onset of structures in these systems.

Any open-surface reaction (such as photochemistry, reactions at gas-liquid interface [4], vaporization and adsorption) yielding or consuming a coloured product will lead to the formation of spatial structures.

The quasi-stationary mosaïc-striped structures often observed in bromate or glycolytic oscillators must be related to convective patterns [5]. Moreover the striations could induce the localization of center waves.

As convective regimes give rise to slow stirring, the experimental value of the quantum yield of photochemical reactions in unstirred open liquid phase largely depends on the layer depth.

This work has been supported by the NATO research grant 024483. M.G. acknowledges a scientific and technical grant from the European Communities Committee.

1 : P. Möckel Naturwiss. 64, 224 (1977)
2 : a) M. Kagan, A. Levi and D. Avnir Naturwiss. 69, 548 (1982)
b) D. Avnir, M. Kagan and A. Levi Naturwiss. 70, 144 (1983)
c) D. Avnir and M. Kagan Naturwiss. 70, 361 (1983)
3 : a) M. Gimenez and J.C. Micheau Naturwiss. 70, 90 (1983)
b) J.C. Micheau, M. Gimenez, P. Borckmans and G. Dewel Nature 305, 43 (1983)
4 : S. C. Müller and T. Plesser, Lecture Notes in Biomath, (1983)
5 : a) M. Orban, JACS 102, 4311 (1980)
b) K. Showalter, J. Chem. Phys. 73, 3735 (1980)

Toroids, Spirals and Other Chemical Waves

Arthur E. Burgess

Department of Chemistry, Glasgow College of Technology
Glasgow G4 OBA, United Kingdom

The spatial structures recorded from sealed tubes (10 mm diameter) were obtained in the Belousov Zhabotinskii reaction medium under similar conditions with initial concentrations: potassium bromate 0.09M, sulphuric acid 0.4M, malonic acid 0.3M and cerous nitrate 1.2mM. Sufficient ferroin dye was included for a visual colour contrast of pastel blue waves in pale orange background to be enhanced by black and white photography (1, 2).

Simple scroll waves are the most frequent early structures with wavelength of nearly 3.5mm and each wave segment propagating vertically to the scroll length at a velocity of ~0.04mm s^{-1}, ambient temperature ~22°C. Minor discontinuities and gas bubble punctures of the waves are smoothed as the structures evolve and show with time a general tendency to increased symmetry.

Waves annihilate upon contact with each other and with the container wall. Waves evolving in an arc form a scroll ring (a toroidal scroll structure). As the ring propagates it forms a series of concentric shells revealing a disc-like and increasingly spherical structure that progressively diminishes in size. The central structure reduces eventually to a point then vanishes. Elongated scrolls contract in length and may subdivide into more symmetric versions.

Structures once established continue for several hours and are quite stable to small thermal and mechanical disturbance. Deliberate and moderate disturbance creates disorder in the propagating waves out of which spatial forms of unusual geometry may emerge. Events have been observed that resemble biological cell division and indicate a fruitful area for further investigation. A general observation is that identifiable patterns most frequently occur near the meniscus (3).

Complex structures occur during the early stages of wave development shortly after pouring or perturbation. Some of the evolving scroll waves may appear to be twisted or linked. They exist in a condition of quasi-stability and undergo transition by division or smoothing into subsequent stages of greater symmetry. In the final stages of wave propagation only simple scrolls and bands remain in the tubes.

1 B J Welsh, J Gomatam, A E Burgess : Nature 304, 611 (1983)

2 A E Burgess, B J Welsh, J Gomatam : to appear in the Proceedings of the Workshop on Modelling of Patterns in Space and Time, ed W Jäger (Springer 1983)

3 B J Welsh : PhD Thesis - CNAA Glasgow College of Technology (1984)

A Model for the Oscillatory Precipitations of Minerals in Chemically Modified Rocks: The Autocatalytic Role of Surfaces

Bernard Guy and Jean-Jacques Gruffat

Département Géologie, Ecole des Mines, 158 Cours Fauriel
F-42023 Saint-Etienne Cédex, France

The chemical transformation of rocks by pervading aqueous solutions may sometimes lead to the precipitation of minerals in the form of recurrent alternations of different "strata" (1).

The model we set forth is based on the autocatalytic role of surfaces (2), (3) : let us consider the case where an external fluid carrying components A and B has dissolved a rock that contained component C, and where different minerals of composition $A_xB_{1-x}C$ may precipitate; x is the molar fraction of A in the solid and we will call y the corresponding fraction in the fluid phase. At the surface of the growing solid phase, we may write four reactions with respective rates v_1 (precipitation of AC on AC), v_2 (AC on BC), v_3 (BC on BC) and v_4 (BC on AC), where the selective autocatalytic effect of each phase is expressed with exponential terms.

For the steady state, $v_2 = v_4$ and we obtain a law $y = g(x)$ which may have two branches : v_1 predominates on the first branch, v_3 on the second ; starting on the first, AC precipitates and the fluid is enriched in B till the moment when the second is reached ; BC then precipitates and so on, provided that the transport is slower than chemical kinetics and imposes the concentration of components around the equilibrium value of the phases defined by the two branches.

We show semi-quantitatively that the spatial wave-length of the structures is related to the amplitude of the chemical change between the two minerals involved : the greater the chemical change, the greater the wave length, a result in good agreement with natural evidences. On another hand, if transport is efficient with respect to kinetics the wave length decreases and the oscillation may disappear, in accordance with the situation where the oscillatory structures are found within chemically modified rock systems, e.g. in places where transport is operated by diffusion rather than convection.

1. Guy B., C.R. Acad. Sc., Paris, 292, II, 413-416 (1981).
2. Chaix J.M., thèse Doct. Univ. Dijon, 190 p., annexes 33 p (1983).
3. Slin'ko M. and Slin'ko M., Catal, Rev. Sci. Eng., 17(1), 119-153 (1978).

Study of the Electrochemical Properties of Colloids by Means of Photo-Ionic Effect

G. Cherbit, M. Poulat, B. Guillermin, and J. Chanu

T.M.I.B.-Université Paris VII, 2 Place Jussieu
F-75251 Paris Cédex 05, France

The photo-ionic detection consists in the measurement of the conduction modification of a solution, due to an optical resonant interaction of a laser beam with the species of the solution [1, 2]. The optical pulse modifies the distribution on the different electronic energy levels, and therefore the steady-state current.

This method is highly specific and sensitive [3,4].

Experimental set-up

A silver colloïd solution [CAREY LEA [5]], with a particle size of about 50 [Å], [6,7], flows through two cylindrical stainless steel electrods, in a quartz cell, so that the electrical field is uniform and about 1 [KV/m], [fig.1].

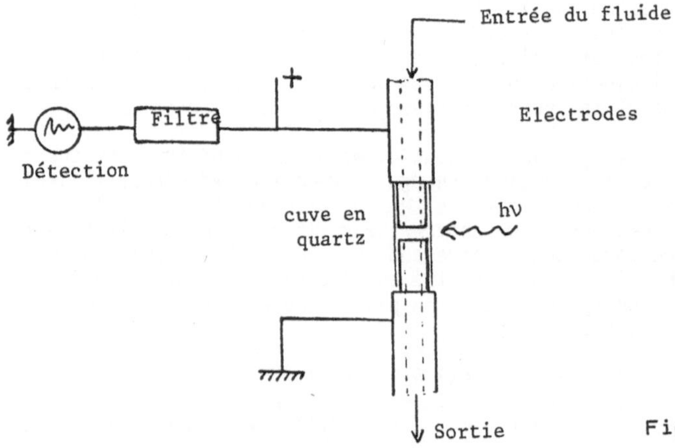

Fig.1

The cell is illuminated by a pulsed dye-laser and the tuning is achieved by a rotating mirror. When the wavelength of the pulse is resonant with the energy levels of the species in the solution, the electrical conduction is modified.[8]

Results

The results we obtained (at 337 [nm] and from 473 to 547 [nm]) do not agree with the silver atomic lines or with the crystal bands of energy. This could be explained as an intermediate state of the aggregate colloïd between the atomic state [spectral lines] and the crystalline state [Fermi levels], resulting from the coupling between the few atoms [about 2000] of the colloïd particle.

References

[1] G.CHERBIT, M.POULAT : Congrès de la Société Française de Biophysique, Orléans, 5-9 Juillet 1982
[2] G.CHERBIT, M. POULAT : Journée de Biophysique, E.N.S. St-Cloud, 16 Novembre 1982
[3] G.CHERBIT, M.POULAT : 33ème Congrès I.S.E., Lyon, 6-10 Septembre 1982
[4] G. CHERBIT,M.POULAT, D. VEKHOFF : J. Phys. n°11, 44, 371-375, Novembre 1983
[5] CAREY LEA : Amer. J. Sci., 37, p. 476, 1889
[6] JOLIVET : Etude des colloïdes d'argent , Université Paris VI 1981
[7] S.M. HEARD & co : J. of Coll. and Int. Sci., 93,545-555,1983
[8] G.CHERBIT, M.POULAT, B. GUILLERMIN : 35th Congrès I.S.E., Berkeley, 5-10 Aout, 1984

Diffusion-Reactions in Solid-Liquid Systems Mathematical and Geological Features

F. Kalaydjian and M. Cournil

Ecole Nationale Supérieure des Mines de Saint-Etienne, 158, cours Fauriel
F-42023 Saint-Etienne Cédex, France

1 INTRODUCTION AND HYPOTHESES

Many processes involving interactions between a liquid solution and a finely divided solid phase occur in geology, especially in the first steps of the metasomatic mineralisations. Growth, dissolution, diffusion and source terms govern mass-transfers of a component in the close or open studied systems. Models have been developed assuming the following hypothese.

Evolution of grain radius R meets Nernst equation : $dR/dt=k(C-Ce)$. Because of small size of grains, the equilibrium concentration Ce of the liquid phase is a R-function according to Ostwald-Freundlich law : $C=Co.\exp(\beta/R)$.

Mass-balance involves Fick's diffusion and source terms written as $Q.(Cs-C)$.

The boundary conditions are : on the external system surface, a Dirichlet condition or a usual mass-transfer law ; on the grain-solution interface, a mass-balance equation of dissolution-growth : $D\partial C/\partial r=(1/V-C).dR/dt$, characterizing a free-boundary problem.

2 RESULTS

For close systems with instantaneous diffusion, steady states are either stable or unstable depending on whether the solid phase consists in a single grain or in N grains (N>1). For open systems and uniform liquid phases, steady states are unstable saddle-points. For close systems with diffusion, results are similar to the first case ones. For open systems with diffusion, steady states are unstable and uniform.

Stochastic Analysis of a Complex Bifurcation

A. Fraikin

Laboratoire de Chimie Générale, Université P. et M. Curie, Tour 55
F-75230 Paris Cédex 05, France

The stochastic aspect of a complex bifurcation arising in a two variables chemical system is studied. The dynamics reduces, in a suitable region of the phase space, to a normal form for which both roots of the characteristic equation vanish simultaneously. In conditions close to this degenerate situation, the normal form can be viewed as a perturbation of an exactly soluble hamiltonian system, of hamiltonian h, which exhibits a homoclinic trajectory, h = 0. BAESENS and NICOLIS [1] have shown that the phase portrait of the dissipative sytem displays two steady states that coalesce, a focus F and a saddle S. Moreover, as one moves in the parameter space, a limit cycle surrounding F, bifurcates from a homoclinic trajectory and then disappears by Hopf bifurcation.

This degenerate situation allows for a local description of the stationary solution of the master equation. Using the general method presented in this volume [2], the Taylor expansions of the stochastic potential U around each extremum F or S are determined up to the seventh derivatives, at the dominant order with respect of the small control parameters. In this way, U can be expressed in terms of a polynomial of h, at least for the calculated terms. The extrema of U consist then of the two steady states F and S, and of a trajectory h = c, where c is fixed. The sign of c, depending on the parameters, determines the domain of existence of the cycle or of the homoclinic trajectory. Note that the stochastic approach assigns to the dissipative system trajectories of the hamiltonian one. By analysing the matrix of the second derivatives of U on these trajectories, the latter appear to be attractive. On the homoclinic trajectory, the first nonvanishing derivatives at S are the fourth ones. Only in this case, this point is not a saddle of the potential surface, which is very flat.

The two kinds of expansions of U around F and S are suitable to describe respectively the Hopf bifurcation and the bifurcation of the limit cycle from the homoclinic trajectory. The results are qualitatively in agreement with those obtained by BAESENS and NICOLIS but our approach does not lead to the exact condition of existence of the homoclinic trajectory.

References

1. C. BAESENS and G. NICOLIS, Z. Phys. B 52, 345 (1983)
2. H. LEMARCHAND and A. FRAIKIN, *Stochastic description of various types of bifurcations in chemical systems*, in this volume. Others references may be found at the end of this contribution.

Relaxation Processes in Spatially Distributed Bistable Chemical Systems: Deterministic and Stochastic Analysis

M. Frankowicz

Institute of Chemistry, Jagellonian University, ul. Karasia 3
30-060 Krakow, Poland

Two simple versions of an inhomogeneous system displaying bistability (2- and 3-box Schlogl model with diffusion) are analysed from the deterministic and stochastic **points of view. The deterministic** bifurcation diagrams are constructed. The results of Monte Carlo simulation of the time-evolution and the stationary probability distributiuon function constructed by an iterative procedure are presented for the 2-box case. It is shown that certain inhomogeneous states which are asymptotically stable in the deterministic case, appear during the stochastic evolution of the system as transient structures. These states can facilitate the transition between homogeneous steady states (there exists a minimum in the dependence of the mean first passage time on diffusion) [1] . A qualitative picture of the transition mechanism for the 3-box case is advanced. The above conclusions have been recently supported by analytical calculations [2] . Transient inhomogeneous structures may be also responsible for the stabilization of the unstable state observed by Blanché [3] . It has to be mentioned that recently compartmental systems have been studied experimentally and various inhomogeneous configurations were observed [4] .

References

1. M. Frankowicz and E. Gudowska-Nowak: Physica 116A,331-344(1982)

2. D. Borgis and M. Moreau, to be published.

3. A. Blanche: Thesis, Université Bordeaux I (1981)

4. I. Stuchl and M. Marek: J. Chem. Phys. 77,2956-2963(1982)

A Lagrangian Formulation of the Dynamics of Chemical Reaction Systems

J.S. Shiner

Institut für Physikalische Chemie, Universität Würzburg
D-8700 Würzburg, Fed. Rep. of Germany

A variational formulation of the dynamics of systems of chemical reactions is presented which utilizes a Rayleigh dissipation function to extend the Lagrangian approach to conservative systems to dissipative systems [1]. The formulation is valid arbitrarily far from equilibrium and is based on macroscopic, deterministic chemical kinetics. It thus stands in contrast to variational principles for dissipative systems which are based on studies of fluctuations and are restricted to the near equilibrium regime [2] or, far from equilibrium, to the vicinity of stable stationary states [3].

The equations of motion are obtained in the standard manner from a Lagrangian, the dissipation function, and equations of constraint, which here express conservation of mass. Since "inertial" effects are absent on the macroscopic level of deterministic kinetics, the Lagrangian (at constant temperature and pressure) is simply the negative of the Gibbs free energy, which is composed of two contributions. The first is the free energy of the internal species of the system; the second is due to external sources which control the chemical potentials of some of the internal species and thus allow the system to be driven away from equilibrium. The key to the formulation is the dissipation function, which is written in the standard fashion as a quadratic form in the rates of reaction $\dot{\xi}_j$:

$$F = \frac{1}{2} \sum R_j \dot{\xi}_j^2 \tag{1}$$

The "resistances" R_j for the reactions are, however, in general not constant but defined by the relation

$$A_j \equiv R_j \dot{\xi}_j \tag{2}$$

where the A_j are the reaction affinities, and A_j and $\dot{\xi}_j$ are expressed in the accustomed manner of chemical kinetics in terms of rate constants and chemical potentials (or activities). Thus, R_j is also expressed in terms of these same quantities.

A particularly important result is that the matrix of phenomenological resistances (linear combinations of the R_j) which appears in the equations of motion relating the independent reaction rates to their conjugate thermodynamic forces in the stationary state is symmetric. However, the symmetry is of an algebraic nature; the more stringent condition of differential symmetry does not hold in general.

1. H.Goldstein: Classical Mechanics (Addision-Wesley, Cambridge 1950)
2. L.Onsager: Phys. Rev. 37, 405 (1931); L.Onsager & S.Machlup: Phys. Rev. 91, 1505 (1953); S.Machlup & L.Onsager: Phys. Rev. 91, 1512 (1953)
3. J.Keizer: Biosystems 8, 219 (1977)

JSS is recipient of an Alexander von Humboldt Fellowship.

Relaxation Phenomena in a Bistable Chemical System

G. Dewel, P. Borckmans, and D. Walgraef

Chercheurs Qualifiés au Fonds National Belge de la Recherche Scientifique
Service de Chimie Physique II, Université Libre de Bruxelles, Campus Plaine, CP 231
B-1050 Bruxelles, Belgium

A theoretical investigation is presented of the relaxation dynamics of a simple model which gives a near-quantitative description of the iodate-excess arsenous acid systems (1). Near the hysteresis limits ($K \gtrsim K_1$, $K \lesssim K_2$, K= input flow rate) and near the critical point K_c an exact expression is obtained for the nonlinear response to a sudden change of the concentration. As a result, the critical exponents characterizing the lengthening of the relaxation time near K_1, K_2 and K_c (critical slowing down) and the long time tails at K_1, K_2 and K_c are derived. Long-lived unstable modes ("dynamic metastability") are also found near the bistability limits ($K \lesssim K_1$, $K \gtrsim K_2$). The duration of these concentration plateaus is very sensitive to the distance to the hysteresis limits (K_1, K_2). They could therefore explain the distribution of the lifetimes of the metastable states (2) (cf.Fig.).

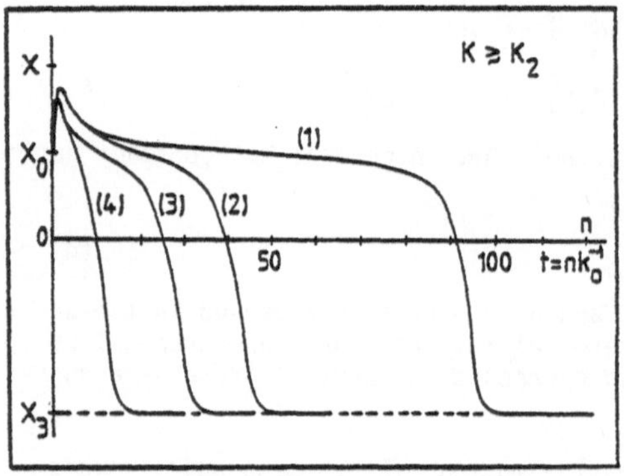

Fig.: Relaxation curves slightly beyond hysteresis limit for the same initial condition but decreasing residence time ((1): K= 1,01 K_2; (2): K = 1,05 K_2; (3): K= 1,1 K_2; (4): K=1,5 K_2).

References:

1. N.Ganapathisubramanian and K.Showalter, J.Phys.Chem. 87, 1098, (1983
2. J.C.Roux, P.De Kepper and J.Boissonade: Phys.Lett. 97A, 168,(1983).

Effect of Stirring on Chemical Bistability in CSTR

P. Borckmans, G. Dewel, and D. Walgraef

Chercheurs Qualifiés au Fonds National Belge de la Recherche Scientifique
Service de Chimie Physique II, Université Libre de Bruxelles, Campus Plaine, CP 231
B-1050 Bruxelles, Belgium

The continuous flow stirred tank reactor (CSTR) offers simple conditions for studying one of the most characteristic features of chemical non-equilibrium systems : bistability [1] . Indeed, such open reactors may exhibit two coexisting stationary states over a range of pumping rates.

It was usually assumed that the stirring is sufficiently effective to destroy any inhomogeneities in the system ("gradientless reactors") and thus that the transition between the steady states takes place at the marginal stability points. It has, however, been shown experimentally that a reduction of the parameters (stirring rate, size of stirrer...) affecting the effectiveness of mixing reduces the width of the bistable region and also affects the amplitude of the steady states [2].

An interpretation of these results is presented in terms of the competition between turbulent mixing and the inhomogeneities generated by the feeding of the reactor. The effects are illustrated (Fig. 1) on a realistic example, the simple kinetic model which has been used to give a near-quantitative description of the bistable region of the iodate-arsenous acid system [3, 4].

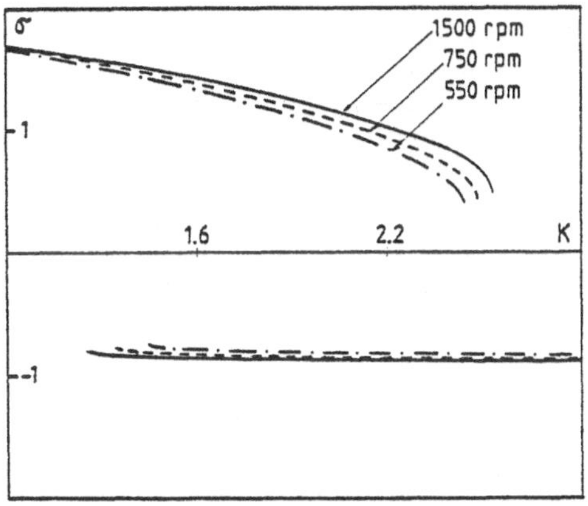

Figure 1 : Steady states for the iodate-arsenous acid system under the conditions of ref. [3] as a function of the flow rate (reduced units) for three stirring rates.

1. I.R. Epstein J. Phys. Chem. 88, 187-198 (1984).
2. J.C. Roux, P. De Kepper and J. Boissonade: Phys. Letters, 97A, 168-170 (1984).
3. N. Ganapathisubramanian and K. Showalter : J. Phys. Chem. 87, 1095-96 (1983).
4. G. Dewel, P. Borckmans and D. Walgraef : (these proceedings).

Oscillatory Combustion of Heterogeneous Flames

L. De Luca, G. Riva, C. Zanotti, R. Dondè, D. Marongiu, A. Molinari, and V. Lapegna

Dipartimento di Energetica-CNPM, Politecnio di Milano, 32 Piazza Leonardo da Vinci I-20133 Milano, Italy

Two different steady reacting modes can be experimentally observed in heterogeneous combustion, depending on the nature of the burning material and operating conditions (pressure, temperature, diabaticity). The first is a steady, time-invariant regression rate of the burning surface (interface between the condensed and gas phases); the second is an oscillatory combustion with regression rate of the burning surface cyclically varying in time. For adiabatic systems, oscillatory combustion regimes can be observed for a more or less wide pressure range above the pressure deflagration limit (PDL). Experimental tests conducted on AP84/CTPB16 composite solid rocket propellant, at 300 K of ambient temperature, revealed oscillatory combustion in the pressure range $0.07<p<0.3$ atm. For $p<0.07$ atm (the PDL), no self-sustained reacting mode was observed, while, for $p>0.3$ atm, the combustion was characterized by a time-invariant burning rate. The nature of the reacting mode can also be affected by adding energy to the system: an external energy source (CO_2 laser) was employed in order to drive the combustion process of the same AP-based solid propellant. Several tests at different pressure levels and radiant flux intensities were performed. In this experimental configuration, reacting modes were detected even for $p<0.07$ atm. It was found that low values of pressure and radiant flux intensity lead to pulsating combustion. By increasing pressure and/or radiant flux intensity, the frequency of the oscillations increases. For sufficiently large pressure and/or radiant flux intensity, time invariant regression rate is obtained.

These experimentally detected combustion modes were analytically predicted following a nonlinear stability analysis of the set of equations governing the combustion process (essentially the energy conservation in the condensed phase with appropriate initial and boundary conditions). This nonlinear analysis accounts for the influence of the properties of the burning material and the ambient conditions (included pressure and diabaticity), allowing to predict PDL and the values of pressure and radiant flux intensity originating oscillatory combustion. Moreover, several numerical checks of the analytical predictions were performed by numerical integration of the basic set of equations under the appropriate ambient conditions. Both the numerical checks and experimental results fully confirm the validity of the analytical predictions.

Mécanisme de la Réaction de Bray

Guy Schmitz

Université Libre de Bruxelles, Avenue F. Roosevelt 50
B-1050 Bruxelles, Belgium

La réaction de Bray est la décomposition du peroxyde d'hydrogène catalysée par le couple iode-iodate. Dans certaines conditions cette décomposition a un caractère périodique: on observe alternativement la réduction de l'iodate en iode suivant la réaction (R) et l'oxydation de l'iode suivant la réaction (O).

$$2\ IO_3^- + 2\ H^+ + 5\ H_2O_2 = I_2 + 5\ O_2 + 6\ H_2O \qquad (R)$$

$$I_2 + 5\ H_2O_2 = 2\ IO_3^- + 2\ H^+ + 4\ H_2O \qquad (O)$$

La somme (R)+(O) donne la réaction globale

$$2\ H_2O_2 = 2\ H_2O + O_2$$

Nous avons étudié la cinétique de la réaction de Bray en suivant l'évolution au cours du temps des concentrations (H_2O_2),(I_2) et (I^-). En particulier nous avons montré que l'étape déterminante de la réaction (R) est la même que celle de la réaction iodate-iodure dans les mêmes conditions.

Le mécanisme proposé comporte trois groupes de réactions:Des réactions faisant partie du mécanisme du système iodate-iodure-iode suivant les travaux classiques de Bray,Skrabal et Liebhafsky; Des réactions de réduction et d'oxydation des dérivés de l'iode par le peroxyde d'hydrogène; Des réactions probables mais n'ayant qu'une importance accessoire telles que l'oxydation de l'iodure par l'oxygène. Les étapes principales du mécanisme proposé sont:

$$IO_3^- + I^- + 2\ H^+ \rightleftharpoons HIO_2 + HIO$$

$$HIO_2 + I^- + H^+ \longrightarrow I_2O + H_2O$$

$$I_2O + H_2O \rightleftharpoons 2\ HIO$$

$$HIO + I^- + H^+ \rightleftharpoons I_2 + H_2O$$

$$HIO + H_2O_2 \longrightarrow I^- + H^+ + O_2 + H_2O$$

$$I_2O + H_2O_2 \longrightarrow HIO_2 + HIO$$

Ce mécanisme permet d'interpréter toutes les caractéristiques de la réaction de Bray. L'intégration numérique des équations cinétiques associées donne des courbes d'évolution des concentrations conformes aux courbes expérimentales.

Process (E2) of the Explodator Model

Zoltán Noszticzius

Henrik Farkas Technical University of Budapest, Institute of Physics
H-1521 Budapest, Hungary

Zoltán A. Schelly

University of Texas at Arlington, Department of Chemistry,
Arlington, TX 76019-0065, USA

Recently a new reaction scheme the "Explodator" core was proposed by us [1,2] as a simple model for chemical oscillations. Limit cycle oscillators can be constructed using that explosive core and different limitary reactions.

$$\{A\} + X \longrightarrow (1+\alpha)\,X \qquad (E1)$$
$$X + Y \longrightarrow Z \qquad (E2)$$
$$Z \longrightarrow (1+\beta)\,Y \qquad (E3)$$
$$Y \longrightarrow \{B\} \qquad (E4)$$

Figure 1. The Explodator core

As a possible chemical realization of our theoretical scheme the halate-driven oscillating reactions - especially the BZ reaction of oxalic acid [3] - were regarded:

$$\{HBrO_3\} + HBrO_2 \longrightarrow 2\ HBrO_2 \qquad \alpha = 1 \qquad (E1)$$
$$HBrO_2 + Br_2 \longrightarrow 3\ HOBr \qquad (E2)$$
$$3HOBr \longrightarrow 1.5\ Br_2 \qquad \beta = 0.5 \qquad (E3)$$
$$Br_2 \longrightarrow \{Br_2\ gas\} \qquad (E4)$$

Figure 2. Chemical realization of the Explodator core in the BZ reaction of oxalic acid. $X \equiv HBrO_2$, $Y \equiv Br_2$, $Z \equiv 3HOBr$. (Only the important chemical components are shown and no limitary reactions are included)

According to the critics of the Explodator the most problematic step in our new scheme is the chemical realization of process (E2). They suggest that (E2) in fact goes through bromine hydrolysis (R1)

$$Br_2 + H_2O \longrightarrow H^+ + Br^- + HOBr \qquad (R1)$$
$$HBrO_2 + Br^- + H^+ \longrightarrow 2\ HOBr \qquad (R2)$$
$$\overline{HBrO_2 + Br_2 + H_2O \longrightarrow 3\ HOBr} \qquad (E2)$$

and bromous acid can react directly only with bromide ions in reaction (R2) and no direct reaction can take place between bromous acid and elementary bromine. They regard (R2) to be more realistic than (E2) because (R2) is analogous to the well-estabilished reaction of (R3) [4]

$$2\ H^+ + BrO_3^- + Br^- \longrightarrow HOBr + HBrO_2 \qquad (R3)$$

Here an oxygen atom transfer occurs from $HBrO_3$ to HBr. However (E2) can also go via an oxygen atom transfer (from $HBrO_2$ to Br_2) according to the following steps

$$HBrO_2 + Br_2 \longrightarrow HOBr + Br_2O \qquad (M1)$$
$$Br_2O + H_2O \longrightarrow 2HOBr \qquad (M2)$$
$$HBrO_2 + Br_2 + H_2O \longrightarrow 3\,HOBr \qquad (E2)$$

Other oxygen atom transfer reactions between the different oxidation states of bromine are also accepted like (-R3) or even proved experimentally as in the case of the disproportionation of bromous acid [5]; thus we have no reason to reject (E2) on a purely theoretical basis. On the other hand the explanation of non--bromide controlled oscillations [6] in the BZ reaction would be rather straightforward if elementary bromine can react directly with bromous acid.

Unfortunately, we have no direct experimental evidence about the rate constants in question [7] . We could give an upper limit for k_{R2} only [5]. (Our estimate for its lower limit is valid only if (M1) is a relatively slow reaction.) In our opinion both elementary Br_2 and Br^- ions can play a control role in the different BZ reactions depending on the $[Br_2]/[Br^-]$ concentration ratio.

In the end, however, we have to stress that the most important feature of the Explodator model is not the chemical identification of its intermediates - they are surely the different oxidation states of the halogens in the case of halate driven oscillators [5,8,9,10,11] - but its structure. That "serial" structure [2] contains two consecutive autocatalytic processes just like some of our previous models [5,8,9,10] or the new revised Oregonator [11,12] .

The authors thank Professors R.J.Field, H.D.Försterling, R.M.Noyes and J.J.Tyson for a helpful correspondence.

References

[1] Z.Noszticzius,H.Farkas,Z.A.Schelly: J.Chem.Phys.80 6062 (1984)
[2] Z.Noszticzius,H.Farkas,Z.A.Schelly: React.Kinet.Catal.Lett. 25 305 (1984)
[3] Z.Noszticzius,J.Bódiss: J.Am.Chem.Soc. 101 3177 (1979)
[4] R.J.Field,E.Kőrös,R.M.Noyes: J.Am.Chem.Soc. 94 8649 (1972)
[5] Z.Noszticzius,E.Noszticzius,Z.A.Schelly: J.Phys.Chem. 87 510 (1983)
[6] Z.Noszticzius: J.Am.Chem.Soc. 101 3660 (1979)
[7] J.J.Tyson in Oscillations and Traveling Waves in Chemical Systems Eds: R.J.Field and M.Burger (Wiley, New York 1984)
[8] Z.Noszticzius,J.Bódiss: Ber.Bunsenges.Phys.Chem. 84 366 (1980)
[9] Z.Noszticzius,H.Farkas in Modelling of Chemical Reaction Systems Eds. K.H.Ebert etal. (Springer,Berlin,1981) p. 275
[10] Z.Noszticzius,A.Feller: Acta Chim.Acad.Sci.Hung. 110 261 (1982)
[11] R.M.Noyes: J.Chem.Phys. 80 6071 (1984)
[12] J.J.Tyson: J.Chem.Phys. 80 6079 (1984)

Formation of CO_2 in the Belousov-Zhabotinsky System

H.D. Försterling, H. Idstein, R. Pachl, and H. Schreiber
Fachbereich Physikalische Chemie, Philipps-Universität Marburg, Hans-Meerwein-Straße
D-3550 Marburg, Fed. Rep. of Germany

FKN-Mechanism

$$2Ce^{4+} + MA + H_2O \xrightarrow{k_1} 2Ce^{3+} + TA + 2H^+$$

$$2Ce^{4+} + TA \xrightarrow{k_2} 2Ce^{3+} + GOA + CO_2 + 2H^+$$

$$2Ce^{4+} + BrMA + H_2O \xrightarrow{k_3} 2Ce^{3+} + GOA + CO_2 + Br^- + 3H^+$$

Calculation of the rate of CO_2-formation

$$d[TA]/dt = k_1 [Ce^{4+}] \cdot [MA] - k_2 \cdot [Ce^{4+}] \cdot [TA]$$

$$d[CO_2]/dt = k_2 [Ce^{4+}] \cdot [TA] + k_3 \cdot [Ce^{4+}] \cdot [BrMA]$$

Integration of the rate equations ($[Ce^{4+}]$ approximated by the mean value $[Ce^{4+}]_m$, $[MA]$ by the initial value $[MA]_o$):

$$\frac{d[CO_2]}{dt} = \left\{ k_1 \cdot k_2 \cdot [Ce^{4+}]_m^2 \cdot [MA]_o + k_3 \cdot [Ce^{4+}] \cdot \frac{d[BrMA]}{dt} \right\} \cdot t$$

Comparison of experimental and calculated values of the rate of formation of CO_2

In a BZ system (0.1 m MA, 0.1 m $HBrO_3$, 10^{-4} m Ce^{4+}) Ce^{4+}, BrMA and CO_2 are measured directly: $[Ce^{4+}]_m = 2.8 \cdot 10^{-5}$ m, $d[BrMA]/dt = 1.67 \cdot 10^{-6}$ $m \cdot s^{-1}$, $d[CO_2]/dt = 2.5 \cdot 10^{-6}$ $m \cdot s^{-1}$. Using the values for Ce^{4+} and BrMA, the value $d[CO_2]/dt = 2.6 \cdot 10^{-8}$ $m \cdot s^{-1}$ is calculated at t = 5400 s (end of the induction period; $k_1 = 0.10$, $k_2 = 0.33$, $k_3 = 0.045$ 1/mol s).

Result

The experimental rate is larger than the calculated rate by 2 orders of magnitude.

Conclusions

The rate of CO_2 formation is strongly affected by the presence of bromate; the discrepancy between theory and experiment is not due to a direct attack of bromate itself, but by oxybromine transients on organic radicals.

Reference

H.D.Foersterling,H.Idstein, R.Pachl and H.Schreiber, Z.Naturforsch. 39a, in press.

Experimental Study of Stirring Effects of a Bistable Chemical System in Flow Reactor Mode

H. Saadaoui, J.C. Roux, P. de Kepper, and J. Boissonade

Centre de Recherche Paul Pascal, Université de Bordeaux I, Domaine universitaire
F-33405 Talence Cédex, France

The bistable chemical system $[\mathrm{C\ell O_2^-} - \mathrm{I^-}]$ [1] is studied in a flow reactor with a controlled stirring rate [2]. The chemicals are fed into the reactor by two inlet ports in the vicinity of the stirrer (propeller). We show that the position of the transition point from the thermodynamic branch toward the flow branch is stirring-rate dependent : the better the mixing, the larger the bistable domain (Fig. 1,2). Furthermore this transition could occur in two steps : after a change in one control parameter the system changes only slightly, stays on this new value (still on the thermodynamic branch) for a while and abruptly jumps on the flow branch (Fig. 3). The autocorrelation time of the pretransitional fluctuations increases when we approach the transition point ; in some cases these fluctuations seem to become periodic.

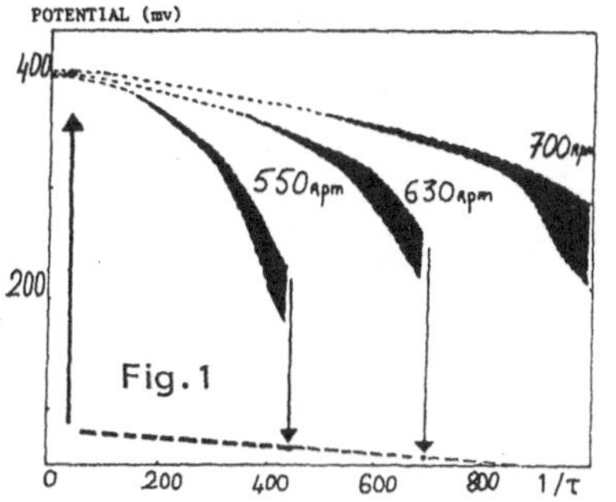

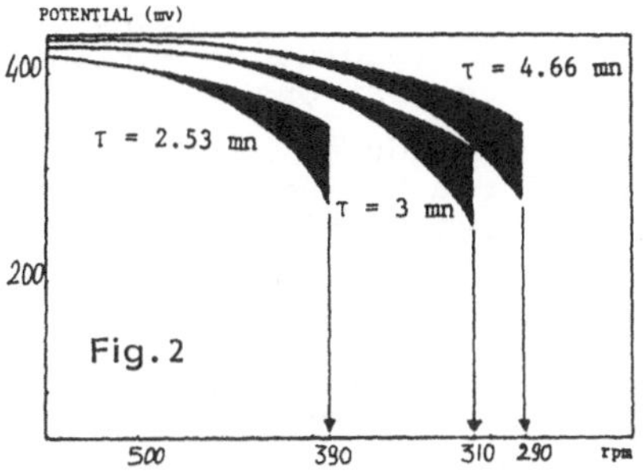

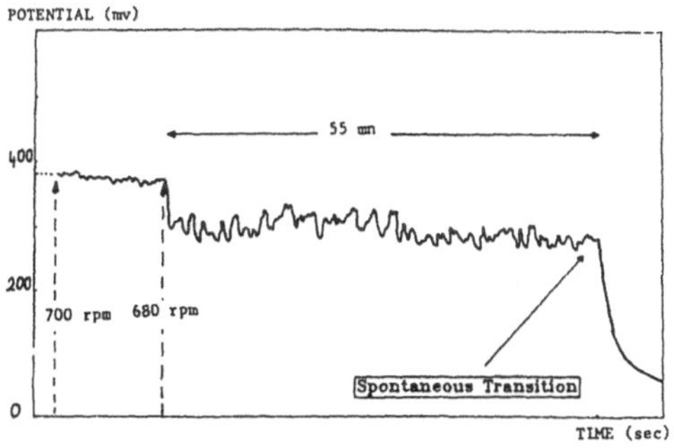

Fig. 1 Bistable domain versus flow rate

Fig. 2 Stirring effect on the transition point

Fig. 3 Measured lifetime of a metastable state

References

1. C.E. Dateo, M. Orban, P. de Kepper, I.R. Epstein, J. Am. Chem. Soc. 104, 504 (1982)
2. J.C. Roux, H. Saadaoui, P. de Kepper, J. Boissonade, in "Fluctuations and Sensitivity in non equilibrium systems", Ed. by W. Horsthemke, D.K. Kondepudi (Springer Berlin 1984), p. 70

Study of Bromide Control in Bromate Oscillators

M. Varga, L. Györgyi, and E. Kőrös

Institute of Inorganic and Analytical Chemistry, L. Eötvös University
H-1443 Budapest, P.O. Box 123, Hungary

The control function of bromide ions in bromate oscillators has been questioned recently, and suggestions have been made to replace the Oregonator model by other ones in order to account for the occurrence of oscillations in silver ion containing BZ systems [1,2].

With the aim in mind to clarify the still existing problems we examined how the characteristics of chemical oscillation /e.g. period time, amplitude/ are altered by the addition of bromocomplex-forming ions [thallium/III/, mercury/II/] to the reacting systems. For, these ions perturb the dynamic bromide concentration conditions prevailing in bromate oscillators [3].

At low/10^{-4}-10^{-5}M/ thallium/III/, respectively, mercury/II/ concentrations a considerable increase in the period time of oscillations, at high /0.01M/ thallium/III/ concentration high frequency oscillations were recorded. Oscillations did not occur at high mercury/II/ concentrations.

All phenomena induced by thallium/III/, respectively, mercury/II/ could be simulated by the Oregonator model if the dynamics of complex equilibria between bromide and the metal ion was considered:

$$M^{n} + Br^{-} \underset{k_{-1}}{\overset{k_1}{\rightleftharpoons}} MBr^{/n-1/} \qquad \text{etc.}$$

The rate constants of complex formation and dissociation were calculated from known relations and considering analogous ligand substitution reactions. The experimentally-obtained curves and those simulated by the Oregonator model show good agreement.

The increase in period time at low metal ion concentrations can be interpreted in terms of the bromide buffering effect of the metal bromocomplexes. The BZ system shows limit cycle behaviour. At high thallium/III/ concentrations, however, the BZ system does not exhibit limit cycle, it approaches rather rapidly to equilibrium.

Our results and those of Ruoff /with Ag^{+}/ [4] clearly demonstrate that even in the presence of bromide-removing ions Oregonator can simulate the experimentally recorded curves, and modification neither in the Oregonator model nor in the skeleton of the chemical mechanism is necessary. This means that bromide ion should not be replaced by an other control intermediate /e.g. HOBr, Br/ and more than one autocatalytic reaction should not be taken into consideration.

1 Z. Noszticzius, J. Am. Chem. Soc., 101, 3660 /1979/
2 N. Ganapathisubrimanian and R. M. Noyes, J. Phys. Chem., 86, 3217 /1982/
3 E. Kőrös and M. Varga, React. Kin. Cat. Lett. 21 521 /1982/
4 P. Ruoff, Z. Naturforsch., 38a, 974 /1983/

Complex Wave-Forms in the Belousov-Zhabotinskii Reaction

Richard M. Noyes
Department of Chemistry, University of Oregon, Eugene, OR 97403, USA
John G. Sheppard
Computer Centre, University of Zimbabwe, P.O. Box MP 167, Harare, Zimbabwe

The Belousov-Zhabotinskii reaction with ethyl acetoacetate (3-oxobutanoic acid, ethyl ester; $CH_3COCH_2COOC_2H_5$; EA) as organic substrate is of interest as the batch reaction has shown both high and low frequency oscillations in the same experiment, (1), and because no bubbles of gaseous products are formed.

We report a preliminary investigation of the BZ-EA system in a continuously stirred tank reactor.

In general, the behaviour of the BZ-EA system was found to be similar to that of the BZ-malonic acid system, but the absence of CO_2 bubbles in the reactor makes the system somewhat easier to control. The concentration of catalyst necessary for oscillations is considerably higher than that needed for malonic acid.

Only a small section of the phase-space has been explored thus far. The BZ-EA system shows burst oscillations (also called "composite double oscillations") of the type found in the BZ-MA system (2). We have also found "doubled carrier wave" oscillations previously found in batch experiments with EA (3).

1. L.F. Salter and J.G. Sheppard: Int.J.Chem.Kinet. 14, 815 (1982)

2. (a) P.G. Sorensen: Faraday Symposia Chem. Soc. No. 9, 88 (1974)
   (b) M. Marek and E. Svobodva: Biophys. Chem. 3,263 (1975)
   (c) P. De Kepper, A. Rossi and A. Pacault: C.R. Acad. Sc. Paris,C283, 371 (1976)
   (d) P.G. Sorensen: Ann. NY Acad. Sci. 316, 667 (1979)

3. L'Treindl and A. Nagy: Coll. Czech. Chem. Commun. 48, 3229 (1983)

On the Possibility of Chaos in Formal Kinetics

Vera Hárs
CHINOIN Chemical and Pharmaceutical Works Ltd., Tô u. 1-5,
H-1045 Budapest, IV, Hungary
P. Érdi
Central Research Institute for Physics, Hungarian Academy of Sciences, P.O.B. 49
H-1525 Budapest 114, Hungary
J. Tóth
Computer and Automation Institute, Hungarian Academy of Sciences, Kende u 13-17
H-1502 Budapest, Hungary

A natural way to construct a formal chemical reaction showing chaotic behaviour would be to transform an equation with chaotic solutions into a kinetic equation, i.e. into a differential equation that may be considered as the induced deterministic kinetic equation of a chemical reaction. Using a necessary and sufficient condition for the form of kinetic equations [1] we show the impossibility of this procedure for the case of the Lorenz-equation. In other words, we show, firstly, that no proper orthogonal transformation turns the Lorenz-equation into a kinetic one. Secondly, no improper orthogonal transformation does this either. [2]

It is also shown that a kinetic equation remains a kinetic one if transformed by a positive definite diagonal transformation /change of scale/ and any natural transformation of time does not change the presence or absence of negative cross-effects.

Based upon this results we strongly believe that impossibility to eliminate negative cross-effect from equations showing chaotic behaviour may prove an important, if not characteristic, property. Nevertheless, it should be mentioned that kinetic equations with apparent /numerically demonstrated/ chaotic behaviour do exist [3] , although most examples contain negative cross-effects [4].

1. Vera Hárs: "On the Inverse Problem of Reaction Kinetics", in Coll. Math. Soc. J.Bolyai 30 (Qual. Theory of Diff. Eqs.) pp.363-379.
2. J.Tóth and Vera Hárs /in preparation, 1984/.
3. O.E. Rössler: Z. Naturforsch. 31a, 259 (1976)
4. O.E. Rössler: Z. Naturforsch. 31a, 1664 (1976)

Diversity and Complexity in Dynamical Behaviour of a Simple Autocatalytic Brusselator-Type Model

P. Tracqui, P. Brezillon, and J.F. Staub

Service de Biophysique, (CNRS LA 163), C.H.U. St-Antoine, 27, rue Chaligny
F-75571 Paris Cédex 12, France

A simple two-variable theoretical model is proposed. It exhibits most of the temporal dynamical behaviours reported for nonlinear chemical systems [1] : oscillations, excitability, multistability. This autocatalytic model, part of our nonlinear model of calcium metabolism [2], has been associated with bone calcification processes (nucleation and crystal growth). It is described by the differential system :

$$\begin{aligned} \dot{x} &= a - k_1 x - (k + y^2)x + k_2 y \qquad & x(t_0) &= x_0 \\ \dot{y} &= b + (k + y^2)x - (1 + k_2)y \qquad & y(t_0) &= y_0 \end{aligned} \tag{1}$$

Using a and k as bifurcation parameters, bistabilities and even tristabilities involving 1 or 2 steady states and 1 periodic nonsteady state have been displayed. These properties are often associated with multiple limit cycles. All the situations we observed are similar to those reported in [3], except for topologies depicted by Fig. 1. According to whether there is a stable limit cycle or not, two main classes of behaviour can be charcterized. In the first case, two kinds of topological configurations are performed, depending on whether the limit cycle surrounds all the singular points or not. Transition between these configurations originates from a secondary bifurcation. Associated with 1 or 3 singular points and for a critical value of the influx a, this bifurcation leads to a switch between two stable limit cycles with very different amplitude and period (period ratio > 3). This kind of behaviour, which could be involved in quasi-periodic or chaotic dynamics, can be related to experimental observations [4] and theoretical studies [5] on the Belousov-Zhabotinsky reaction.

Moreover, system (1) can be connected to a Brusselator-type reaction scheme in which both chemicals are involved in symmetrical steps except for the autocatalytic process.

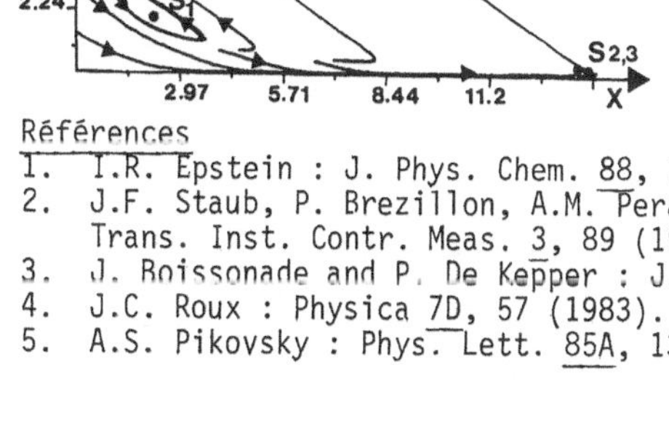

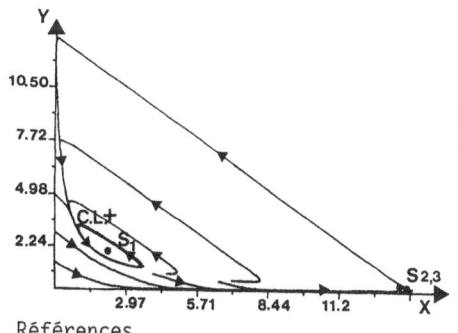

Fig. 1 A phase portrait example for which the stable limit cycle C.L$^+$ surrounds only one of the three singular points S_i . S_3 is a stable steady state.
(k_1 = 0.15 , k_2 = 3. , b = 0.
k = 0.02075 , a = 2.23)

Références

1. I.R. Epstein : J. Phys. Chem. 88, 187 (1984).
2. J.F. Staub, P. Brezillon, A.M. Perault-Staub and G. Milhaud : Trans. Inst. Contr. Meas. 3, 89 (1981).
3. J. Boissonade and P. De Kepper : J. Phys. Chem. 84, 501 (1980).
4. J.C. Roux : Physica 7D, 57 (1983).
5. A.S. Pikovsky : Phys. Lett. 85A, 13 (1981).

Designing and Excluding Periodicity

J. Tóth
Computer and Automation Institute, Hungarian Academy of Sciences, Kende u. 13-17
H-1502 Budapest, Hungary
P. Érdi
Central Research Institute for Physics, Hungarian Academy of Sciences, P.O.B. 49
H-1525 Budapest 114, Hungary
Vera Hárs
CHINOIN Chemical and Pharmaceutical Works Ltd., Tõ u. 1-5
H-1045 Budapest IV, Hungary

1. Designing reactions with prescribed linearized form

From the investigations of HANUSSE [1], TYSON and LIGHT [2] and PÓTA [3] it is known that in two-component bimolecular systems there is only one oscillator: the Volterra-Lotka model. The following question has arisen: is it also true that this model is the unique simplest one among all the models with the same linearized form around their own stationary state ? The answer to this question being yes does add something new to the result cited above.

It has also been shown by the same method that there is essentially only a single model different from the explodator [4] with the same linearized form. This model

$X + 2Y \rightarrow 3Y \qquad A + X + Y \rightarrow 2X + Y \qquad Y \rightarrow B$

seems to be worth investigating [5].

Finally, a general procedure has been sketched on how to look for reaction models with a prescribed linearized form in a certain class: methods of mathematical programming are proposed to be used.

2. Excluding periodicity

The theorems of Bendixon and Dulac are generalized for the case of n-dimensional systems with first integrals. It is shown as an illustration that the classical Michaelis-Menten reaction has no oscillatory solution. As another illustration we show that reactions with M components composed of M-1 atoms have no oscillatory solution either [6].

1. P. Hanusse: C.R.Acad.Sci.Ser. C274, 1245 (1972)
2. J.J. Tyson and J.C. Light: J.Chem.Phys. 59, 4164 (1973)
3. Gy. Póta: J.Chem.Phys. 78, 4164 (1973)
4. Z. Noszticzius, H. Farkas and Z.A. Schelly: J.Chem.Phys. 80, 6062 (1983)
5. J. Tóth and Vera Hárs /in preparation 1984/
6. J. Tóth: "Bendixon Type Theorems with Applications", in Coll. Math. Soc. J.Bolyai /in preparation/.

The Effect of Inhomogeneity of the Residence Time in a Chemical Flow Reactor

C. van den Broeck

Dept. Nat., Vrije Universiteit Brussel, Pleinlaan 2, B-1050 Bruxelles, Belgium

1 Introduction.

Let us consider a chemical reaction of the radioactive type $X \xrightarrow{\lambda} A$. In many experimental situations, the chemical reaction rate λ depends on the position of the molecule X, due to the presence of a concentration or temperature gradient or of an external field. The position of the molecule changes due to diffusion and convection. The question arises: what will be the effective reaction rate λ^*, describing the decrease in the concentration of X ? Another way to state the problem is to ask for the average residence time of a molecule $\overline{t_R} = 1/\lambda^*$. In the present contribution, we will consider a one-dimensional reactor.

2 The 2-Box Model : Exact Results.

Consider a molecule X which can exit from a system consisting of two boxes. The exit rates are λ_1 and λ_2 respectively and the transition rates between the boxes $k_1^+ = k_1$ and $k_2^- = k_2$. For the decoupled case ($k_1 = k_2 = 0$), one obviously has $\overline{\lambda} = \lambda_1 p_1 + \lambda_2 p_2$, where p_1 and p_2 are the initial probabilities to be in box 1 and 2. Henceforth, we will assume stationary initial probabilities. For the case of coupled cells, one then finds (one easily verifies that $\lambda^* \leqslant \overline{\lambda}$) :

$$\lambda^* = \frac{(\lambda_1 \lambda_2 + \lambda_1 k_2 + \lambda_2 k_1)(k_1 + k_2)}{\lambda_1 k_1 + \lambda_2 k_2 + (k_1 + k_2)^2} \leqslant \overline{\lambda} \qquad (1)$$

3 The N-Box Model : Approximate Results.

We consider the generalization of the above model to the case of a linear array of N boxes. The transition rates from box i to i ± 1 are denoted by $k_i^{\pm}$ and $p_i = \mathcal{N} k^+ \ldots k_{i-1}^+ \cdot k_{i+1}^- \ldots k_N^-$ is the stationary probability distribution. One finds, asymptotically, for small $\overline{\lambda} = \Sigma \lambda_i p_i$:

$$\lambda^* = \overline{\lambda} - \sum_{r=1}^{N-1} \frac{\left[\sum_{i=1}^{N} \sum_{j=1}^{r} (\lambda_i - \lambda_j) p_i p_j \right]^2}{k_r^+ p_r} \leqslant \overline{\lambda} \qquad (2)$$

The equality sign $\lambda^* = \overline{\lambda}$ only holds for the homogeneous case ($\lambda_1 = \ldots = \lambda_N = \overline{\lambda}$). It is straightforward to obtain from (2) the result for a chemical flow reactor in the continuum limit.

4 Conclusion.

In all the cases considered, the effect of inhomogeneity is to decrease the effective reaction rate λ^*, as compared to the weighted average $\overline{\lambda}$.

Dynamical Aspects in the Noise-Induced Freedericksz Instability

F. Sagués[1 2] and M. San Miguel[1]

Universitat de Barcelona, Departament de Fisica Teòrica[1] and Departament de Quimica Fisica[2], Diagonal 647, E-08028 Barcelona, Spain

The role of the internal and external fluctuations in certain dynamical aspects associated with the so-called Freedericksz transition corresponding to a nematic liquid crystal under twist deformation is analyzed [1]. The external fluctuations come from the applied magnetic field viewed as a random control parameter. Owing to its nonlinearity we are faced with very interesting features concerning the appropiate formulation of the model via a Langevin or a Fokker-Planck equation [2]. Moreover, the global (internal) fluctuations of the system are taken into account through a time-dependent Ginzburg-Landau approach (TDGL).

Let us specify our system by considering a nematic layer enclosed between two parallel plates perpendicular to the z-axis. A magnetic field is applied along the y-axis. We restrict ourselves to sufficiently weak fields and deformations. The amplitude, $\theta(t)$, of the lowest Fourier mode of the deformation angle $\phi(z,t)$ obeys a stochastic differential equation (SDE) written in terms of a reduced magnetic field h

$$\partial_t \theta(t) = -U'(\theta) + \xi(t) \quad , \quad U(\theta) = (1-h^2)/2\, \theta^2 + h^2/8\, \theta^4 \qquad (1)$$

in which both an internal gaussian white noise $\xi(t)$ of intensity ε and an external Ornstein-Uhlenbeck noise, $w(t)$ appear

$$h = h_d + w(t) \quad ; \quad \langle w(t)\rangle = 0 \, , \; \langle w(t_1) w(t_2)\rangle = \frac{D}{\tau} e^{-\frac{|t_1 - t_2|}{\tau}} \qquad (2)$$

The dominant effect of the external noise corresponds to a systematic contribution obtained in a "deterministic" limit: $D \to 0$, $\tau \to 0$, D/τ finite [1,2]. The SDE is in turn expressed in terms of a modified potential $\bar{U}(\theta)$

$$\partial_t \theta(t) = -\bar{U}'(\theta) + \xi(t) \quad ; \quad \bar{U}(\theta) = \frac{(1-\alpha^2)}{2}\theta^2 + \frac{\alpha^2}{8}\theta^4 \, , \quad \alpha^2 = h_d^2 + \frac{D}{\tau} \qquad (3)$$

For $\alpha^2 < 1$, the single-well potential $\bar{U}(\theta)$ is flatter at $\theta = 0$ than the deterministic potential $U_d(\theta)$. $U_d(\theta)$ corresponds to $U(\theta)$ for $h = h_d$. This obviously implies a slowing down in the relaxation towards $\theta = 0$. For $\alpha^2 > 1$ the effect is the opposite one since the bistable potential has deeper wells than those of $U_d(\theta)$. Likewise the maximum of $\bar{U}(\theta)$ at $\theta = 0$ for $\alpha^2 > 1$ is more pronounced than the corresponding one of $U_d(\theta)$. As a consequence, the time spent by the system before leaving the unstable state when a field is switched on from $h=0$, is smaller than in absence of field fluctuations.

References

1) W. Horsthemke and R. Lefever: Noise-Induced Transitions (Springer-Verlag, Berlin 1984).

2) F. Sagués, M. San Miguel and J.M. Sancho: Z.Phys. B55,269(1984).

Variety of Evolutions to Stationary Periodical Structures

Andrzej Lech Kawczyński

Institute of Physical Chemistry, Polish Academy of Sciences, 01224 Warsaw, Poland

The system (1) with S-shaped $f(u,v)=0$ and bifurcations from one stationary state to trigger, and from trigger to another stationary state (in the subsystem u,v) induced by $\varphi(p)$

$$
\begin{aligned}
u_t - D_u u_{xx} &= \varepsilon_1 f(u,v) \\
v_t - D_v v_{xx} &= \varepsilon_2 (g(u,v) + \varphi(p)) \qquad (1)\\
p_t - D_p p_{xx} &= \varepsilon_3 (\alpha u - \beta p) \qquad \varepsilon_1 \gg \varepsilon_3 \gg \varepsilon_2
\end{aligned}
$$

gives different types of evolutions to stationary periodical structures for proper initial conditions and values of parameters.

For N-shaped $\varphi(p)$ the generation of the structure occurs by induction of the next pulse by the former one. The structure is stabilized by the fact that at each maximum (minimum) of the distribution the subsystem u,v has only an attractor on the upper (lower) branch of $f(u,v)=0$, respectively [1] [2].

For $\varphi(p)$ with a maximum only, the generation of the structure can also occur by introduction of the next pulse by the former one, but the structure is stabilized by the self-adjustment distribution of p which controls the threshold for excitation of u [3].

For the same type of $\varphi(p)$ but other values of parameters the structure is generated by a division of the pulse of u behind the front of the running wave. The structure is stabilized in the same way as in the previous case [4].

Examples of chemical models describing coupled enzymatic reactions together with results of numerical calculations are given in [2] [3] [4]. Amplitudes and periods of these structures do not depend essentially on the size of the system nor on initial conditions.

References

1. A.L. Kawczyński, A.N. Zaikin: J.Non-Equilib.Thermodyn. 2, 139 (1977)
2. A.L. Kawczyński, J. Górski: Pol.J.Chem. 57, (1983)
3. J. Górski, A.L. Kawczyński: Pol.J.Chem. to appear
4. J. Górski, A.L. Kawczyński: Pol.J.Chem. to appear

Transient Bimodality: A New Mechanism of Stochasticity in Explosive Chemical Reactions

F. Baras

Faculté des Sciences de l'Université Libre de Bruxelles, Campus Plaine, C.P. 226
B-1050 Bruxelles, Belgium

M. Frankowicz

Institute of Chemistry, Jagellonian University, ul. Karasia 3
30-060 Krakow, Poland

We study the stochastic model of thermal explosion in a chemical system. The principal assumptions are : neglecting the consumption of the reactant, Semenov boundary conditions (uniform reactant temperature, Newton heat losses at the boundary), and the Arrhenius temperature dependence of the rate constant. We use the Semenov-Frank Kamenetskii approximation (large activation energy) [1]. The Frank-Kamenetskii parameter, δ , is our control parameter (ratio of the characteristic heat tranfer time to the adiabatic induction time). There exists a critical value of δ , δ_c , such that for $\delta > \delta_c$ the temperature tends asymptotically to a finite stationary value (subcritical regime). In supercritical regime ($\delta < \delta_c$) the ignition takes place. As in our simplified model there is no steady state corresponding to the combustion regime. Ignition is manifested as a divergence of temperature in finite time.

To analyse the stochastic aspects of this problem, we use Fokker-Planck equation obtained from the master equation (whose transition probabilities were derived from thermodynamics arguments, see [2]). The Fokker Planck equation was solved numerically. Also, a series of simulations using Langevin equation was carried out.

The principal results are :
1. In the supercritical region the probability distribution function displays transient bimodality.
2. The distribution of ignition times displays long time tail.
3. The mean ignition time calculated from the stochastic model is significantly shorter than the deterministic ignition time.

References

1. Frank-Kamenetskii : Diffusion and heat transfer in chemical kinetics (Plenum Press, New York-London 1969)

2. G. Nicolis, F. Baras and M. Malek Mansour : in these proceedings

Complex Periodic and Chaotic Behavior in the Belousov-Zhabotinskii Reaction

K. Coffman, W.D. McCormick, and H.L. Swinney

Physics Department, University of Texas, Austin, TX 78712, USA

J.C. Roux

Centre de Recherche Paul Pascal, Université de Bordeaux I, Domaine universitaire F-33405 Talence Cédex, France

In recent years the authors and co-workers [1] have investigated the dynamics of the Belousov-Zhabotinskii (BZ) reaction in a stirred-flow reactor. The experiments [2] have revealed a series of alternating periodic and chaotic regimes as a function of flow rate. With increasing reactor residence time (τ) the regimes are labeled P_1, C_1, P_2, C_2, etc. where P_1 indicates large amplitude oscillations, P_2 one large followed by one small amplitude oscillation, P_3 one large followed by two small oscillations and so on. Detailed investigation of C_1 has shown that it, in turn, is a complex series of chaotic states separated by periodic windows which may be understood in terms of an underlying one-dimensional map. Here we wish to report very briefly on three new features that we have discovered in the dynamics of this system:

1. Experiments show that the major periodic regimes P_1, P_2, P_3 etc. all lose stability to chaos by the mechanism of period doubling; and that this happens in both directions, that is, towards either longer or shorter residence times.
2. The periodic windows in the chaotic regimes lose stability either by period doubling (towards longer τ), or by a tangent bifurcation and intermittency (towards shorter τ). Figure 1(a) shows the third iterate of the return map of an intermittent state, for τ just less than a value in the period three window in C_1. The same behavior has been seen just outside a prominent period five window in C_2.
3. The return maps describing the chaotic dynamics in C_2, C_3, etc. are still one dimensional, like those in C_1, but more complex. Figure 1(b) shows an example from C_2. One implication of the multivalued nature of this map is that it now takes two consecutive values of the amplitude to predict the next amplitude.

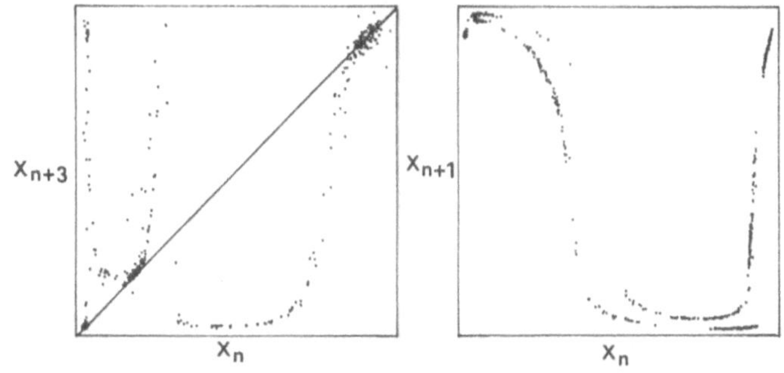

Fig. 1(a) Third iterate of return map, showing tangency and intermittency.
(b) Multivalued (folded) return map in C_2.

References

1. J.S. Turner, J.C. Roux, W.D. McCormick and H.L. Swinney, Phys. Lett. 85A (1981) 9.
2. For further references and background see the review "Observations of Order and Chaos in Nonlinear Systems", H.L. Swinney, Physica 7D (1983) 3.

Periodic Operation of a Stirred Flow Reaction with Limit Cycle Oscillations

W. Geiseler and T.W. Taylor

Institut für Technische Chemie, Technische Universität Berlin
D-1000 Berlin 12, Fed. Rep. of Germany

The oscillating flow reaction of bromate, bromide, and cerous ions, referred to as a minimal oscillator [1,2] and well understood in terms of elementary reaction steps [3], was investigated computationally in an isothermal CSTR where periodic perturbations were imposed on either the total flow rate or one of the reactant inflow concentrations. Both the perturbation amplitude and frequency were varied over wide ranges. The investigations were primarily conducted with sinusoidal perturbations; however, square pulse and saw-tooth periodic functions were also applied.

The model calculations, based on the widely accepted NFT mechanism [4], predict a variety of interesting non-equilibrium phenomena including a) Harmonic entrainment (H), b) Subharmonic entrainment (S) of order 1/n (n being an integer), c) Regular transient behavior (T), d) Chaotic oscillations (C), and e) Low amplitude harmonic entrainment (LH). These dynamic patterns, generally occurring after a short transient time following the switch from constant to periodic operation of the reactor, are not restricted to discrete perturbation frequencies, but exist within certain ranges or bands of frequencies. In Fig. 1 typical frequency bands of distinct dynamical behavior are shown for the CSTR where the inflow concentration of bromide ions was sinusoidally varied at distinct values of the perturbation amplitude. Similar results were obtained when the flow rate was periodically perturbed or other periodic functions were applied.

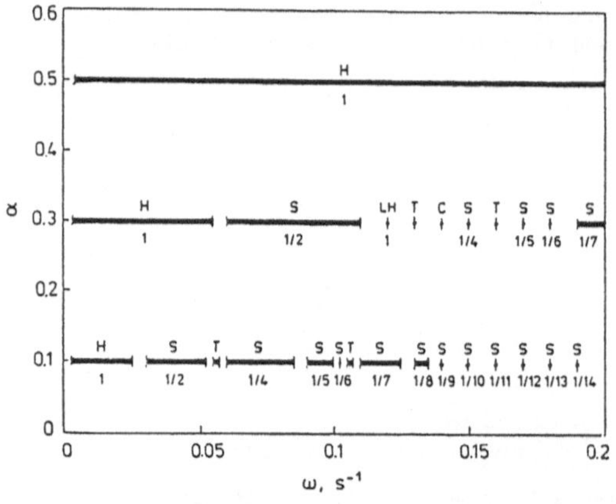

Fig. 1 Entrainment bands in the plane of perturbation amplitude α vs. perturbation frequency ω. Operating conditions: $[BrO_3^-]_0 = 0.150$ M, $[Ce^{3+}]_0 = 0.0003$ M, $[H^+]_0 = 0.75$ M, and $k_0 = 0.005\ s^{-1}$. Periodic perturbation: $[Br^-]_0 = [\overline{Br^-}]_0 (1 + \alpha \sin \omega t)$ with $[\overline{Br^-}]_0 = 0.00041$ M. Natural frequency $\omega_0 = 0.00548\ s^{-1}$.

The model predictions are in excellent agreement with presently known experimental results [5] and show surprising parallels to the work of Dulos [6]. The theoretical behavior of the system, studied under a variety of experimental conditions, suggests many feasible experiments which are interesting in themselves, but may also serve as an additional probe of the complex NFT mechanism used.

References

1. W. Geiseler: Ber. Bunsenges. Phys. Chem. 86, 721 (1982)
2. M. Orban, P. DeKepper, and I.R. Epstein: J. Am. Chem. Soc. 104, 2657 (1982)
3. K. Bar-Eli and W. Geiseler: J. Phys. Chem. 87, 3769 (1983)
4. R.M. Noyes, R.J. Field, and R.C. Thompson: J. Am. Chem. Soc. 93, 7315 (1971)
5. H. Lachmann: Personal communication
6. E. Dulos: Synergetics, Vol. 12, p. 140, Springer-Verlag (1981)

Index of Contributors